FRUIT AND VEGETABLE PRODUCTION IN WARM CLIMATES

NATURAL RESOURCES INSTITUTE
CENTRAL AVENUE
CHATHAM MARITIME
CHATHAM
KENT ME4 4TB

First African edition published 1987
This international edition published 1990

Published by *Macmillan Publishers Ltd*
London and Basingstoke
Associated companies and representatives in Accra,
Auckland, Delhi, Dublin, Gaborone, Hamburg, Harare,
Hong Kong, Kuala Lumpur, Lagos, Manzini, Melbourne,
Mexico City, Nairobi, New York, Singapore, Tokyo

British Library Cataloguing in Publication Data
Rice, Robert P.
    Fruit and vegetable production in warm climates
    1. Tropical regions. Crops : Fruit
    Production 2. Tropical regions. Crops.
    Vegetables. Production
    I. Title   II. Rice, Laura Williams
    III. Tindall, Harold Donovan
    634'.0913

ISBN 0−333−46850−3

ISBN 0−333−51255−3 (cased)

Printed in Hong Kong

Contents

Acknowledgements

The invaluable assistance of all who contributed to the preparation of the text material is sincerely acknowledged by the authors.

Particular thanks are due to Joanne Clark for the excellent drawings which have been used in this book and to Susan Carr for her contribution in compiling the data on pests and diseases of vegetables. The assistance of Eugene Memmler and Vocational Education Productions in supplying photographs is also gratefully acknowledged.

Robert P. and Laura W. Rice
Crop Science Department
University of Zimbabwe
Harare
Zimbabwe

H. D. Tindall
Emeritus Professor
Cranfield Institute of Technology
Bedford
England

Preface

This book has been designed to meet the need for a text which covers both fruit and vegetable production in a single book, making it an ideal text for college, technical school, and secondary school courses in horticulture. The aim has been to present the basic principles of fruit and vegetable crop production together with practical techniques relating to all operations from plant establishment to harvesting. It is assumed that students using this text will already possess a background in general agriculture; therefore an attempt has been made to discuss only those principles and techniques which are specific to horticultural crops. The authors have also attempted to discuss crops within the framework of production systems used in warm climates and to include the large number of indigenous crops, especially vegetables. At the same time, since many exotic fruits and vegetables are produced both for local consumption and for export, these have also been included. A short chapter has also been included dealing with the production of temperate fruit, the production of which is increasing in importance and on which information is not readily available.

A special feature of the text is the extensive appendix which will make the book useful to students, once they have completed their studies, and to extension personnel in the field. The compilation of the tables in the appendix has involved an extensive review of the literature and has resulted in the presentation of data which in many cases have never been compiled before. Since students are often limited in the number of texts they can purchase, *Fruit and Vegetable Production in Warm Climates* will provide them with both a general text and a field reference of long-term usefulness.

It is hoped that this book will be of lasting value to horticultural students and the many extension workers who are striving to improve the nutrition and income of their people through teaching methods of producing fruits and vegetables.

Conversion Factors for Imperial and Metric Units

Length
1 kilometre, km   =0.621 miles, mi
metre, m   =1.094 yard, yd
centimetre, cm   =0.394 inch

Area
kilometre^2, km^2 =0.386 mile^2, mi^2
kilometre^2, km^2 =247.1 acre, acre
hectare, ha (0.01 km^2)   =2.471 acre, acre

Volume
metre^3, m^3 =0.00973 acre-inch
litre, l   =0.0284 bushel, bu
litre, l   =1.057 quart (liquid) qt

Mass
metric tonne, t   =1.102 ton (English)
1 kilogramme, kg   =2.205 pounds, lb
gram, g   =0.035 ounce (avdp), oz

Yield or Rate
metric tonne/ha, t/ha   =0.446 ton (English)/acre
kg/ha   =0.892 lb/acre

Temperature
Celcius = 5/9 (F−32)   (9/5°C) + 32 = Fahrenheit
−17.8°C   =0°F
0°C   =32°F
20°C   =68°F
100°C   =212°F

Section I — Fruits

CHAPTER 1

Fruit Tree Propagation

Fruits are important in the tropics and subtropics due to their carbohydrate and vitamin contribution to the diet. Most fruits contain large quantities of sugars and are high in vitamins such as vitamin A and C which are not abundant in the staple food of many warm areas. In addition, since fruits are normally eaten fresh rather than cooked the vitamin content is not diminished in preparation. Because fruits are palatable they are often consumed in large quantities thus contributing markedly to nutrient intake. Fruits may also be economically important since they can produce large yields in small areas and yet sell for high prices when compared to crops such as maize, groundnuts, or rice. This combination of high yield and high price means that in heavily populated areas, farmers can make good incomes even from small farms.

The production of high-quality fruits which provide all the benefits described above occurs only if the farmer starts with healthy plants of cultivars which are:

(1) adapted to the environment in which they are to be grown; and
(2) genetically capable of producing good fruit.

Unlike most cereal crops and many vegetables, fruits are not generally sown directly in the field where they are to grow. Instead, they are started or propagated in nurseries and are then transplanted to the field when they are large enough to survive the relatively harsh conditions in the orchard. The success or failure of an orchard is therefore largely dependent on how well the propagating and growing of fruit trees has been done in the nursery.

Selection and preparation of the nursery site

Nurseries vary widely in their size and scope. In some countries, a single nursery may propagate and grow all the fruit plants required for the entire country. In other countries, the government or private companies may operate several nurseries, each producing plants for the geographical region in which it is located. An individual grower generally does not produce his

3

own plants, although if nursery plants are not available, it may be necessary to do so.

No matter what the size and purpose of the nursery, it should be located in an area which is climatically suitable for the species being grown. Nurseries require deep, well-drained soil for growing plants in beds or in the field as well as for mixing with organic matter for container-grown plants. A reliable water supply is another necessity since nursery plants will require some watering even in wet climates. For the production of a fairly wide range of fruit plants in the tropics, moderate elevations of 1000 to 1500m are desirable. These elevations are high enough to permit some deciduous fruits to be grown but are not so high that tropical fruits cannot grow. In addition, the lower amounts of rainfall which tend to occur at these elevations help minimize pest and disease problems.

The nursery should be located as close as possible to fruit-producing areas and should be readily accessible by all-weather roads for the transport of the plants. Remember that fruit plants, especially those growing in containers, are heavy and bulky and that planting will most commonly occur during the rainy season when roads are at their worst.

After the site has been selected, propagation structures, buildings, windbreaks, and nursery beds should be laid out so that plants are moved a minimal distance from the time they are propagated to the time they are finally sold or transplanted to the orchard. Figure 1-1 shows typical movement of plants in a nursery. If this pattern of material and plant movement is considered in the nursery layout, there will not be wasteful movement of plants and materials.

After the layout is complete, areas to be used for propagation beds and field areas for the production of field stock (plants grown in the ground as

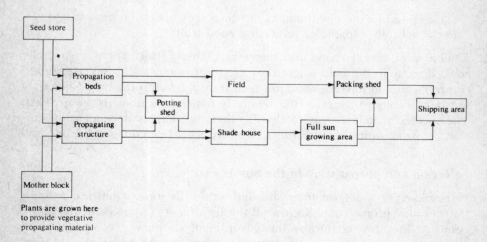

Figure 1-1
Typical pattern of plant and material movement in a nursery

opposed to those grown in plastic plant bags or other containers) should be tested for nematodes and, if necessary, fumigated with methyl bromide, EDB (ethylene dibromide), or some other fumigant. Even if nematodes are not a problem, the fumigation of nursery beds is desirable as a control for soil-borne disease and insects. Where chemical fumigation is not possible, a decrease in soil pathogens may be obtained by covering the moist beds with a sheet of clear polyethylene for 30 to 60 days during hot weather. The increased temperatures which will be created under the sheet during sunny weather will aid in soil pest control although chemical fumigation is more effective.

Propagation structures

Propagation structures are desirable in a nursery because they permit the nursery worker to control the environment. Structures range from simple shade houses to complex greenhouses with automatic controls and accordingly, they vary in the extent to which they control the environment.

A simple shade structure can be constructed by using poles to support a roof of wire mesh upon which a thin layer of thatching grass is tied to give filtered sunlight beneath (Figure 1-2). This type of structure (especially if also provided with grass walls) can be used to propagate many species of fruit crops and can also be used for growing newly transplanted stock. Poles

Figure 1-2
A typical shade structure constructed from local materials—treated poles
and thatching grass

Figure 1-3
A propagating frame
(Reproduced by permission of Saunders College Publishing)

should be treated with a wood preservative to prevent rotting and termite damage. Creosote may be used below ground level but should not be used above ground as it gives off fumes which are toxic to plants. Copper-containing preservatives such as copper naphthenate are ideal if they are available. If these materials are not obtainable, soak the bottom of the poles in paraffin (kerosene) so that the poles are protected to 15 cm above the soil level.

If a more complex structure is required, propagating frames or hot beds can be constructed. A propagating frame (Figure 1-3) is a low structure made of brick, wood or other material, and has a removable top made of light-admitting plastic or glass. The top is slanted towards the equator to allow maximum light penetration and to allow water to run off. At locations on the equator, the direction of slant is unimportant. This structure allows for the regulation of temperature and humidity by opening or closing the top.

A hot bed, which may be useful in high-altitude areas, is of identical construction except that a heating cable is buried in the rooting medium to provide heat in the root zone. If electricity is not available a 15 cm thick layer of fresh manure can be placed below the rooting media. Heat from the rotting of the manure will warm the root zone. If propagating frames or hot beds are used for propagation, an attendant should be nearby at all times to regulate the temperature since these structures can overheat rapidly. For rooting of cuttings the best environmental control device is an intermittent mist system which can be mounted in a hot bed, shade structure, or green-house. A mist system (Figure 1-4) is composed of nozzles mounted above the cuttings and a timing mechanism. The nozzles spray the cuttings every few minutes with a fine mist so that the cuttings are wet and transpiration is minimized. The timing mechanism can be either a time-clock or an 'electronic leaf', a device which senses when the leaves of the cuttings are dry and then turns on the mist nozzles. When a mist system is used, heating cables should be placed in the rooting medium to offset the cooling effect of the mist due to evaporation from the leaves and medium.

Potting soils

One of the common problems which occur in nurseries where plants are grown in containers is that poor potting soils or potting mixes are used. When soil is removed from the field and placed in a container its properties change drastically, thus field soils are seldom suitable for use as potting soils unless they are heavily amended.

A good potting soil must perform a number of functions well if high quality nursery stock is to be produced. Firstly, a mix must adequately support the plant. This means that it must be heavy enough to prevent the plant from falling over, without being so heavy that handling and shipping costs are unduly increased.

Secondly, the mix must provide a reservoir for air, moisture, and nutrients. Roots must be adequately supplied with both air and moisture, however these two compete with each other to occupy the spaces between the particles of the mix. The tendency is for water to occupy small spaces while air occupies larger spaces. For this reason the size of the spaces between soil

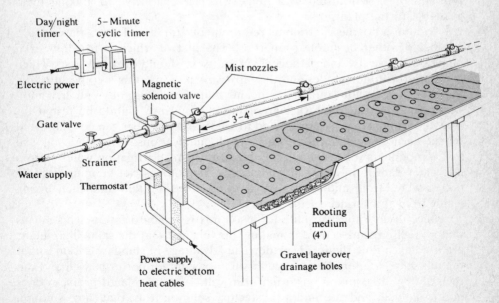

Figure 1-4
The parts of an intermittent mist system with an electric soil-heating cable

mix particles is very important. In the ground, most clay-containing soils will aggregate or clump together in such a way that there is space for both water and air. When these soils are transferred to a pot, however, the aggregated particles are usually destroyed and only very small spaces remain. In addition, the shallowness of the pot reduces drainage and tends to cause the water-logging of the soil. Hence the soils do not contain sufficient air for good growth of plants. For this reason field soils are almost never good potting soils unless they are amended to increase their aeration. If the roots of container plants tend to be near the surface, near the sides of the planting pockets, or near drainage holes with only minimal growth in the centre of the pot, the potting soil is low in air and you can be sure that your plants are growing much more slowly than is desirable and that the likelihood of root rot diseases is greatly increased.

The soil mix must, however, also hold large quantities of water since the root system is confined to a small container. If the soil mix does not, the nurseryman will have to water frequently, and even then the plants may be water stressed resulting in reduced growth and quality. This means that the soil mix must contain water absorptive materials. Many types of organic matter are very efficient in this regard.

Thirdly, the soil mix must supply all the nutrients the plant needs — both those needed in large quantities, such as calcium, magnesium, nitrogen, phosphorus, and potassium, and also those needed in small quantities, such as iron, zinc, and manganese. These nutrients have to be added to the soil mix and the pH adjusted to a range which is conducive to making them available to the plant.

In addition to these absolute requirements for a soil mix, there are a number of other desirable properties. The first of which is freedom from nematodes, insects, and pathogens. No nursery should ever sell stock which is infested with nematodes and the only way to ensure this is to be sure that the media does not contain them and that the propagules and irrigation water are likewise free. All ingredients in a soil mix should be tested for freedom from nematodes or treated to kill them. Soil pests can be eliminated by chemical or heat sterilization, pesticide applications, or by choosing soil mix components which are naturally free of them. If irrigation water is from a river or dam it may contain nematodes. Nematodes can be minimized by taking water from mid-stream or by using a settling tank and then taking water from the surface of the tank.

A second desirable attribute for a soil mix is standardization. This means that a precise recipe is used in making the mix and that the same ingredients are used each time the mix is made. The advantage of standardization is that the mix can be tailored to particular situations until it performs well and can then always be counted on to perform well. If a soil-related problem does occur one year and you find a satisfactory solution to it, that same solution should work in succeeding years. A standardized potting mix allows the grower to standardize fertilizer and watering programmes and to produce reliably good quality plants.

A soil mix should also be stable enough to ensure that its properties do not change drastically during the period of time that the plant is in the nursery. Materials which tend to compact or organic matter which tends to decompose rapidly are usually not satisfactory ingredients in soil mixes because although they may perform satisfactorily initially their performance will change drastically after a period of time has passed.

What are some of the materials which might be suitable for use in commercial soil mixes? *Pine bark* is an excellent material which meets the criteria described above. This could comprise 50 to 75 per cent of mixes for many plants. The bark should be passed through a hammermill and screened until the size is appropriate (usually 0.5−1.0cm) and then allowed to sit for a few weeks before use. *Groundnut hulls* are also an excellent material when ground or composted and can be used in a manner similar to bark. They may, however, contain nematodes and should be sterilized prior to use. *Coarse sand* is a good material to add weight and to improve the drainage of mixes as it is inexpensive and readily available. Particle size is, however, very important and should range between 0.5 and 2mm. The sand should be screened to eliminate smaller particles which will tend to retard aeration and drainage. *Soil* may also be used, however, it has some very definite disadvantages. Firstly, it is very likely to be nematode-contaminated and will require treatment. Secondly, it is difficult to ensure its uniformity and thus the standardization of the soil mix. The best soils are usually sandy loams taken from the top 15cm below a grass cover. Be sure that herbicides have not been used recently. If soil is used, it should rarely comprise more than 50 per cent of the mix.

In addition to these major ingredients, essential nutrients must also be added and the pH adjusted to about pH 6.0−6.5. The amount of nutrients will vary depending on the soil mix and fertilizer programme. However, the following may be regarded as a starting point in determining, through experimentation, what is best.

For each cubic metre of soil mix (75 per cent bark/25 per cent sand) add:

3.5kg dolomite (supplies magnesium and calcium, and counteracts acidity)

100g potassium nitrate

100g potassium sulphate

1000g single superphosphate

10g zinc sulphate

10g iron sulphate

10g manganese sulphate

Alternatively, use a mixture containing micronutrients to supply the necessary items.

Note that the above mix is adapted from work done at the University of California (Baker, 1957) and should be regarded as a starting point only. Be sure that your mix is thoroughly mixed to ensure the even distribution of the nutrients. Soil mixes are best made when the materials are slightly moist.

Propagation techniques

The propagation or reproduction of fruits is either sexual or asexual. Sexual propagation is by planting seeds resulting from the fertilization of ovules and often an exchange of genetic material through cross-pollination. Fruit trees reproduced in this way are not identical to their parents and in most fruit species propagation by seed is not desirable. However passionfruit, tree tomatoes, and paw paws (papayas) can be propagated well from seed. Sexual propagation is most commonly used for growing rootstocks onto which selected, high-yielding cultivars will be grafted.

Asexual or vegetative propagation does not involve an exchange of genetic material. The offspring therefore are genetically identical to the parent plant. Since most fruits are highly heterozygous, and do not produce offspring similar to the parent, vegetative propagation is preferred for most species.

Sexual reproduction by seed

For fruits propagated by seed, the seed should be chosen from vigorous, healthy plants which have desirable characteristics. By selecting seed from the best plants, the chances of producing good-quality seedlings are increased. If seed is not to be planted immediately, it should generally be stored at 0–5°C and at 50 per cent or lower relative humidity. Seed which is not carefully stored will lose viability rapidly in warm, humid areas.

Many seeds will germinate soon after planting with no special treatment. However, others require treatments such as scarification and stratification before they will germinate and grow normally. Scarification is the scraping of the seed coat to help water penetration and is not normally needed for fruit seeds. Stratification, which is the process of chilling seeds under moist conditions at temperatures slightly above freezing, is needed for the germination of seeds of many temperate tree fruit species. The optimum length of stratification varies with species. An easy way to stratify seeds is to alternate a layer of seeds with a layer of soil or humus in a box, moisten and then enclose it in a plastic bag. The container should then be placed in a refrigerator for the time required for the species or for about 40 days. After stratification the seeds should be planted immediately or they may enter a secondary dormancy period.

Seeds can be planted in nursery beds or in individual containers. The soil in which they are grown should be sterilized, if possible, prior to planting to control soil-borne pests. Methyl bromide is the most convenient material for soil sterilization, and is most commonly available in small pressurized cans. When the can is punctured the chemical escapes as a gas and is toxic to most living organisms.

The procedure for fumigation starts with preparing the soil in the bed for planting, including the addition of any manure or compost. Special care should be taken that any lumps or clods are broken and that the soil is moist but not waterlogged. If soil is being sterilized for filling containers, it should be spread to a depth of 30cm on top of hard ground or concrete. Then the procedure for treating both nursery beds and soil is the same. After the soil has been prepared a shallow pan is placed on the surface in the centre, and the end of the special applicator tube is anchored in the container. The soil is then covered with a polyethylene sheet with the edges sealed by placing soil on top. The can of methyl bromide is next placed in the applicator and the operator, standing upwind, releases the gas which flows into the evaporating pan under the sheet (Figure 1-5). Depending on weather conditions, the soil should remain covered for 24 to 72 hours; it should then be uncovered and aired for the same period. After airing, planting can begin with no further waiting. The user of methyl bromide should take extreme care as it is very toxic to humans and can kill within a very short time. If available, a respirator should be worn. It is always wise to stand upwind when using methyl bromide so that, if the gas escapes, it will blow away from the user.

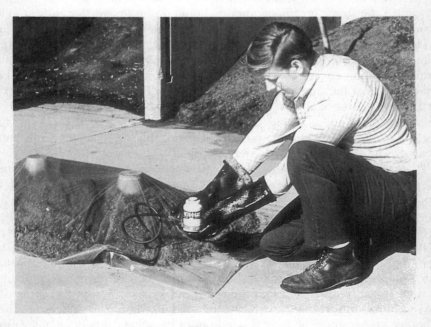

Figure 1-5
The procedure for sterilizing small quantities of soil with methyl bromide
(Reproduced by permission of Vocational Education Productions)

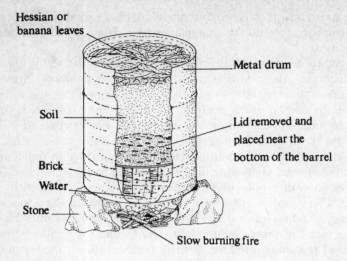

Figure 1-6
A simple soil sterilizer made from a used drum

Small quantities of soil may be sterilized satisfactorily by use of steam. The soil is heated in a drum as shown in Figure 1-6. The top of the drum should be removed and many small holes should be made in it. This is then placed in the drum on three supports about 15 cm above the bottom. The drum is put on cement blocks or stones at sufficient height for lighting a fire underneath. The bottom of the drum is filled to a level of 10 cm with water. The soil is placed on the perforated piece to one inch below the top of the drum, and then covered with banana leaves or sacking. A small wood fire should be kept burning under the drum until a temperature of at least 80°C at the top of the soil has been maintained for at least 30 minutes. If a thermometer is not available, a sweet potato can be placed 15 cm below the surface of the soil. When the potato is cooked the soil has been sufficiently heated. The soil can be used after it cools to air temperature.

If sterilization is not possible, the soil should be tested for the presence of nematodes by sending a moist soil sample containing small roots to a nematode testing laboratory. Only nematode-free soil is suitable for propagation.

As a guide, seeds should be planted at a depth two to three times their diameter and should be covered firmly with soil. The soil should then be kept moist but not wet until the seedlings have emerged. It is important that the soil does not dry out completely at any time during the germination period. It is also important that the soil is not kept too wet or 'damping off' (a soil-borne fungal disease which kills seedlings before or just after they emerge) may result. The typical symptom of damping off is the death of the stem at ground level and the subsequent wilting and falling over of the seedling. Where damping off is a problem, watering should be decreased.

After the seedlings have formed true leaves, they can be lightly fertilized every two weeks and should be watched for the development of diseases and insect pests.

Asexual propagation

Apomictic seeds   Growing plants from seeds is commonly regarded as being solely a sexual propagation technique and, in most cases, this is true. However, in some fruit trees such as certain kinds of citrus and mango, seeds are formed which do not result from pollination but are formed instead from the ovary tissue of the mother plant. Thus the plants which grow from these seeds are genetically identical to the mother plant and are called apomictic seedlings.

There are two disadvantages of propagating plants using apomictic seeds:

(1) It is often difficult to distinguish between seedlings growing from apomictic seeds and those growing from sexually-produced seed. Since the two types of seeds are mixed in the fruit, some variability of offspring may result.
(2) Plants grown from seed, including apomictic seed, often bear fruit only after they are six to eight years old, due to being in a juvenile, non-fruiting phase until that time. Plants which are not grown from seed can bypass the juvenile phase if vegetative propagules (parts of the plant capable of forming roots and shoots) are taken from adult portions of the mother plant. This results in earlier bearing.

Apomictic seeds are planted and grown in the same manner as sexually-produced seed. They are often used for the production of rootstocks of mangoes and other fruits since they are less variable than ordinary seedling rootstocks. The long juvenile period is in this case not a problem since adult shoots or buds are grafted onto them.

Cuttings   Cuttings are an important way of propagating a wide variety of plants due to the ease with which many plants grow from cuttings. In the case of mulberries, grapes, pineapples and some other plants all that is required is to stick a branch or offshoot in the ground during the rains. Other species require the special environmental conditions provided by hot beds and intermittent mist. While cuttings are most often made from stem sections, some plants, such as breadfruit, will reproduce from root sections.

Stem cuttings are classified according to the age of the wood from which they are taken (Figure 1-7). *Hardwood* cuttings are often slow to root. However, they are not as sensitive to environmental conditions as are other types of cuttings. They are usually 12 to 20 cm in length and can be stuck directly into ground beds or containers. *Softwood* cuttings are made from new growth. They are succulent and very sensitive to the environment, however they often root quickly. *Semi-hardwood* cuttings are from wood

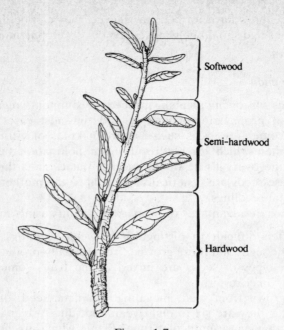

Figure 1-7
A stem showing the source of soft-wood (new growth),
semi-hardwood (bark just beginning to form), and
hardwood (fully developed bark) cuttings

which has started to mature and has some bark present. However, they do not have fully-developed bark as do hardwood cuttings. Semi-hardwood cuttings are intermediate in environmental sensitivity and speed of rooting.

The type of cutting which is used will depend on the species to be propagated and the propagation structures and cutting type available. The Appendix lists the types of cuttings preferred for each species.

Whichever type of cutting is used, the procedure for rooting the cuttings is similar. Cuttings are pruned from a vigorous, healthy mother plant using a sharp knife or secateur (hand pruner). Cuttings should be kept cool and moist while being transported to the propagation shed. The cuttings are then prepared by first removing the leaves from the bottom 4−6cm to avoid leaves rotting in the rooting medium and contaminating it and second by dipping into rooting hormone. Next the cuttings are placed upright in the rooting medium (Figure 1-8).

Rooting hormones are growth regulators which encourage the formation of roots. The most common is IBA (indolebutyric acid) which is available either as a powder or as a liquid. Hormones are available in different concentrations. The more difficult the species is to root, the higher the concentration used. For small nurseries, powdered formulations are most convenient.

The rooting medium or soil is the material into which the newly formed roots will grow. A sterile, inert material such as sand is preferred. Other materials which can be used are sterilized compost, perlite, and vermiculite or combinations of these materials. In many areas, sand or sand mixed with sterile compost is used.

Since semi-hardwood and softwood cuttings are very sensitive to the environment, conditions that minimize transpiration and encourage rooting should be provided. Remember that stem cuttings have no roots and are therefore not efficient at water uptake. If water loss from leaves is too rapid, cuttings will wilt and die. Water loss is minimized by a high humidity and cool temperatures around the leaves. Rooting is speeded by warm temperatures in the rooting media. The use of hotbeds or intermittent mist systems with heating cables provides this environment best.

Root cuttings are sections of roots at least 1 cm thick and 7−10 cm long which are cut from the mother plant and then placed in moist soil to grow a shoot. Roots are placed either horizontally or vertically about 3 cm below the soil surface and kept moist until shoots develop. Generally speaking, plants which sucker readily from the base are most likely to grow successfully from root cuttings.

Layering   Layering is a propagation technique which is used when only a few offspring are desired or when the plant is difficult to propagate by other techniques. Basically it consists of wounding a plant stem and then covering the wounded part with the rooting medium. When roots form, the stem and attached roots are cut from the mother plant and planted. A large number of species can be propagated in this way since the 'cutting' is still attached to the mother plant during rooting and can obtain water and nutrients easily during the process.

The three most common layering techniques are mound (or stool), tip, and air layering (marcottage). In mound layering, soil is mounded over the base of a shrubby plant and roots are allowed to form. After rooting, branches are cut off and planted and new branches are allowed to form on the mother plant. The following year the same process is repeated. This process is commonly used to propagate clonal apple rootstocks. For mound layering a single-stemmed plant the process is slightly different (Figure 1-9). The plant is cut off 3 cm above the ground to force it to produce several new shoots. After the shoots are 8 to 12 cm tall, the plant is mounded with soil and when the shoots have rooted they are cut from the parent plant.

Tip layering is used to propagate plants whose lower branches can be bent to contact the soil without breaking. The stems are bent to the ground and a section of bark and phloem is removed. This part of the stem is then covered until rooting occurs (Figure 1-10). Air layering involves the wounding of the stem, covering the wound with a moist rooting medium and then wrapping around with a polyethylene sleeve to hold the medium in place and keep it moist (Figure 1-11). Roots grow into the medium and the stem is

(a)

(b)

Figure 1-8
Steps in the preparation of cutting:
(a) cutting after removal from the parent plant;
(b) removal of bottom leaves;

(c)

(d)

(c) coating base of stem with rooting hormone;
(d) insertion of cutting in the rooting medium
(Photographs by Alan Thomas)

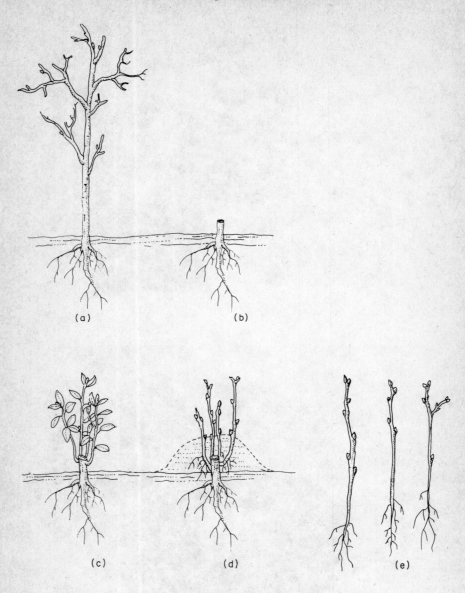

Figure 1-9
The steps in propagation using mound layering:
(a) one-year-old plant growing in the field;
(b) top is removed 3 cm above ground;
(c) new shoots grow from the cut stem and when they are 8−12 cm tall soil is
mounded around them to half their height and added periodically until it is 15−20 cm
deep;
(d) at the end of the season shoots have formed roots at their base;
(e) rooted cuttings are removed and planted in the nursery

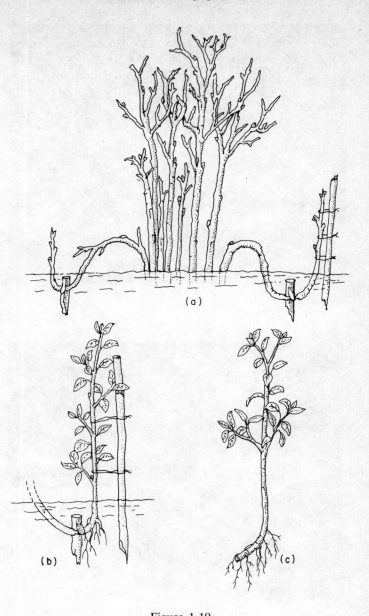

Figure 1-10
The steps in tip or simple layering:
(a) shoots are bent to the ground during the rainy season and
a second bend is made just below the tip. The shoot is
girdled at the second bend and anchored with a stake;
the injured area is then covered with soil;
(b) roots will form on the buried portion of the shoot just above the injured area;
(c) the rooted shoot is cut from the parent plant and replanted

(a)

(b)

Figure 1-11
The steps in air layering or marcottage:
(a) a 2cm strip of bark and phloem is removed from the stem of the plant at a point
just below the desired rooting point;
(b) rooting hormone may be placed on the wounded area;

(c)

(d)

(c) a moist, sterile material is placed around the injured area and covered with polyethylene to retain moisture;
(d) after roots have developed the stem is severed just below the roots and planted in the nursery
(Reproduced by permission of Vocational Education Productions)

then cut and planted. Air layering can be done anywhere on the plant, even on tall branches, as long as the diameter of the branch does not exceed 1.5 cm.

The process of wounding or girdling the stems which are to be layered prevents the movement of sugars and auxin from the top portions of the plant to the base, and they accumulate just above the girdled stem section. The presence of sugars and auxin encourages the formation of roots at this point. Rooting hormones can also be applied to the girdled section of stem.

Grafting   Grafting is a technique by which two plants are joined together and eventually grow together to become one. For example, a stem from one plant can be joined to a stem from another plant or the root from one plant can be joined to the stem of another. The upper part of the graft is called the scion and the lower part is called the stock or rootstock. Grafting is an efficient way of propagating fruit trees as it is quick and very limited numbers of the mother stock are required.

For grafting to be successful, several conditions must be met:

(1) the scion and the stock must be compatible, that is, they should be closely related;
(2) the grafting technique must be one that is successful with the species being grafted;
(3) the grafting operation must be performed well and at the right time of the year;
(4) the environment must be suitable to encourage the graft to heal.

Graft compatibility means that the scion and stock are capable of joining together properly. As a rule only closely related plants can be grafted together, for example, cultivars of the same species or species of the same genus. Accordingly, *Citrus sinensis* (sweet orange) can be grafted on to *Citrus limon* (lemon) because they are both in the same genus, but *Citrus sinensis* cannot be grafted on to *Malus pumila* (apple) since they are in different genera. In some cases, even closely related plants cannot be grafted together due to infection by a virus or other factor. When in doubt about whether a graft is likely to be compatible it is best to rely on previous experiments and published information because an unsuccessful graft may show up only after several years.

There are many grafting techniques used for fruit tree propagation. The most common ones are whip, approach, cleft, T-budding or inverted T-budding, chip budding and modified Forkert budding. Some species are more likely to graft by one technique than another and therefore the recommended grafting technique for a particular species should be used (see appendix). Most grafts are performed slightly before or during a period of active growth. When the plant is actively growing the bark is said to be 'slipping' and the bark and phloem layers can be easily separated from the xylem at the cambium layer. This is due to cell division in the cambium

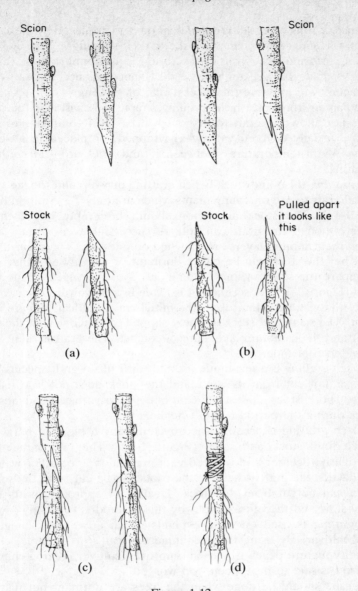

Figure 1-12
The steps in whip and tongue grafting:
(a) a long sloping cut 3−6cm long is made at the top of the stock and a matching cut is made at the base of the scion;
(b) a second downward cut is made in both the scion and the stock starting one-third from the tip of the first cut;
(c) the stock and the scion are joined with the tongues interlocked and cambium layers matching;
(d) the graft is then tied in place with sisal, polyethylene tape, or other water-retaining material

layer. Since healing of a graft is dependent on cells produced by the cambium, most rapid healing will occur when the bark is slipping.

Care in performing the grafting operation is also important as the neater the graft is made the greater will be the chance of success. As with most skills practice will improve the success rate of grafting.

Finally, an environment which encourages rapid cell division in the cambium but does not allow the graft to dry out will be most conducive to success. Grafts are usually covered with wax, wrapped, or placed in a very high humidity. Warm temperatures, between 13 and 32°C, are preferred but are not essential.

Whip grafting (also known as bench grafting or whip and tongue grafting) is a useful graft to use on young plants which are only 7–15mm in diameter (Figure 1-12). The stock and scion should preferably be of the same diameter although the graft can still be successful even if they are not, providing the cambium layers are matched on one side. The scion should be dormant but the stock can be either dormant or actively growing.

To prepare the stock, a long sloping cut, 3–6cm long, is made through the stem (Figure 1-12). A second cut is made in a downward direction about a third of the way from the top of the initial cut. Identical cuts are made in the scion. The scion and stock are next slipped together so that the tongues interlock and the cambium layers are in contact. The graft is then wrapped with a waterproof tape.

Approach grafting is a technique used for difficult-to-graft species because the scion is not cut from its roots and the stock does not lose its leaves. Therefore, both stock and scion continue photosynthesis and absorption processes during the period of graft healing.

Approach grafting is accomplished by removing a piece of bark 2–6cm long from both stock and scion (Figure 1-13). The two parts are then pressed tightly together and covered with grafting tape or other waterproof material. After the graft is healed, the roots can be cut from the scion and the shoot can be cut from the stock. Healing of approach grafts is often relatively slow and therefore output by this method is relatively low.

Cleft grafting is used on mature plants which are to be changed to a different cultivar. By using this technique one can, for example, change a poor-quality mature peach tree to an improved cultivar. A tree changed by this method is said to have been 'top worked'.

Cleft grafts should be done when the trees are dormant but just before new growth begins. Limbs from the stock plant are cut crosswise and then split vertically down the centre (Figure 1-14). Scions of one-year-old wood are cut so that they have long tapering wedges at the base. One is inserted at each side of the central split so that the cambium layers of stock and scion meet. The graft is then covered with a waterproof grafting compound to prevent drying. After healing, one of the two scions is gradually removed so that nutrients are channelled into the remaining scion. If both scions remain, a weak crotch will result and will be prone to break under heavy fruit loads.

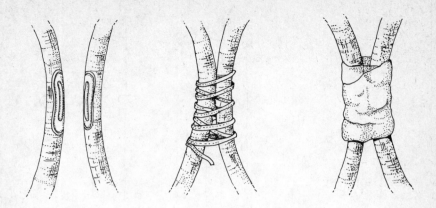

Figure 1-13
The steps in approach grafting:
(a) a thin layer of bark is shaved off to expose the cambium of both the stock and the
scion;
(b) the stock and scion are joined so that the cambiums match;
(c) the graft is tied and covered with grafting wax or other waterproof material;
(d) after the graft has healed the scion is separated from its roots by cutting just
below the graft

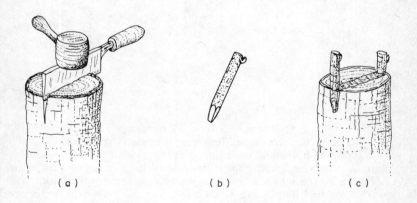

(a)   (b)   (c)

Figure 1-14
The steps in cleft grafting:
(a) the stock is prepared by splitting the stub of a cut branch to a depth of 5—8cm;
(b) the scion is prepared by cutting the base into a long, gradually tapering wedge
with the outside edge slightly thicker than the inside edge;
(c) the split in the stock is held open with a wedge and the scions inserted so that the
cambium layers match;
(d) the wedge is withdrawn and the entire union including the tips of the scions is
covered with grafting wax

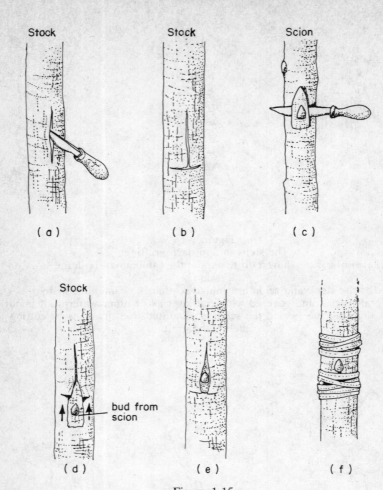

Figure 1-15
The steps in inverted T-budding:
(a) a vertical cut about 3 cm long is made in the stock;
(b) a 2 cm long horizontal cut is made at the base of the vertical cut;
(c) beginning 2 cm above a bud a slicing downward cut is made to 1 cm below the
bud; a horizontal cut is made at the base to remove the bud;
(d) the bud is inserted by pushing it upwards under the bark flaps on the stock;
(e) the two horizontal cuts are matched; and
(f) the bud union is tied with a waterproof material

Veneer grafting is a technique whereby a thin shallow slice about 5 cm
long is cut from the stock and a second cut at a 45 degree angle to the first is
made to remove the slice. The method is similar to that illustrated in Figure
1-16b except that the vertical cut is parallel with the stem and longer. The
scion is prepared with a similar long and short cut to match that made in the
stock and inserted so that the cambium layers match. The union is then

carefully wrapped. If the scion has leaves present, steps should be taken to reduce transpiration.

Budding is an adaptation of grafting where the scion consists of a single bud. Most budding techniques are easy and if done properly the chance of success is high. T-budding and inverted T-budding (Figure 1-15) are the most common techniques, however they can be done only when the bark is slipping. Chip budding is less commonly done, but dormant plants can be chip budded.

In *T-budding*, a vertical cut about 2 cm in length is made through the bark to the cambium, followed by a small crosswise cut at the top of the initial cut. In the humid tropics the inverted T-technique is used in which case the cross-cut is made at the bottom of the vertical cut (Figure 1-15). Since the bark is slipping, the flaps formed by the cuts can be readily separated from the underlying xylem. The bud (scion) is removed from the mother plant by making a cross-cut slightly above (T-bud) or below (inverted T-bud) and then, from a point just above or below the bud, slicing beneath the bud until the original cross-cut is reached. The resulting sliver of bark containing the bud is inserted under the bark flaps already prepared in the stock so that the straight sides on the top or bottom of both the scion and stock touch. The area is then wrapped with a waterproof tape but the bud is left showing. After the bud has healed, the top of the stock is cut off just above the bud, thus forcing the new bud to grow.

Chip budding (Figure 1-16) is more difficult than T-budding but is useful when dormant stock must be budded. A 45° angle downward cut is made in the scion just above a bud. An inward cut is then made just below the bud to meet the first cut and remove a 'chip' of wood containing the bud (Figure 1-16). Identical cuts are made in the stock and the 'chip' is inserted so that the cambium layers match. The bud area is then wrapped with the bud left exposed. After healing, the top of the stock is cut off to force the bud into growth.

Modified Forkert budding is a technique adapted from Forkert budding, which was developed in Malaya for the propagation of rubber plants. A horizontal cut is made at right angles to the stock and the bark peeled back to a length of about 2 cm. The manner of the cut is similar to that used to remove a bud in Figure 1-15c. After the bark has been peeled back, two-thirds of it is cut off leaving a stub. A bud is removed from the bud stick in the same manner as in Figure 1-15c, except that it is trimmed to fit as closely as possible into the wound made in the stock. After trimming to fit, the bud is put in place on the stock using the bark stub for support. It is then wrapped until the bud has healed.

Division   Division is a simple and reliable technique for the propagation of any plant with a clumping habit. Bananas and plantains are propagated solely by division. Division is accomplished by separating a clump into sections, each of which contains a growing point and a portion of the root

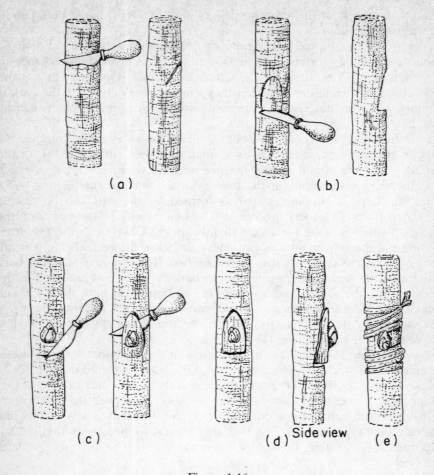

Figure 1-16
The steps in chip budding:
(a) a cut is made in the stock at a 45° angle going about a quarter of the way through the stock;
(b) a second cut is made going inward until it meets the first cut;
(c) the bud is removed from the budstick using identical cuts;
(d) the bud is inserted in the stock; and
(e) wrapped with sisal, polyethylene or other waterproof material

system (Figure 1-17). The sections are then replanted either in the nursery bed or directly into the field where they will soon re-establish themselves.

Micropropagation

Micropropagation is the use of very small plant parts to propagate plants using sterile conditions and controlled environments. Micropropagation techniques are often called *tissue culture* and have a number of advantages over more conventional propagation methods. Firstly, they allow for the rapid multiplication of plants in a very small space. For example, it is possible to produce thousands of plants from a single mother plant in only a few square metres of space within a six month period; whereas by using cuttings, many mother plants and hundreds of square metres of space are required to accomplish the same thing. Thus tissue culture is ideal for situations where a large number of plants are needed but where the availability of mother stock is the limiting factor. Secondly, tissue culture can be used to 'clean-up' and maintain disease-free clones, especially when combined with techniques such as heat treatment. Where an existing clone has been contaminated by viruses, these can be eliminated using tissue culture and heat treatment techniques, and new plants can be cultured from the plants in which the virus has been eliminated. The mother plants can then be maintained *in vitro* (under sterile conditions growing in a nutrient medium) in the tissue culture laboratory so that if at any time more virus-free plants are needed they can be quickly cultured from the mother culture. In many cases, tissue cultured plants are also more uniform and of better quality than those produced by other propagation methods. Tissue culture is also useful for the exchange of clonal material where phytosanitary regulations normally necessitate long quarantine periods.

Micropropagation is not, however, the panacea that some think. It also has disadvantages. Tissue culture requires a laboratory with expensive equipment and well trained technicians to operate it. Herbaceous plants are generally easiest to culture, and whereas the research has been done to know how to culture many species, the techniques will vary from species to species and research has to be carried out on each one. Tissue culture laboratories have to be operated carefully and efficiently because errors are multiplied many times due to the rapidity of reproduction in tissue culture systems. The contamination of cultures and mutation are constant concerns and quality control systems have to be utilized to ensure that a consistent, good quality product is produced.

Despite these advantages, many countries do have tissue culture laboratories and are finding them to be valuable additions to their capabilities in the area of plant propagation.

The techniques of tissue culture are complex and varied but all involve the use of a small part of the mother plant, which is removed under clean conditions, surface sterilized by immersion in a disinfectant solution, and then placed in a test tube or jar containing a sterile nutrient medium. The medium is commonly composed of macro- and micronutrients, vitamins, sugar, and hormones, along with agar to solidify the medium so that it will support the plant. The test tube is sealed and placed under artificial light (usually fluorescent) and allowed to grow. Periodically the mass of tissue is divided and placed on separate media and allowed to resume growth. This process is repeated until the desired number of plants is obtained. The type of growth which occurs is controlled by manipulating hormones in the

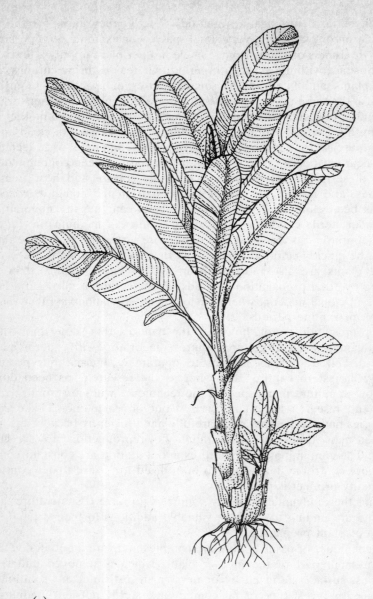

(a)

Figure 1-17
The steps in division of a banana:
(a) a sharp tool is used to cut the rhizome between the parent plant and the offshoot;
(b) the offshoot is then separated and replanted

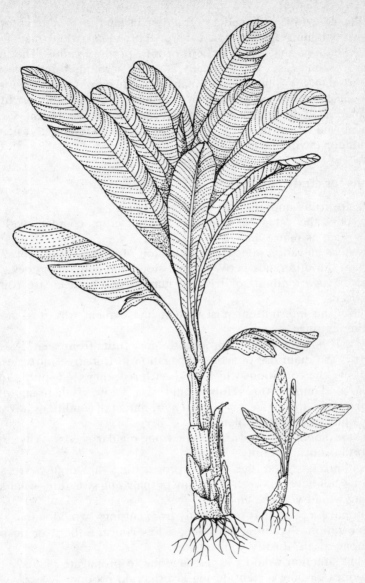

(b)

media. Plants grown on a media high in auxin tend to produce roots; high cytokinin containing media tend to produce shoots; while media with a balance of the two may produce both or may produce callus. The multiplication process increases geometrically and this accounts for the very rapid increase in plants. When sufficient plants have been obtained they must be gradually acclimatized to the harsher environment outside the test tube and then planted in sterile potting soil in a shaded, humid environment. Gradually they are moved out of the protected area until they are able to tolerate the harsh outdoor environment.

Questions for study and review

1. Why are fruits important in the average diet?
2. How does the value per unit area of fruit crops compare with staple crops such as maize and rice?
3. How does elevation affect the production of tropical fruit crops? Why?
4. Explain the difference between field stock and container stock.
5. Describe two methods of treating nursery soil to decrease soil-borne pathogens.
6. Describe an intermittent mist system and explain why it is an aid in rooting cuttings.
7. Why is it not advisable to propagate most fruits from seed?
8. What is the main use of seed propagation in fruit tree nurseries?
9. What are the conditions which are best for storing seeds in the tropics?
10. What is damping off? How would you know if damping off was occurring in the nursery? What environmental conditions favour the development of the problem?
11. How are apomictic seeds different from regular seeds? Why are they important horticulturally?
12. If you are assigned the job of propagating 400 orange trees from cuttings but have not been given any propagating structure, what kind of cutting would you use and why?
13. If you have a plant that roots easily from cuttings, would you use a high concentration of rooting hormone, a low concentration, or no rooting hormone at all? Justify your answer.
14. In what situation would you use layering to propagate plants?
15. Why does girdling a stem during layering aid rooting?
16. You have noticed that in your orchard one lemon tree has very large, juicy fruits and another has very strong roots. How could you obtain a new tree with both of these valuable characteristics? Explain in detail.
17. Matching the cambium is very important in grafting. Based on your knowledge of the function of cambium, explain why cambium matching is stressed.
18. Give three examples of viable uses for a tissue culture laboratory in your country.

CHAPTER 2

Establishing the Orchard

Since an orchard continues to fruit for many years, the planning associated with selection, layout, land preparation, and planting is important if the orchard is to produce at maximum capacity. Mistakes made during this phase will cause problems for many years to come.

Site selection

Each species of fruit has specialized environmental requirements which must be met for good growth and production. Of these, soil, topography, climate, water availability, and drainage must be considered when selecting an orchard site. In addition, labour availability, transport, and nearness to markets must be considered if a commercial orchard is to be profitable.

Soil

Orchard crops are deep-rooted and will therefore require deep soils. Before planting the orchard, holes 1–2 m in depth should be dug randomly throughout the orchard site to detect if plinthite (ironstone, laterite) or other impervious layers are present. Some indication of soil drainage can also be obtained from these holes. If plinthite layers are present or if water in the holes does not soak into the ground overnight, another location should be selected.

If deep holes reveal that the land is suitable, soil from the top 30–45 cm should be sampled and analysed for pH and levels of macronutrients present. This analysis can generally be arranged through the extension service. At the same time, samples of topsoil can be sent to a nematode laboratory for tests which will reveal the presence or absence of parasitic plant nematodes. The information from these tests should be used as a guide when deciding which species are likely to succeed. Nematode samples should be collected from moist soil and should contain some plant roots. Sampling for nematodes

33

is particularly necessary if vegetables or fruits have previously been grown on the site. As with all crops, it is unwise to replant a species of fruit in the same location where it was grown previously unless a rotation has been followed.

Topography

An orchard can be planted on all but the steepest terrain but flat or gently sloping sites are preferred. If the land is steep, erosion may be a problem and special steps such as box ridging or terracing will be required. In addition, orchard operations such as picking and spraying will be more difficult. When orchards are planted at high elevations where frosts occur, a gentle slope is preferred as this will decrease the possibility of any cold injury by allowing the cold air to flow downwards away from the plants.

Climate

The climate of an area will significantly affect the growth of all fruit crops. For this reason a careful assessment of such components of climate as minimum and maximum temperatures, rainfall levels and distribution and, to a lesser extent, wind should be made.

In the tropics, temperature is most affected by elevation. The higher the elevation, the cooler the climate. Thus a knowledge of the approximate elevation of the site will be necessary in deciding which fruit and nut crops are likely to succeed. In some areas, vastly different climates exist within a few miles of each other due to elevation changes. In Lebanon, for example, the hot climate of the coast favours the cultivation of such crops as bananas and citrus while in the mountains apples and peaches will grow. In many parts of West Africa, on the other hand, most fruits are grown in lowland areas and deciduous fruits are rarely cultivated.

Rainfall quantity and timing is another important climatic component to consider since it will determine whether irrigation is required to grow a particular crop and even if a particular crop will be economical. Mangoes, for example, do not fruit well when flowering occurs during the rains. In areas where rainfall and mango flowering coincide, mangoes would not be a profitable commercial crop. Where there is insufficient rainfall or where the rains are separated by a long dry period, irrigation is necessary for many fruit crops, particularly in the early stages of growth.

Wind is a component of climate which, when it is strong or frequent, is likely to damage plants or, at the least, increase the water consumption of the orchard. If wind is likely to be a problem, windbreaks of *Eucalyptus*, *Casuarina*, or other fast-growing trees are planted between the orchard boundary and the source of wind. The best way to assess the climate in a particular location is to look at fruit crops which are currently grown in the area and to talk with area farmers. If weather statistics are available these are most useful and should be utilized.

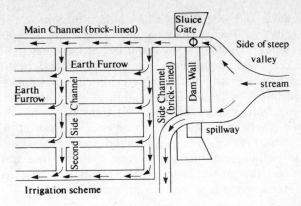

Figure 2-1
The design of a furrow irrigation scheme

Water availability

An orchard should be located near a source of water for irrigation (if needed), spraying, and washing of fruit prior to marketing. Irrigation systems for orchard use can range from complex techniques which require expensive pipes and a pump to relatively simple systems utilizing streams and changes in elevation. In general, furrow and basin systems of irrigation work well in the orchard and are relatively inexpensive.

Where the orchard is lower than a nearby stream, water can be diverted when needed to the orchard. Where this is not possible, a pump is used to direct the water to the highest point in the orchard from where it will flow by gravity through irrigation ditches to the rest of the orchard (Figure 2-1). Sprinkler irrigation can also be used but this is expensive, enhances disease problems, and will also increase weed control problems due to the inability to place water only in the root zone of the trees. Drip irrigation is an ideal system where the relatively expensive equipment required can be installed and maintained.

Drainage

Most orchard crops require well-drained soils. If water stands around the roots after a rain or if the water table is near the surface, the orchard will not succeed. In areas with a distinct rainy season, it is wise to observe the orchard site after a heavy rain. If drainage is poor, it can be improved by digging drainage ditches to carry off the surplus water but this adds to the cost of both establishment and maintenance of the orchard.

Transport and market location

The potential fruit-grower should remember that fruits are highly perishable and must be sent to market as soon as possible after harvest. Unlike staple crops such as maize and groundnuts which store for months, fruits will often spoil within a day or two of picking unless they are refrigerated. Since refrigeration is uneconomical in many countries, produce has to be transported quickly to market. Before the first fruit tree is planted, the location and extent of the market should have been determined. Questions such as: 'Will the fruit be consumed by the local population or will it be exported?' should be asked. If the fruit is to be sold locally, transport is less likely to be a problem than if it has to be sent to a large city or exported by air or sea. When road transport is required, roads should preferably be passable all the year round but definitely during the harvest period. Since transport is expensive, the closer the grower is located to his market, the greater will be his profit and the less spoilage loss during transport.

Labour availability

Orchard operations tend to be labour-intensive, particularly during harvest, and a supply of reliable labour is therefore a necessity.

Layout

Prior to ordering and planting trees, an orchard plan should be drawn on paper to show the location and cultivar name of each tree, irrigation fixtures (if applicable), trellises or tree supports, orchard roads and paths, packing shed and any other permanent features of the orchard. If the plan is drawn properly, the number of plants, stakes, and so on which will be needed can be determined and obtained before the start of planting.

If the slope of the land is steep, contour planting is desirable to minimize erosion. On flat or gently sloping land, trees are generally planted in one of three patterns: (i) square (ii) quincunx, or (iii) hedgerow. In a square planting (Figure 2-2a), trees are planted equidistant from each other at the spacing recommended for mature trees. Quincunx planting is the same as square planting except that an additional tree is placed in the centre of each square (Figure 2-2b).

Quincunx planting has the advantage of producing higher yields during the early years of production but the additional central tree must be cut down when the trees begin to form a mature canopy. A hedgerow orchard consists of trees planted closely together to form solid thick rows (Figure 2-2c). Hedgerow planting is best for dwarf deciduous trees and requires special pruning and training techniques. The primary advantages are high yield and low labour requirements per hectare.

There is now a growing trend in orchard planting to plant dwarf trees where available and to use a high plant population per hectare. The advantages of this are much higher early yields, and less expense in picking the fruit, and spraying and pruning the trees. To some extent this is offset by higher costs for both trees and labour, which together can be termed planting costs. In deciduous fruits, dwarfed trees are obtainable through the use of dwarfing rootstocks or through 'genetic dwarfs' — cultivars which are naturally dwarf. In other fruits dwarfing rootstocks can sometimes be used (for example, 'Flying Dragon' rootstock in citrus), but for many species standard sized trees must be used because of the unavailability of dwarfing rootstocks. Another trend in deciduous fruit production is the use of trellises to control the shape of the tree and to increase the penetration of light into the canopy, thus increasing yield. One of these systems called the tatura system employs an X-shaped trellis. Depending on the degree of sophistication of the farmer, the planting stock available, and the capital available for the establishment of the orchard, these systems should be investigated during the planning stages of the orchard.

Land preparation

Land preparation prior to orchard planting consists of clearing, terracing (if needed), installation of irrigation pipes or ditches (if needed), fencing, planting of windbreaks, and digging planting holes. The extent of the clearing operations necessary prior to planting an orchard depends on whether the land has been previously cropped. If the land has been recently cropped all that is necessary is to level any furrows or ridges, slash or mow down annual weeds, and dig out or kill perennial weeds and brush. If the orchard site has not been cultivated before all trees and shrubs should be ringbarked a year before the planned planting date and removed a year later. Ringbarking involves cutting through the phloem and cambium of the trunk leaving only the xylem. Ringbarking is required due to the presence of *Armillaria mellea* in many soils. This disease survives by attacking the roots of trees and assimilating the carbohydrates stored there. Ringbarking causes the trees to deplete the carbohydrates in their roots thus killing the *Armillaria* fungus. If the fungus survives, for example in the roots of trees which are simply 'chopped down' it will grow from the roots of the cleared trees to the new fruit trees, causing their death.

Since fruit trees are deep-rooted, it is impossible to prepare the soil to the full depth of their ultimate root penetration. In most cases, it is sufficient to prepare the planting holes well. Cultivating the soil of the whole of the orchard is not done except in cases where cover crops are to be grown and incorporated into the soil. Where the entire area is ploughed, a liberal application of superphosphate should be mixed with soil. Since young fruit trees do not have sufficient roots or large enough canopies to prevent erosion, a ploughed orchard can be expected to erode seriously. For this reason the ground between rows of trees should be grassed and mowed or slashed periodically. Only the area within 1−2m of the tree itself is kept completely free of vegetation, but this should be mulched.

After clearing, terracing or ridging should be done if warranted by a slope which will encourage erosion. On slopes of 10−15 per cent, well-maintained

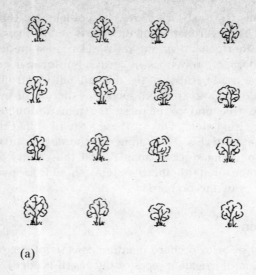

(a)

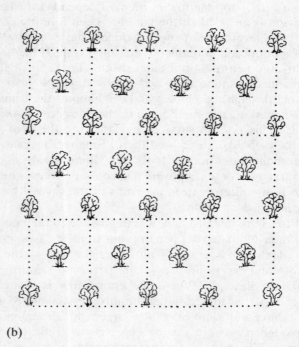

(b)

Figure 2-2
Orchard planting patterns:
(a) square, (b) quincunx, and on opposite page (c) hedgerow

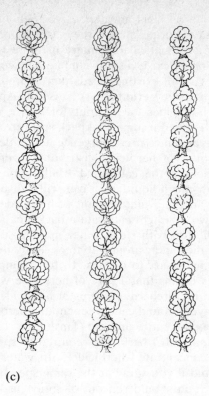

(c)

terraces are necessary to prevent soil erosion. Generally, combinations of ridging, terracing, and cover crops are necessary in the tropical orchard if severe erosion is to be prevented.

If irrigation is necessary, the installation of pipes or the digging of ditches is done prior to planting. Since gravity flow irrigation systems depend on proper grading and orchard layout, the irrigation system to be used must be considered in conjunction with layout planning and soil conservation practices such as terracing.

Fencing is an important feature of an orchard since young trees are likely to be eaten by cattle and goats if they are not protected. In addition, when trees begin to fruit, theft may become a problem. Fencing will help to alleviate both of these problems and, if properly constructed, can also serve as a support for passionfruit and other vines.

Weed control

Controlling weeds in the orchard is very important since research has shown repeatedly that weed competition around the base of the tree will seriously slow the growth of the tree. Traditionally, weed control in orchards has been done manually. However, there are now many excellent herbicides

available which can be used as an alternative to hoeing where labour is
unavailable or in the case of perennial weeds which are difficult to control.
In most cases, weeds are controlled a metre or more from the base of the
tree. Where weeds already exist, a spray containing a combination of a post-
emergence herbicide (to kill existing weeds) and a pre-emergence herbicide
(to prevent new weeds from germinating) is useful. Appendix Table 3 lists
some recommended herbicides for a variety of fruits.

The herbicides are applied using a backpack sprayer or, if it is available, a
tractor-mounted sprayer. Since pre-emergence herbicides are applied on the
basis of a given amount of herbicide distributed over a given area, it is
important that the sprayer be calibrated. Calibration allows the user to
determine exactly how much liquid a sprayer will deliver under actual spray
conditions and so to add the right amount of herbicide to the liquid.

An example of how one might calibrate a backpack sprayer is as follows:
Fill the sprayer to the top with clean water making sure that the pressure
chamber is filled, and then spray around 10 trees. Refill the sprayer, measuring
the amount of water necessary to do this. If, for example, it took 1 litre to
refill the sprayer, this means that 1 litre of herbicide will spray 10 trees. If
there are 100 trees in the orchard, 10 litres of herbicide will be required. If
an area of $1\,m^2$ is sprayed around each tree, enough herbicide to spray $100\,m^2$
will be needed to spray the entire orchard ($1\,m^2$/tree $\times$ 100 trees). Therefore,
simply prepare the amount of herbicide required to spray $100\,m^2$. This will
involve diluting the concentrated herbicide with water as directed. If you
calibrated correctly and if you spray at the same speed and pressure as you
did when calibrating, you should run out of spray just as you are finishing
the last tree.

Herbicide applications should be applied periodically when the regrowth
of weeds begins. When herbicides are used, the soil should generally not be
disturbed. Be sure that the herbicide chosen will control the weeds present
and that it is safe to use around the species of tree in the orchard, then apply
it according to the directions on the label.

Windbreaks

Wind can have a definite harmful effect on fruit production and, where it is
a serious problem, windbreaks must be used to decrease the wind velocity.
It is well known that windbreaks can result in a 45–60 per cent reduction in
wind velocity as well as reducing transpiration by as much as 65 per cent. At
the same time, windbreaks will shade and compete with fruit trees so they
must be used judiciously.

Ideally, windbreaks should be planted from two to three years before the
orchard is established so that they are sufficiently large to provide immediate
protection to the newly planted orchard trees. Where this is not possible,
protection of young trees with grass fences or even maize sown between the
rows is sometimes necessary. The choice of the windbreak species depends

on the location of the orchard. The species chosen should, however, be fast-growing, disease and pest free, tall and narrow, and strong and dense enough to block the wind. Species of *Eucalyptus* such as *E. globulus*, *E. globulus* 'compacta', and *E. rostrata*, are common windbreak trees: *Casuarina equisetifolia*, pitanga cherry (*Eugenia uniflora*), and Madras thorn (*Pilhecellobium dulce*) are also used.

Planting

Prior to planting, planting holes are staked out, and large holes 0.6−1.0m in both width and depth are dug; topsoil should be placed in one pile and subsoil in another. The hole should then be refilled with a mixture of 50 per cent topsoil and 50 per cent well-rotted manure, compost, or other decomposed organic matter. In soils where phosphorus is lacking, superphosphate should also be added as the hole is refilled. The hole should be allowed to settle for several weeks and then planting can begin.

Bareroot plants are light in weight and are easy to transport. However, they must be kept cool and the roots must be kept moist and protected from the sun. When the bareroot trees arrive, they will be wrapped in banana leaves, plastic, or some other moisture-conserving material. After they are unwrapped, the roots should be immersed in water for a few hours before being planted. If plants cannot be planted soon after delivery a trench can be dug in a shady location and the roots covered with moist soil.

In planting, some of the soil is removed from the previously prepared hole and a cone-shaped pile of soil is placed in the centre (Figure 2-3). At this stage, any posts or stakes should be inserted firmly into the planting hole; later insertion may result in damage to the root system. The root system is trimmed to remove damaged or diseased roots and is then spread in the hole. Roots should not be kinked or bent, but should be arranged as they would have grown naturally. The hole can then be refilled, taking care that air spaces do not remain around the roots by firming the soil at regular intervals during the operation. The tree should be planted at the same level as it was growing; a slight stain on the bark of the trunk usually shows where the soil line was in the nursery. The last step is to prune off half to a third of the top shoots to compensate for the loss of roots which occurred in the digging-up process. Failure to prune will result in weak growth during the first year and an increased chance of plant death.

Container-grown plants should be removed from their pots and any circling roots should be cut off (Figure 2-4). Soil is removed from the prepared holes and the plants are then set and covered to the same depth that they were growing in the container.

The most critical period in the life of the orchard is the first year after planting. During this time it is important that plants are not allowed to dry out and that they are kept free from insect and disease infestations. Newly planted trees are particularly prone to attack by termites (white ants), and it

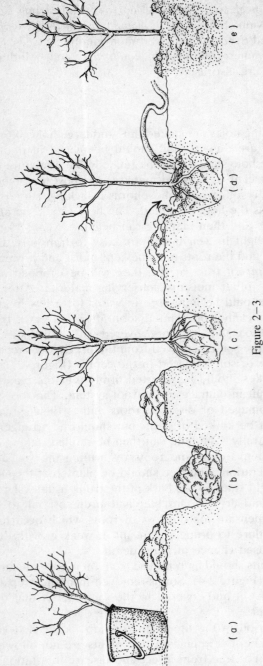

Figure 2–3

The steps in planting a bareroot tree: (a) soak the plant in a bucket of water; (b) form a cone of soil in the
bottom of a hole made large enough
to take the roots easily; (c) spread the roots evenly over the cone of soil; (d) add soil gradually firming as
added and water deeply; (e) form a
watering basin

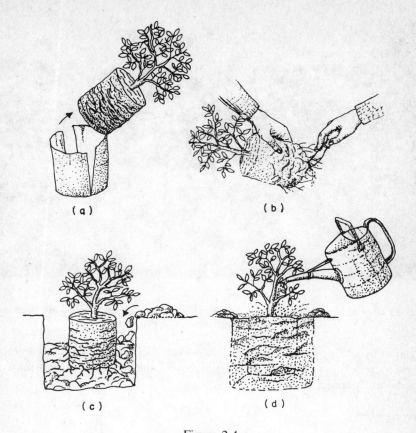

Figure 2-4
Steps in planting a container-grown plant:
(a) remove plant from container by slitting sides;
(b) if circling roots are present cut them in several places or loosen them;
(c) place plant in hole which is at least twice the diameter of the root ball at the same
depth at which it was growing in the container;
(d) refill with soil and water well

is usually wise to spray the base of the plant and the surrounding soil with a persistent insecticide such as dieldrin or chlordane to prevent damage. Mulching is a beneficial practice both to prevent weed growth and to conserve soil moisture. Mulches can be of any organic material including dried grass, banana leaves, chopped plant refuse, and so on. To be most effective, mulches should be at least 10cm thick and should be placed in an area 1–3m in diameter around the tree but not touching the trunk itself. Commonly a 2m wide strip centred on the tree row is mulched and the area between is grassed (Figure 2-5). This grassy strip is mowed or slashed so that it does not grow tall enough to impede orchard operations. Fertilizer is not generally required during the first season if the planting hole has been correctly

Figure 2-5
Orchard floor management. Notice the mulched trees and the grassed area between trees which minimizes soil erosion

prepared by adding organic material; however, the addition of small amounts of nitrogen fertilizer is beneficial to vigorous species such as passionfruit, banana and papaya.

Intercropping

When an orchard is newly established, there is a great deal of space wasted between the small trees. This is an ideal situation in which to intercrop annual crops between the rows of trees. If the orchard is irrigated, it becomes an ideal location for growing high-value crops which require irrigation, such as vegetables. Care should, however, be taken not to grow tall crops too close to the trees since they will then shade them. Likewise, no crop should be planted so close to the trees that they will seriously compete with them for essential nutrients, or that tree roots will be damaged during cultivation operations.

When trees become large, the grassed areas between rows can furnish grass for cattle-feed or for mulching. In addition, prunings from fruit trees can be used for firewood.

Questions for study and review

1. Discuss ways of checking for soil drainage in a prospective orchard site.
2. Why should a windbreak be used only when absolutely necessary?

3. Explain how distances to market affect profit in fruit growing.
4. What are the advantages and disadvantages of quincunx planting?
5. What is ringbarking? Why should trees be ringbarked instead of simply chopped down in a future orchard?
6. What can be done to prevent erosion in an orchard?
7. List several reasons why an orchard should be fenced.
8. What are the characteristics of a good windbreak plant?
9. How should the roots of bare-root trees be prepared before planting?
10. What are the possible consequences of failure to prune a bare-root tree at planting?
11. How can you decide how deep to plant trees in an orchard?
12. What can be done to prevent termites (white ants) from eating plants in the orchard?

CHAPTER 3

Citrus Fruit

Crop description

Citrus trees (Rutaceae family) are aromatic, broad-leaved, evergreen trees native to the subtropical regions of eastern Asia. Trees vary in size from the 3–5m tall lime up to 10m for grapefruit cultivars. Leaves are leathery, dark green on the upper side and light green on the under surface. The ovate leaves are dotted with glands and vary in having broadly-winged petioles in grapefruit, narrow-winged petioles in oranges and limes, and wingless petioles in most lemon cultivars (Figure 3-1). Flowers are perfect, regular, and have sepals, petals, and stamens in multiples of five. They may be borne either singly or in terminal or axillary cymes. The flowering period is concentrated during warm periods with regular rainfall or may continue throughout the year if temperatures are warm and irrigation is practised. The fruit is a berry with a leathery pericarp which has numerous oil sacs in its tissue. The primary species of cultivated citrus are the sweet orange (*Citrus sinensis*), the lemon (*C. limon*), the grapefruit (*C. paradisi*), the pummelo (*C. maxima*), the lime (*C. aurantifolia*), and the mandarin, also known as tangerine or naartchie (*C. reticulata*). Various hybrids include the tangor (tangerine × sweet orange) and the tangelo (tangerine × grapefruit).

Economic importance and distribution

Citrus are now the major fruit of subtropical regions. The main centres of production in the world are southern Africa, Israel, the United States, Brazil, Spain, Japan, Italy, and Mexico. Of these, the United States is the largest producer. Citrus can be produced economically in the tropics although fruit quality is sometimes inferior in colour to that produced in subtropical regions.

46

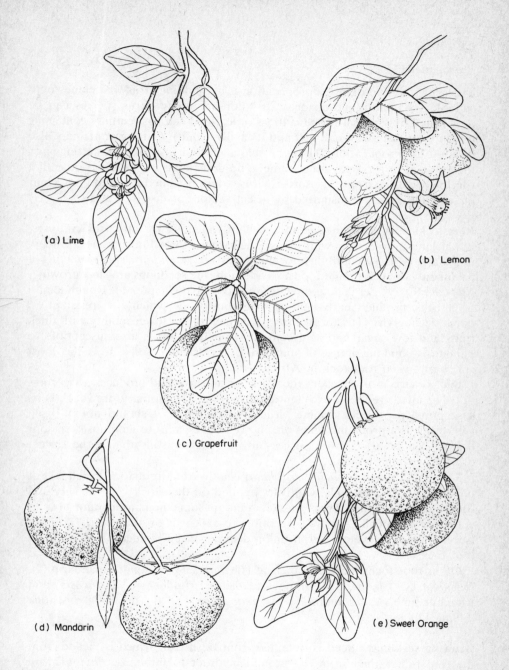

Figure 3-1
The foliage and fruit characteristics of (a) lime, (b) lemon, (c) grapefruit,
(d) mandarin, and (e) sweet orange

Culture

Propagation

The usual method of citrus propagation is by budding virus-free clones onto
suitable (compatible) rootstocks. In addition, many citrus species can be
propagated by semi-hardwood cuttings or leaf-bud cuttings (cuttings containing
a section of stem with one leaf and its axillary bud). Cutting grafts may also
be used for propagation. They are scions grafted to rootstock cuttings and
then placed in a propagating frame. The graft heals at the same time that
rooting occurs. Except for psorosis virus, transmissible diseases do not occur
in seedlings unless transmitted by insect vectors or diseased budwood.

Rootstocks Sweet orange is an excellent rootstock for all citrus cultivars,
producing vigorous trees with thin-skinned fruits of good quality. The primary
drawback is its susceptibility to gummosis (*Phytophthora* spp.), making it
useful only on well-drained soils. In addition, the seedlings are slow growing.
They are 70–90 per cent apomictic (see apomixis, Chapter 1). Rough lemon
rootstocks produce early-bearing, vigorous trees which are resistant to
quick decline virus (tristeza). Fruits are generally of lower quality with thick
rinds and low sugar and acid content. Rough lemon is also susceptible to
gummosis, making it useful only on freely-draining soils. It is the most
commonly used rootstock in Africa.

Sour orange is an excellent rootstock for all citrus and produces a vigorous,
deeply-rooted tree with excellent quality fruit. The primary drawback is its
susceptibility to quick decline virus, and therefore it should not be used
where this is a problem. Trees are, however, resistant to gummosis, so sour
orange can be used more successfully in heavier soils than either sweet
orange or rough lemon.

Trifoliate orange (*Poncirus trifoliata*) is a dwarfing rootstock which is best
on medium-textured soils. Cultivars budded on this stock yield heavily and
fruit quality is good. Trees are resistant to some nematodes and also to
gummosis although they are susceptible to exocortis or 'scaly bark' disease.
Eureka lemons develop a graft union disorder when budded on *P. trifoliata*
rootstocks.

Other rootstocks include grapefruit (*C. paradisi*), Cleopatra mandarin (*C.
reticulata* cv. Cleopatra), citranges (*Poncirus trifoliata* × *C. sinensis*) and
Rangpur lime (*C. aurantifolia* × *C. reticulata*), and the dwarfing rootstock
'Flying Dragon'.

Growing seedlings Seed is obtained from fresh fruit which is picked from
the tree rather than from the ground in order to minimize *Phytophthora*
infection. The easiest way to extract the seed is to cut shallowly through the
rind all the way around the fruit and then twist the fruit apart. The pulp can
then be squeezed through a sieve and the seeds washed. Seeds must not be

allowed to dry out or germination will be sharply reduced. After extraction, seeds should be placed in water at 47°C for 10 minutes to kill any spores of *Phytophthora citrophthora* and *P. parasitica* which may be adhering to the seedcoat. After treatment, the surface of the seeds should be rapidly dried to prevent drying of the inside, by spreading on absorbent paper and fanning. The seeds can then be treated with a seed fungicide such as thiram or captan and either planted or stored. To store citrus seeds, place them in sealed plastic bags and keep at temperatures between 6° and 10°C. Seeds stored under these conditions will survive up to six or eight months.

Seeds are planted in containers (Figure 3-2), or more commonly into seedbeds. The soil should be at least 30 cm deep, light in texture, uniform, and well-drained. Disease and nematode problems will be less severe if citrus have not been grown in the soil previously and if the soil is slightly acid. Soil sterilization is desirable when nematodes are present although, due to the death of some beneficial fungi (mycorrhizae) during the sterilization process, the seedlings may then be slower growing.

Citrus seeds are planted 3–5 cm deep and covered with sand or other porous material which will not compact. Germination is most rapid (10 days) when the soil temperature is between 26° and 32°C although seeds will still germinate, albeit more slowly, providing the temperature is above 12°C.

Before and after seedling emergence the soil should be given frequent light waterings to keep the surface moist. When the plants are 10 cm high, the frequency of watering should be decreased but quantity increased. In

Figure 3-2
Seedling citrus trees. Planting pockets should ideally be elevated above the ground to minimize contamination by pathogens and nematodes

many areas, the seedlings will require protection from the sun and wind until they are well established. Careful attention should be paid to controlling insects, mites, diseases, and weeds in the seedbed. When seedings are 20−60cm high (usually, within six months to a year depending on the species) they should be lifted and transplanted either to individual containers or to rows in the nursery. Seedlings are removed from the moist seedbed by using a spading fork, care being taken to damage the roots as little as possible. During this operation, plants should be carefully inspected and those which are smaller than average or more vigorous or have bent or badly twisted roots are discarded. Seedlings (now called liners) are planted at the same depth at which they were growing, in rows 1−1.5m apart with plants about 30cm apart in the row.

After the liners have reached pencil thickness but prior to their reaching 1.25cm in diameter, they are ready to be budded. Budding can occur at any time when the bark is slipping, and T-budding is most commonly used. In high rainfall areas, the inverted-T method is normally used so that the union will shed rain and be less susceptible to rot.

When selecting budwood it is important that the budwood source be free from virus or virus-like diseases. Budwood should be selected from the upper portion of vigorous, heavily-producing trees which are apparently disease-free. The best budwood is that taken from wood below the last flush of growth. Budwood is cut just prior to the actual budding process, and leaves are removed. Buds can usually be unwrapped in three to six weeks. Trees are then cut about half-way through above the bud and the tops bent over to the ground to force the bud into growth.

Climatic requirements

Climate is one of the most important factors affecting the profitability of a citrus orchard. Temperature, rainfall, and wind are the most important components of the climate affecting citrus production. Extremely hot or extremely cold temperatures are damaging to citrus. Frosts damage trees by injuring flowers, young leaves, and fruit.

In most parts of the tropics freezing temperatures are not likely to be a problem, but high temperatures may be detrimental at low elevations. High temperature injury occurs at temperature above 42°C and is most severe during flowering or if cool temperatures are followed by hot. Damage is in the form of flower and leaf drop. The mean temperature is also important since it will affect the types of citrus and even the cultivars which can be successfully grown in an area. For example, while lemons will succeed where mean temperatures are cool, grapefruits will produce poor-quality fruit in the same area.

Wind can cause serious damage to citrus trees and fruit. Hot, dry winds will often scorch trees by drying young leaves. Winds of high velocities will scar fruits and cause fruit drop. Where wind is a problem, windbreaks can be effective in reducing the velocity below damaging levels.

For good citrus production, trees will require rainfall or irrigation throughout the year. Where dry seasons occur, planning for irrigation is required. Water requirements vary according to climate and soils from as little as 45 cm to as much as 270 cm per year.

Soil requirements

Citrus trees will grow in a wide variety of soils but they grow best in soils of a medium texture and moderate depth, with good drainage and high fertility, and which are not high in soluble salts or too alkaline. A slightly acid pH is preferred by most citrus. If soils do not meet these criteria, citrus may still be grown although more care in cultural practices will be required.

Planting and spacing

The planting techniques for citrus are the same as those used for other orchard trees. Trees can be planted at any time of year provided that water from rainfall or irrigation is provided regularly. Trees may be planted bareroot, from containers, or 'balled and wrapped'. Balled and wrapped trees are lifted from the field with a ball of soil attached and wrapped immediately in hessian, plastic, or other material. Whatever the type of tree, extreme care must be taken to ensure that the roots do not dry out. Citrus are not as tolerant of root damage as many kinds of deciduous trees.

Spacing within the orchard depends largely on the scion and rootstock combination. As a general guideline, oranges, tangerines (naartchies), and grapefruit are spaced 7−9 m apart, and lemons 6−8 m apart. Limes can be planted 5 m apart. To increase early yield, trees can be spaced more closely, but this will result in mutual shading and subsequent yield reductions as trees mature. Pruning or removal of trees can partially alleviate this problem.

After planting, trunks should be protected from sunburn in most tropical areas. Maize stalks or grass tied around the trunk work well provided that they do not shade the upper part of the tree so that photosynthesis is decreased. Low-growing branches should be left on trees since the foliage protects the trunk from sun scorch. Leaves growing along the trunk should not be removed as these provide energy for the growth of a sturdy trunk.

Fertilizer requirements

Citrus trees respond vigorously to high levels of soil fertility although they are also highly sensitive to excess soluble salts in the soil; therefore fertilizer must be used with good judgement. Unfortunately fertilizer programmes vary widely throughout the world and standardization is achieved only through practical experience and leaf analysis. Leaf analysis is not, however, practical for most small growers.

Citrus trees generally benefit from nitrogen, phosphorus, potassium, iron, magnesium, manganese, and copper fertilization. Of these, nitrogen and perhaps potassium are the only nutrients likely to be needed routinely on adult trees. Table 3-1 can be used as a general guideline for trees up to eight years old. Nitrogen should be divided into three applications, applied either during or just before periods of active growth. Phosphate can be applied at any time but must be mixed with the soil and potassium should be applied in two applications. Animal manure applied to the surface or worked in lightly is also beneficial. Most fertilizers are sprinkled on the soil surface in an area one metre greater in diameter than the canopy of the tree and prior to rainfall or irrigation. Micronutrients or trace elements are applied only when deficiency symptoms exist and can be applied to the soil or as foliar applications (see Appendix).

Table 3—1
Suggested rates for fertilization of young citrus trees
(kg/tree/year)

Age	Nitrogen (elemental)	Phosphorus		Potassium	
		Super-phosphate	Double super-phosphate	KCl	K_2SO_4
1	0.06	0.20	0.075	0.20	0.25
2	0.12	0.40	0.150	0.20	0.25
3	0.18	0.60	0.250	0.20	0.25
4	0.25	0.80	0.350	0.25	0.30
5	0.34	1.00	0.450	0.50	0.60
6	0.42	1.25	0.500	0.75	1.00
7	0.50	1.50	0.600	1.00	1.25
8	0.59	2.00	0.800	1.50	2.00

Source: Adapted from *Farming Bulletin*, E.4/1978, by S. F. Plessis.

Diseases and pests

There are a number of insects which attack citrus but the severity of damage varies with location and predator populations. Among the most serious are scales, including but not limited to the Australian bug (*Icerya purchasi*), soft brown scale (*Coccus hesperidum*), soft green scale (*Coccus aethiopicus*), and wax scale (*Gascardia* spp.). Many species of mealybugs and fruitflies such as the Mediterranean fruitfly (*Ceratitus capitata*) and Natal fruitfly (*Pterandus rosa*), are damaging. Mites, especially citrus red mite (*Panonychus citri*), thrips (several species), aphids (especially black citrus aphid, *Toxoptera citricidus*, and brown citrus aphid, *Toxoptera aurantii*) can reduce yields. The false codling moth (*Cryptophlebia leucotreta*), fruit-piercing moths (*Achaea* spp., *Anomis* spp., *Calpe* spp., *Ophideres* spp., *Sphingomorpha* spp.), and citrus psyllid (*Trioza erytreae*) make up the remainder of the serious pests.

In small citrus plantings, most of the insects listed above do not cause serious damage due to the presence of predators which feed on them and thus provide control. In fact, most scales and mealybugs will not become a serious problem unless wide-spectrum pesticides such as parathion have been used. Insecticides should not therefore be used to control scales and mealybugs unless it has been determined that biological controls are not working. If this is the case, ants which 'cultivate' scales and mealybugs should be controlled by spraying the base of trees with a residual insecticide. The scales and mealybugs themselves can be sprayed with malathion, diazinon, dimethoate (not on rough lemon), and azinophos. The addition of a light oil to malathion sprays greatly increases control.

Fruitflies lay eggs just under the epidermis of the fruits and larvae tunnel into the fruit. Due to the oil present in citrus rinds, oviposition is seldom successful, but infection by fungi such as green mould (*Penicillium digitatum*) causes fruit decay. Control of fruitflies is by destruction of infected fruit and removal of host plants from around the orchard. Baiting within the orchard with sugar, treacle, protein hydrolysate, or dried fish solids containing malathion, mercaptothion or trichlorfon as toxicants is also effective. Baits are sprayed over half of each tree in large droplets, beginning when fruits are about half developed or at any time when flies or damage are noticed.

Mites produce small white specks on leaves and webbing when severe (Figure 3-3). Infestations are most likely to be high during dry, dusty weather. Control by the introduction of predator mite species is most effective; however, where this is not possible, control is achieved by spraying with light petroleum oil and acaricides such as aramite, tetradifon, dicofol,

Figure 3-3
A severe mite infestation
(Bruce Coleman photograph by Adrian Davies)

demeton, and dimethoate. Spraying should be used only when infestations are severe enough to cause economic damage.

Aphids are sucking insects which damage new growth and transmit tristeza virus. As with scale insects, ant control is the first step in controlling aphids. Spraying is used for control only when infestations are severe and then spot spraying is best. Endosulfan is recommended as it is least likely to affect predators although other insecticides such as methomyl, monocrotophos, dimethoate, azinophosmethyl, and demeton-S-methyl may also be used.

False codling moths lay eggs on the fruit and larvae burrow inward causing the decay and collapse of tissue. Control is difficult and expensive and consists of traps baited with virgin female moths which attract and trap males, and weekly removal of all infected fruits from and under trees.

Adult fruit-piercing moths feed on mature fruits of citrus, mangoes, and other fruit crops by inserting their mouth parts into the fruit. Secondary infection usually occurs and fruits rot and drop prematurely. Because the adults are nocturnal and are rarely seen and since larvae feed on different species from the adults, control is difficult. Damage can be minimized by picking fruit as soon as it is ripe, or even in a slightly underripe condition.

The citrus psyllid is a serious pest in the cooler areas of the tropics. Its feeding causes galls to develop on leaves and causes serious problems by transmitting citrus greening disease. In areas where citrus greening is not present, control measures are often not necessary. However, in greening infected areas, root drenches with dimethoate or sprays of dimethoate and a quarter to a half per cent light/medium mineral oil are used to kill psyllids before they can transmit the disease.

Over 15 species of nematodes attack citrus, the most serious of which are the burrowing nematode (*Radopholus similis*) and the citrus nematode (*Tylenchulus semipenetrans*). Control is difficult, so the orchard site should be tested prior to planting. Citrus should not be replanted where it has grown previously, and only nematode-free nursery stock should be used. Roots of infected nursery stock should be dipped in hot water at 45°C for 25 minutes prior to planting. Pre-planting soil fumigation decreases nematode populations, but damaging levels are likely to be reached again in four to five years. In heavily infected soils the resistant rootstock *Poncirus trifoliata* should be used. Nematicides such as aldicarb, DBCP, ethoprop, furadan, and phenamiphos can be soil incorporated and then irrigated for control when existing trees are infected. These materials are highly toxic and should be used with extreme care, following the recommended safety precautions.

Citrus are susceptible to many diseases. Perhaps the most serious are those caused by virus and virus-like organisms. These are transmitted by aphids, psyllids and other sucking insects, as well as by vegetative propagation from infected trees. Control is therefore by correct propagation practices and by controlling the insect vectors.

The tristeza virus complex causes two disease symptoms of economic importance. (a) Stem pitting is found primarily on grapefruit, limes and citrons. Symptoms are a gradual decline, poor vigour, and small fruit. External bark grooving and pitting of wood are common (Figure 3-4). (b) Quick decline occurs in conjunction with certain rootstock/scion combinations. Where the virus is present, sour orange should not be used as a rootstock. Select instead rough lemon, sweet orange, or Cleopatra mandarin as a rootstock. Symptoms are rapid yellowing and death of foliage accompanied by branch dieback and eventually death.

Psorosis or scaly bark disease can be identified by bark lesions which only develop six to ten years after infection. Early symptoms resemble pimples which gradually enlarge until the bark scales and falls off (Figure 3-5). If trees are not treated, yield is reduced and trees become worthless within 10 to 20 years after the appearance of symptoms. The only practical treatment is to cut out the lesions and scrape away all infected tissue. Since psorosis is transmitted almost entirely by propagation from infected trees, the disease can be largely avoided by careful propagation practices.

Citrus greening disease is a serious problem in middle altitude areas. It is transmitted by the citrus psyllid (Figure 3-6) and by propagation from infected trees. The mycoplasma-like causal organism is heat-sensitive and the citrus psyllid is not a serious problem in hot areas, so the disease is never severe in hot, lowland areas. Symptoms include blotchy mottling of the leaves which resembles zinc deficiency, abnormally upright twiggy trees, fruit drop, sour fruit and a failure of fruit to colour properly upon ripening. Control is difficult and expensive but the disease can be prevented by controlling psyllids through sprays and insecticidal soil drenches (see section on insects). Infected trees can be cured by pressure trunk injections of tetracycline hydrochloride, but this is expensive.

In soils which are poorly drained, forms of gummosis and root and crown rots may be a problem. Prevention is by growing in well-drained areas, not planting too deeply, and using sour or trifoliate orange rootstocks where drainage is inadequate. If the diseases do develop, lesions can be cut out and the infected area painted with a copper-containing paste.

Integrated pest management

A useful technique to keep in mind when dealing with disease and pest problems is integrated pest management. This technique involves controlling pests and diseases as a unit, keeping in mind the interactions between environment, natural predators, and pesticides. When this system is used, the grower watches his crop carefully for pest problems and allows natural controls to work when possible. Spraying is reserved for situations where natural controls are not working and the pest population has reached economically damaging levels. When spraying is used, those pesticides which are least likely to kill beneficial insect species are used. For example, if aphid populations become severe enough to require spraying, endosulfan may be used rather than parathion since endosulfan is less toxic to beneficial insects.

Figure 3-4
Stem pitting caused by Tristeza virus

Figure 3-5
Porous or scaly bark disease

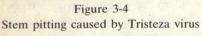

(Reproduced by kind permission of Jürgen Kranz, Heinz Schmutterer and Werner Koch from *Diseases, Pests and Weeds in Tropical Crops*)

Figure 3-6
Larvae and adult citrus psyllids
(NHPA photograph by Anthony Bannister)

Weed control

Weeds should be controlled in an area 2 m in diameter around the base of each tree. Weed control can be accomplished through cultivation or with herbicides (see Appendix). After the weeds have been killed, a thick mulch will prevent new weed growth and will also conserve moisture. The area between rows is best grassed and mowed so that erosion is minimized.

Pruning

Citrus require little pruning and, in fact, pruning will reduce yield. Early tipping of main branches may, however, be required to develop a balanced tree form. In later stages of growth, only branches which touch the ground should be removed, otherwise lower branches should be allowed to remain to protect the trunk from sunscald. Other pruning consists of removing diseased or crossed branches and removing vigorous suckers which occasionally grow up through the centre of the tree or from below the graft union.

Harvesting and handling

Maturity of citrus fruit is indicated by colour changes in the rind. In cool tropical areas good colour will develop. However, in the lowland tropics fruit may remain green at maturity. Though the lack of colouring decreases the attractiveness of the fruit, flavour is unaffected. If degreening is desired, fruit can be treated with ethylene after harvest to destroy the chlorophyll in the rind and allow the underlying colour to show.

Citrus fruits are harvested by clipping or pulling fruit from the tree. Mandarins, lemons, and limes are clipped while others are pulled with a slight twisting motion. Fruit should be handled carefully to prevent bruising and pickers should wear gloves to minimize damage to the rind.

After picking, fruit should be graded, washed, and kept as cool as possible until consumed or sold. Damaged fruit should be discarded or used immediately. Ideal storage temperature is between 4 and 6°C.

Cultivars

Sweet orange (Citrus sinensis)

The sweet orange is the most important citrus fruit, both for eating fresh and for making juice. There are three main types: (a) cultivars with normal fruits; (b) cultivars with red flesh (called blood oranges); and (c) navel oranges with a second row of small carpels at the fruit apex and no seeds. The most widely grown types are the Valencias which require a hot climate and the Navels which do best at higher elevations in the tropics.

Lemon *(Citrus limon)*

Lemons are used primarily for processing into lemonade and squash. They are widely grown in the tropics but grow best at moderate elevations. The most common cultivars are 'Rough Lemon', 'Eureka', 'Lisbon', 'Villa Franca', and 'Meyer'. The cultivar 'Meyer' produces a more thick-skinned, sweeter fruit than the other cultivars. There is some indication that it is also better adapted to low elevations than many other cultivars.

Lime *(Citrus aurantifolia)*

Limes are commonly cultivated in the tropics both for their juice and as a flavouring for other foods. Like the lemon, they are not generally eaten fresh. They are the only truly tropical citrus and are well-suited to low elevations. Trees are small and bushy and can therefore be spaced as close as 5m × 5m.

There are two groups of limes: (a) diploid limes called Key limes, Mexican limes or West Indian limes, and (b) triploid limes including 'Tahiti', 'Persian', and 'Bearss'. Diploid limes are small, seeded, thin-skinned and juicy while triploid limes are the size of lemons, seedless and have a slightly thicker rind. Triploid limes grow better at high elevations than do diploid limes. Limes are especially perishable so they must be handled with care. They will store for four weeks at 4°C.

Grapefruit *(Citrus paradisi)*

Like limes, grapefruits are adapted to low elevations where the climate is hot. They will not have good quality where the weather is cool. Grapefruit trees are larger than most other citrus so they should be planted 8–10m apart. Important cultivars are 'Duncan' (seedy), 'Marsh Seedless', 'Thompson' (seedless, pink flesh), 'Foster' (seedy, pink flesh), and 'Hohn Garner' (similar to 'Duncan' but with only five seeds).

Pummelo *(Citrus grandis)*

Pummelos or shaddocks are grapefruit-like fruits which are important primarily in South-east Asia although they are also grown to some degree in other areas. They are a variable group ranging from fruits which are practically inedible to those whose quality rivals the grapefruit. Fruits are large, thick skinned, and have firmer flesh and less juice than the grapefruit. Most cultivars are yellow although some have a pink to red flesh. Pummelos are less cold tolerant than grapefruit and like grapefruit tolerate high temperatures well. Trees tend to be large (5–10m) though size varies from clone to clone. Culture is similar to the grapefruit.

Mandarin (*Citrus reticulata*)

The citrus grouped under the general heading of mandarins include tangerines or naarchis. Basically, any orange-like fruit with a loose skin is included here. These fruits are very popular for eating fresh due to their attractive flavour and easy peeling. In addition, they are higher in vitamin A than other citrus.

Trees are planted 7–8m apart as they are fairly small. Unlike other citrus, cross-pollination will increase the yield although the fruits will also have more seeds. Fruits spoil quickly and will not retain their quality if left on the tree once they are ripe. Overripe mandarins may rot, become dry, or lose flavour. If they are plucked from the tree the rind is damaged, so fruit must be harvested by clipping and carefully packed. Mandarins are likely to be damaged by high temperatures, so are best grown at moderate elevations.

Questions for study and review

1. Under what situation would you use each of the following rootstocks: rough lemon; sweet orange; sour orange; trifoliate orange?
2. You have received a shipment of rough lemons and you wish to grow the seeds for rootstocks. Describe how you will extract, treat, and store the seeds for a few months until you can plant them.
3. What is a liner?
4. You have moved to an area with very hot temperatures and have been assigned to start a citrus orchard. Which species of citrus will you plant?
5. Describe the techniques used for fertilizing citrus.
6. How would you treat an infestation of scale in a citrus orchard?
7. What is degreening and why is it sometimes needed on citrus grown in the tropics?
8. Why are navel oranges the most popular oranges for fresh eating?
9. What is the recommended method of weed control in citrus?
10. Why should branches which originate below a graft union be removed?
11. What is a blood orange?
12. Which type of lime would you plant at higher elevations?
13. How can a pummelo be distinguished from a grapefruit?

CHAPTER 4

Major Tropical Fruits and Nuts

The major tropical fruits are those economically important species which are native to the tropics and are therefore adapted to tropical climates. Unlike the deciduous fruits discussed in Chapter 6, the culture of tropical fruits is not confined to a relatively few climatically suitable sites, but is spread over most of the tropics.

Avocado *(Persea americana)*

The avocado (Figure 4-1) is native to Central and South America and a member of the Lauraceae family. It is an evergreen tree which grows up to 20 m tall and can be equally wide. Leaves are simple, ovate, and spirally arranged. Flowers are perfect and are borne in axillary panicles. The fruit is generally pear-shaped with a large, round to egg-shaped central seed. The flesh is buttery in texture, contains a high percentage of oil and protein, and exceeds even the banana in calorific value.

Avocados are divided into three ecological races:

(1) The West Indian race is native to the lowland areas of Central America and produces large fruits (1–2 kg) with a relatively low oil content. Trees are very frost-sensitive.

(2) The Mexican race is native to high-elevation areas in Mexico. The fruit is small with a thin skin and large seed. The flesh is high in oil content and the leaves are anise-scented. Though of little commercial importance in itself, the Mexican race is the most cold-tolerant and has been crossed with other races to obtain cold-tolerant hybrids.

(3) The Guatemalan race is native to the highlands of Guatemala and produces large fruits weighing up to 1 kg. The fruit has a brittle, hard skin and a small seed. The oil content and cold tolerance are intermediate between those of the other two races.

Hybrids between the races are common and are commercially important.

60

Figure 4-1
Avocado (*Persea americana*)

Economic importance and distribution

The avocado is becoming an increasingly important crop in many tropical and subtropical areas of the world. According to 1976 statistics, Mexico is the largest producer, followed by the United States and Brazil.

Culture

Propagation For the small-scale production of avocados, seedling trees are often grown. Alternatively, desired cultivars are sometimes grafted onto seedling rootstocks; however, for commercial production, they should be grafted onto rootstocks tolerant to avocado root rot (*Phytophthora cinnamoni*), which is the most serious disease of avocados. Trees grown on susceptible rootstocks will eventually succumb to the disease in most locations, limiting the productive life of the orchard. At present, the following tolerant rootstocks are being used in many areas: Duke-7, G-22, and G-755.

Since the tolerant rootstocks do not come true from seed, and since they are very difficult to root, the recommended method for propagation is the nurse-graft technique. Basically scions of the tolerant cultivar are grafted onto seedling rootstocks (nurse roots) and are encouraged to root while a scion from a desired fruiting cultivar is grafted to it. Eventually the nurse root dies off leaving the tree rooted onto the tolerant stock. The procedure is as follows:

(1) Avocado seeds are obtained from healthy fruits picked from the tree rather than from the ground.
(2) Seeds should be soaked in hot water at 40−52°C for 30 minutes as a treatment against the seed-borne fungal disease *Phytophthora cinnamoni*.
(3) The brown husk around the seed should be removed (if present) and a thin slice cut from the pointed end.
(4) Seeds are planted with the cut end slightly above the soil surface and germination occurs about four weeks later.
(5) As soon as the seedlings are large enough to handle, a scion from the tolerant rootstock cultivar is grafted onto the seedling using splice, cleft (saddle or wedge) or T-buds.
(6) The graft is carefully wrapped and placed in a warm, partly shaded environment.
(7) After the graft has healed, which takes from four to six weeks, the plant is placed in a completely dark room for about a month, which allows the scion stem to elongate without chlorophyll (etiolate). The etiolated stem will root easier than stems containing chlorophyll.
(8) The plants are then moved to the light, an incision is made just above the graft union and an IBA-containing rooting hormone applied. At this time a wire can be tightly twisted around the stem just above the graft union. The wire will girdle the stem and eliminate the need to cut off the seedling rootstock.
(9) The planting pocket is filled with a rooting media and kept warm and moist until rooting occurs.
(10) After rooting occurs, if a girdling wire has not been used, the nurse graft is cut off and the plant is left with the tolerant rootstock.
(11) During or after the rooting process, the desired scion cultivar is grafted onto the tolerant rootstock, using the techniques mentioned above.

If this process is utilized, and if grafting is done early, a marketable tree can be produced in about nine months.

Cultivars Though the seedling trees are important for local consumption, export avocados must be uniform and of good quality. Thus a number of cultivars are grown, including 'Booth 7', 'Booth 8', 'Choquette', 'Collinson', 'Hall', 'Lula', 'Mayapana', 'Nabal', 'Taylor', 'Tonnage', and 'Walden' in lowland areas and 'Ettinger', 'Fuerte', 'Hass', 'Nabal', and 'Puebla' at higher elevations. In general, cultivars of the West Indian race are most likely to perform well in hot, humid areas while Mexican and Guatemalan cultivars are better adapted to high elevations.

Environment and soils Avocados grow best in areas with warm, frost-free climates although the Mexican race will tolerate some chilling without damage if the trees are not flowering. Flowers of all races are sensitive to low temperatures and a frost during blooming may result in substantial crop loss.

Avocados are tolerant of a wide range of soil types and conditions provided that they are well drained and not excessively saline. Optimum growth will occur in fertile, well-drained soils. Under no circumstances should avocado culture be attempted where drainage is poor since trees planted in badly drained soils are likely to succumb to avocado root rot within two years of planting.

Fertilizers Avocados respond well to fertilizer applications, especially to phosphorus in the early stages of growth, and to nitrogen and potassium when older. The ability of trees to absorb nutrients is affected by soil pH and by conditions which affect the health of the roots. The ideal pH for avocados is 5.5–6.5. Waterlogged conditions and cultivation which destroys surface roots are especially detrimental to nutrient uptake.

The ideal way to determine the optimum amount of fertilizer to use is by analysis of leaf samples in a laboratory to determine the quantities of fertilizer elements contained in the leaves. Where this is not possible, phosphorus should be added at tree planting and nitrogen added once or twice yearly at rates approximating 1.5–1.7 kg of elemental nitrogen per tree per year. Potassium, though important, is less likely to be deficient in tropical soils than nitrogen.

Of the micronutrients, zinc is most likely to be deficient in acid soils. Zinc deficiency causes leaf mottle and the production of exceptionally small leaves on terminal growth. This deficiency can be corrected by using zinc sulphate sprays of 1.5–2.5 kg per 400 l of water (plus a surfactant and hydrated lime to neutralize the solution) applied once or twice yearly whenever there is an abundance of mature foliage. Iron deficiency is occasionally a problem and can be corrected with soil applications of iron in a form called chelated iron which prevents it from changing to an unusable form in the soil. Iron sulphate can also be used but is less effective.

Pests and diseases A number of insects attack the avocado but in most countries they do not present a serious problem. The most serious problem

worldwide in avocado production is root rot caused by *Phytopthora cinna-*
momi. This fungal disease should preferably be prevented by planting avocados
only in well-drained soils, as the disease becomes serious only when roots
are growing in water-saturated soil and by using resistant rootstocks. If the
disease does develop, ethazol will provide some control if applied before the
disease is advanced, but this treatment is expensive. Either metalaxyl or
fosetyl-aluminium are even more effective. The disease can also be retarded
by heavy mulching.

In West Africa and Zaire, Cercospora spot or blotch caused by *Cercospora*
purpurea is a serious problem. This disease causes angular sunken spots on
leaves and stems and cracked, brown spots on fruits. These spots serve as
infection sites for anthracnose. The disease is most severe under warm moist
conditions and all avocado cultivars are susceptible. Control is by sprays of
copper or benomyl applied at 10-day intervals during rainy periods or at
28-day intervals during dry weather throughout the periods when fruit is on
the tree.

Stem end rots caused by the fungi *Diplodia natalensis*, *Phomopsis* spp.,
and *Dothiorella* spp. are serious in hot regions and cause loss of fruit after
harvest. Control is through refrigeration of fruit immediately after harvest
since these rots are serious only when temperatures approach 30°C.

The crop cycle

Avocado plants are established 8×5m apart in the field, despite their
eventually much larger size, in order to obtain maximum yields when the
orchard is young. As the plants grow and become crowded it is necessary to
thin out some of the trees of most cultivars. Close planting of orchards is
particularly beneficial with cultivars which produce fruit at an early age.

Grafted or budded avocado trees usually begin to bear at three to five
years of age while seedlings often require five to seven years. Though
avocados are sometimes self-pollinated, the largest crops are produced when
another cultivar is provided for pollination. This is due to the possible non-
receptivity of the stigmas at the time the flowers are releasing pollen. Based
on when pollen is released, avocados are divided into type A pollenizers
and type B pollenizers. Both should be present if economical crops are to be
produced.

Fruits mature from nine to 18 months after flowering, depending on
cultivar and climate and will store well on the tree. Unlike most fruits,
avocados do not normally ripen on the tree although ripening will proceed
very rapidly after natural drop or after picking. Removal of the fruit from
the tree causes ethylene production which, in turn, triggers the climacteric
rise in respiration which corresponds with ripening.

Determining the time of maturity for picking is sometimes difficult as
there are few or no external signs of maturity. Probably the best way to
determine maturity is to pick a fruit which appears to have reached full size

and allow it to ripen. If the ripe fruit is of good quality, then fruits of similar size can be cut. If immature fruits are cut they will shrivel before ripening and will not attain a saleable quality. Oil and water content are other indices of maturity since the percentage of oil in the fruit increases with increasing maturity, and water decreases.

Harvesting and handling

Mature avocados should be harvested by cutting from the tree rather than by picking as picking damages the fruit and provides a site for decay organisms to enter. After cutting, fruits should be handled carefully since they are very sensitive to bruising. Cold storage will increase storage life but it may also result in chilling injury and loss of flavour. Unless definite information is available on the storage behaviour of a cultivar, it is best not to store avocados below 5.5°C. For export, fruit should be maintained below 7°C but above 5.5°C. Depending on the stage of maturity (the longer the fruit remains on the tree, the faster it will ripen after picking) and storage conditions, fruit will store for between two and six weeks after picking. Wrapping fruits in cellophane plastic increases storage life.

Yield of avocados under intensive, modern cultivation ranges from 28−48 kg/tree. A good yield of 'Fuerte' or 'Bacon' cultivar would be 550−1100 kg/ha; 7700−13,200 kg/ha would represent an appreciable yield from 'Hass' or 'Zutano'.

Banana (*Musa* spp.)

The banana (Figure 4-2) is a member of the Musaceae family and native to South-east Asia. The plant is herbaceous and fast-growing and produces short-lived stalks (pseudo-stems) which arise from underground, corm-like rhizomes. They vary in height from 1.5−8 m and are generally divided into the starchy type called plantains and the dessert type known as bananas. Leaves are oblong with a rounded tip, reflexed and are often shredded along the veins by the wind. The pseudostems, which are actually leaf petioles, are tightly wrapped around each other. The shoot thus formed produces a terminal inflorescence and dies gradually after fruiting. The hanging inflorescence is elongated, producing female flowers in nodal clusters (hands) above and male flowers at the tip. Although the inflorescence is positively geotropic (grows towards the ground), the individual fruits (fingers) are negatively geotropic (grow away from the ground) and point upwards along the stalk. Fruits generally form parthenocarpically (without pollination) and are therefore seedless.

Economic importance and distribution

Bananas are the most important fruit in world commerce and are probably the best-known tropical fruit. Africa is the major producer of bananas and

Figure 4-2
Banana (*Musa* spp.)
(Photograph by Rex Parry)

plantains for local consumption, followed by India, Malaysia, Taiwan and the Philippines. Honduras, Costa Rica, and Panama are leading producers for export.

Culture

Propagation   Bananas are propagated vegetatively, usually by lifting and replanting a sucker and its attached rhizome. Several types of planting material may be used, including:

(1) Peepers — young, above-ground suckers which have not yet formed leaves.
(2) Sword suckers — suckers which bear narrow, sword-like leaves and are planted when they have reached a height of up to 75 cm.
(3) Water suckers — small suckers bearing broad leaves.
(4) Maiden suckers — large suckers which have broad leaves but which have not yet flowered. After digging, they are cut back to 15 cm above the corm and the central meristem is destroyed. They are usually planted on their sides about 30 cm deep.
(5) Sections of large rhizome — after a shoot has fruited, the top is removed and the rhizome is then dug up and divided into sections with each section containing a bud. Each section usually weighs 2 kg or more and is planted about 30 cm deep with the eyes pointing downwards.

The type of propagating material used does not generally affect yield; however, the time required from planting to fruiting will vary. To obtain a crop in which all of the plants fruit at the same time, uniform planting material should be used.

Cultivars   There are a number of important cultivars grown, although the majority of bananas grown in villages are of unidentified origin. 'Gros Michel' is generally considered to produce the best quality bananas and is grown to some extent in West Africa, Central America and South-East Asia where it is called 'Pisang Ambon'. This cultivar is tall and is therefore susceptible to wind damage but its primary disadvantage is its susceptibility to Panama disease *Fusarium cubense*.

Bananas of the Cavendish subgroup are by far the most important, particularly 'Dwarf Cavendish' (Basrai in India). The Cavendish bananas are of good quality but they do not transport well unless the hands are cut from the stalk and packed in cardboard cartons. The primary advantages of the 'Dwarf Cavendish' type are short height which makes it resistant to wind damage, and resistance to Panama disease. Other Cavendish cultivars, such as 'Williams' produce larger fruits than 'Dwarf Cavendish' on a tall plant and are adapted to cooler regions. 'Robusta', also called 'Poyo', is a tall Cavendish type and is now one of the leading cultivars.

Environment and soils   Bananas will grow in most regions of the tropics and subtropics where temperatures are between 5°C and 42°C, although the best production occurs where rainfall is abundant and is fairly evenly distributed throughout the year. In areas with either a pronounced dry or cool season, yields are reduced and production tends to be seasonal.

Bananas will grow in a wide range of soil types, provided that drainage is moderate to good. Optimum production will occur in soils which are high in organic matter, retain moisture well and contain abundant amounts of nitrogen and potassium. A pH between 6.0 and 7.5 is normally required.

Fertilizers   Bananas are heavy feeders and yields can be increased dramatically by fertilization. Phosphorus and potassium should be applied in the planting hole and thereafter potassium is applied once or twice yearly as surface dressings. Nitrogen should be applied in small amounts at frequent intervals throughout the year. The addition of manure in the planting holes and as a side dressing is beneficial nutritionally and also contributes to improving the water-holding capacity of the soil. The practice of chopping banana residues into small pieces and spreading them in a thin layer around the plants is also beneficial since this organic material is high in potassium as well as acting as a mulch which conserves soil moisture.

Diseases and pests   Bananas are subject to a number of insect pests and diseases, the most serious of which is Panama disease, caused by *Fusarium* *cubense*. This is a soil-borne fungus which attacks the rhizome first, causing a characteristic purple colouring. The leaves then turn yellow and die. There is no economical method of control, although 2% carbendazim injected into the rhizome has been effective in India, so infested soil should be planted with resistant cultivars such as those of the Cavendish subgroup.

Leaf spot or Sigatoka disease caused by *Mycosphaerella musicola* is the most serious disease of Cavendish bananas. It is most severe in humid, high-rainfall areas. The symptoms are yellow spots on the leaves which later expand and merge together with dead centres which are parallel to the lateral veins. Eventually, the leaf margins die and photosynthetic area is lost. Premature ripening of fruit may also occur. Control is through applications of 15−20l per hectare of pure mineral oil applied with knapsack mist sprayers. Fungicides such as copper compounds and maneb can be mixed with the oil. Oil applications are effective but they may also be phytotoxic if applied at too heavy a rate or during hot, dry weather.

Bunchy top virus is a problem in certain areas of Central Africa. This disease causes the production of brittle leaves and short petioles and eventually kills the plant. Control is through sanitation, propagation by virus-free material, and elimination of the banana aphid which is the vector.

The most serious insect pest of bananas is the banana root borer *Cosmopolites sordidus*. The adult is a dark brown to black weevil about 1 cm in length. Eggs are laid in damaged or rotting stems or in holes made by the adult in living stems. The larvae tunnel into the rhizomes, causing withering of the newest leaves and weakening the stem. Control is by application of the following combination of sanitation methods and by trapping. Infected stems should be cut as near as possible to the ground and the remains chopped into small pieces so that they will decay rapidly and thus will not be suitable for egg laying. Trapping is carried out by laying 30 cm sections of pseudostems or corms on the soil near the base of the plants, preferably in a shady area. The sections should have been split in half lengthwise and placed with the cut side facing down. Adult weevils will gather beneath the stem to feed and to lay eggs and can then be collected every 48 hours. If trapping is not effective, treatments with a residual soil insecticide may be necessary.

Nematodes can be a serious problem, particularly the root knot (*Meloidogyne* spp.) and the burrowing nematode (*Rhodolphus* spp.). Heavy infections cause yellowing of leaves, depressed yields, and poor anchorage of plants. Control is by rotation, pre-plant fumigation, and by treatment of propagating material by placing the bases in hot (62−65°C) water for 10 minutes, or dipping in non-phytotoxic nematocides.

It is a common practice to cover banana bunches with a thin, blue, open-ended plastic bag to protect the fruit in the field.

The crop cycle

Preparation of land to be planted with bananas should include the removal of existing vegetation, terracing if the land is steep, or box ridging if the land is reasonably level. Since bananas do not have an extensive, soil-binding root system, the prevention of soil erosion is a major concern. Planting stations should be 2–3m apart or double rows can be made with planting stations 1.5m apart and 2.5m between the pairs of double rows. Planting holes should be dug to a depth of 60–90cm, and the soil should be amended with manure or other decomposed organic matter.

After planting, the bananas will begin to form suckers from the base. The removal of surplus suckers is an important operation which is often neglected. Ideally, a clump of bananas should have only three shoots at any one time: one which is fruiting, one which is flowering, and one which has not yet flowered. After the fruit has been harvested, the shoot which bore the fruit should be removed and another sucker allowed to develop. If this system is followed, a continuous cycle of fruit production will be maintained and bunch size will be large. If a plant is allowed to develop into a large clump without thinning, yield will be seriously reduced.

After five years, the yield from a normal banana planting will often begin to decline. Replanting of nematode-free propagules in a new location may be advisable, depending on the yields and nematode status of the existing planting.

Harvesting and handling

A banana will flower and produce fruit between six and 18 months after planting. The fruiting stalk is harvested when the fruits are still green but after the ridges have begun to become rounded and the topmost hands have become light green. For long distance shipment they are harvested while still dark green but after each finger has attained full size. The quality of fruit allowed to ripen on the plant is usually lower than that ripened off the plant, due to fruit splitting and lower sugar content.

As with most fruits, ripening occurs in response to ethylene production and applying ethylene will enhance ripening. It is a practice in East and Central Africa to ripen fruits by smoking them in a covered pit. The ethylene in the smoke and the warmth enhance both the speed and uniformity of ripening. Alternatively, bananas will ripen naturally if hung in a shady, cool location.

Since bananas are highly perishable fruits, great care must be taken to prevent bruising during the picking and transportation operations. It is common practice to transport bananas while they are still attached to the stalk; however, as mentioned previously, Cavendish types are often severely bruised by this treatment. It has become common practice in some areas to pack individual hands of bananas for export in cardboard boxes lined with polyethylene or paper in order to prevent bruising. Where whole stalks of bananas are to be transported, trucks should be cushioned with layers of straw, foam rubber, or other material. If the bananas are to be stored, they should be cooled to between 11.7°C and 13.3°C. Bananas are very susceptible to chilling injury so lower temperatures should be avoided.

Collar rot is the most serious post-harvest disease of bananas and is caused by several different fungi. The disease appears as rotting of the hands, beginning at the point of attachment of the stalk. As a preventative measure, it is recommended that hands for export be first washed in clean water and then dipped in a solution of benomyl (50 ml per 100 l of water) or thiabendazole (50 g per 100 l of water) before being dried and packed.

Yields of bananas as high as 31,700−61,200 kg/ha have been reported under intensive commercial growing conditions. Yields would more normally be 6800 kg/ha−15,900 kg/ha, and 22,700−34,000 kg/ha would be considered very good.

Yield will vary enormously with plant density. At a density of 2500 plants/ha yield of the first crop may be double that of bananas planted at 1200 plants/ha; however, the bananas will be smaller and the second crop will decrease by up to 40 per cent.

Cashew *(Anacardium occidentale)*

The cashew (Figure 4-3) is a member of the Anacardiaceae family and native to the American tropics from Mexico to Brazil and to the West Indies, but it has since become naturalized in many lowland tropical areas. The tree is a spreading fast-growing evergreen up to 12 m in height. Leaves are leathery and ovate with prominent veins. Flowers are borne in terminal

Figure 4-3
Cashew *(Anacardium occidentale)*

inflorescences which consist of a mixture of male and hermaphrodite flowers. These are scented and are cream-coloured on opening, later turning red. The fruit is a kidney-shaped nut on the base of the large receptacle called the 'cashew apple' which is thin-skinned and edible.

Economic importance and distribution

The most important product of the cashew tree is the nut, which is used in confectionery. In addition, the cashew apple is consumed locally and is rich in vitamins A and C. The shell of the nut yields phenol-containing oils which are used for preserving and waterproofing and, after distillation, for oilproof brake linings, inks, cements, and so on.

Cashew is an important export crop in southern Asia, Brazil, India, and parts of Africa, and is grown locally in many other parts of the lowland tropics at elevations of up to 1000 m. Since the crop must be harvested by hand, production is dependent on inexpensive labour for harvesting; the overall requirements for growing the crop, however, are low and plants will grow in relatively dry, infertile soils.

Culture

The majority of the world's cashews are propagated by seed despite the fact that there is a great deal of variability in yield between seedlings. Vegetative propagation is possible and is occasionally used. Yields can be dramatically increased by the selection of high-yielding plants and subsequent vegetative propagation.

In propagation by seed, three seeds are planted 5–10 cm deep in prepared planting holes in the orchard and then only the most vigorous seedling is allowed to remain after germination. Seedlings can also be raised in the nursery in plastic planting pockets or similar containers and are then transplanted. Seedlings are particularly sensitive to root damage so should be transplanted with care.

Viability of seed can be determined by placing the seeds in water; only those that sink are reliably viable. Additional selection for denser seeds, wh   d to be more vigorous, can be done by planting only those seeds   hen placed in a solution of 700 g of sugar dissolved in 5 l of
      ds will store in an airtight container for a year without

      agation is accomplished by layering, budding, grafting,
      Layering and approach grafting are most successful if
      r to the pre-flowering flush of growth. Whip grafting
      ful if it is done immediately after the fruit has ripened.
      s difficult to root although ringing the cutting 40 days
      om the parent plant has sometimes improved rooting.
      en from 1–2-year-old shoots whose stems are still light-
co   somewhat flexible.

Cultivars   Due to a lack of breeding and selection work, no cultivars are widely available. However, the following cultivars may be used SD 80, SD 114, SD 349, SD 365, SD 414, F 896.

Environment and soils   The cashew requires high temperatures and a lack of rainfall during the flowering and harvesting periods in order to produce optimum yields. Where rain occurs during flowering, the flowers are damaged by fungal infections. Rainfall during ripening also causes discolouration of the fruits. Apart from these requirements the cashew is quite tolerant of environments which are generally unfavourable for other fruits and nuts. The tree will withstand 6 months of drought without showing adverse effects and will survive with as little as 750 mm per year of rain. The drought tolerance of the cashew is thought to be due to its wide-ranging lateral root system and this may be reduced if trees are planted too closely together. The only soil requirement is that drainage must be good. If this requirement is met, the cashew will yield well on soils of very low fertility and varying pH.

Fertilizers   Complete fertilizers should be applied two or three times each year.

Pests and diseases   The most serious insect pests of cashews are *Helopeltis anacardii* and *H. schoutedeni*. These bugs suck the sap from the fleshy parts of the plant and in the process their saliva kills cells. This damage results in the production of black lesions, dieback of new shoots, and 'witch's broom', which is the growth of many short twigs at one point on a branch. Small populations of these insects can cause considerable damage. Control is by sprays of lindane, carbaryl, and phenthoate. Regular treatment of young trees may be necessary to prevent stunting.

A stem borer, *Mecocorynus loryses*, is sometimes a problem in East Africa. Chemical sprays are not very effective so control is by regular inspection of trees and mechanical removal of eggs, larvae, and pupae.

Thrips, leaf miners, and a scale (*Pseudoaonidice trilobitiformis*) are occasional pests on young trees. The bug, *Pseudotheraptus wayi*, normally a serious pest of coconuts, has been reported to attack cashews in Kenya. Damage is similar to *Helopeltis* damage.

The only disease of any significance is anthracnose or dieback caused by the fungus *Gloeosporium* spp. Control is by pruning out infected twigs and spraying with a copper-containing fungicide.

The crop cycle

Cashews are seeded in the field or transplanted at a spacing of 12 m × 12 m or 15 m × 15 m. Closer planting is possible although thinning will be necessary in most cashew-growing areas to prevent moisture stress. If plants are spaced at 6m × 6m they will need thinning in about five years, and at a spacing of 9m × 9m thinning will be necessary in about seven years. The main advantage of close planting is high early yields resulting in a quick financial return. However, thinning is laborious and expensive and these factors may partially offset this advantage, depending on the local cost of labour and the value of the fuelwood obtained. Cashew trees may also be grown randomly between other crops such as mangoes and coconuts. Smallholders often plant them as shade trees around the village, gaining the additional advantage of their fruits and valuable nuts. The maintenance of cashews is minimal, consisting mainly of controlling weeds until the trees are large and dense enough to shade them out, and spraying young trees to control insects and diseases should they be a problem.

Flowering and fruiting may be expected to begin between two and four years after planting for seedling trees and from one to two years after planting for vegetatively propagated trees. Yields of mature trees vary from 1000 to 4500kg of unshelled nuts per hectare.

Harvesting and handling

Harvesting is done by picking up the fruits after they have fallen to the ground and immediately removing the nut from the apple by twisting. The small amount of apple tissue which adheres to the nut should then be cut away. During dry weather, collection should be made at weekly intervals while the nuts are still grey in colour. Less frequent collection will result in the development of brown, rotten seed coats which will greatly lower the value of the nut. During wet weather, collection should be made daily. After collection, the nuts should be sun-dried until the kernels rattle in their shells. They are then transported to the processing plant.

Processing is designed to remove the cashew nut shell liquid (CNSL) which is an irritant and will contaminate the nut and blister human skin if not handled properly. The traditional method of treatment is to roast the nuts over an open fire; this removes the CNLS, which is a valuable source of natural phenols.

Another technique is to roast the nuts in vats of hot (180−185°C) oil which removes the CNSL from the nut and allows for its collection and sale. Several other techniques have been developed which include cooling to fix the CNSL, and other methods.

Coconut *(Cocos nucifera)*

The coconut (Figure 4-4) is a member of the Palmaceae family and a widely distributed crop of coastal tropical regions. Virtually every part of the plant has a use and the wide variety of products obtained includes copra, from which oil is extracted, coconut meat, shells for carving and conversion to charcoal, coir or coconut fibre (used for ropes, mats, and mattress stuffing),

Figure 4-4
Coconut palm (*Cocos nucifera*)
(Tropix Photo Library)

toddy (a sweet beverage), palm sugar, and palm cabbage (the terminal bud eaten as a vegetable). Leaves are used for thatching, and the trunks for building.

The botany of the coconut is different from that of most other fruit crops. The mature plant is a monocotyledonous tree consisting of a root system which arises from the bole (base of the trunk), a 40−60 cm diameter trunk, and a crown comprising a growing point surrounded by leaves.

Roots are 8−10 mm in diameter, largely unbranched, and lack root hairs. Absorption occurs only in a small area immediately behind the root cap, the remainder of the root being impervious. Roots average about 6 m in length although they may grow to 20 m. They are tolerant of saline conditions and will survive for some time immersed in moving water although eventually they will rot. Moving water is essential for root survival since the roots will be deprived of oxygen in stagnant water.

The trunk is of generally uniform thickness throughout its length and does not increase in diameter because it does not have a cambium. Another result of the trunk containing no cambium is that wounds cannot heal. However, due to the scattered arrangement of the vascular bundles within the trunk, the plant can withstand considerable damage. The trunk is covered by leaves and is therefore not visible for the first several years, but its height then increases rapidly. If a tree is defoliated and the apical bud is weakened, the trunk produced at that point will be narrow but will resume its normal thickness when vigour is regained .

The crown of a mature palm consists of the apical bud, 25−35 open leaves, and flowers and fruits which are borne in the axil of each leaf. Leaves are pinnate, 4.5−6.0 m long and spirally arranged. Petioles are clasping and grooved to channel water into the crown. The inflorescence is monoecious, 1−2 m long, and has 200−300 male flowers and one or more female flowers at the base. The fruit consists of a fibrous outer skin called the 'husk' surrounding the shell within which the 'meat' and liquid endosperm or 'coconut milk' are located.

Economic importance and distribution

The coconut is found in tropical regions, generally within 22° north and south of the equator and most commonly near seacoasts. It is important both for local use and for export, primarily of the oil-containing 'copra', the dried meat of the coconut. Major producing countries include the Philippines, Indonesia, Malaysia, Singapore, Borneo, New Guinea, Mozambique, Sri Lanka, and India. In addition, many small islands depend on coconut production as their main industry although, due to the size of the islands, the amount exported is relatively small.

Culture

Propagation Coconuts are grown from seeds which are collected from the ground or picked from the trees. Viable seeds can be identified by checking that the contents still include liquid when vigorously shaken. Seeds are planted 20 cm apart in shallow trenches with the narrow side of the coconut pointing downwards. Trenches are then filled so that the top of the seed barely protrudes. About four to five months after planting any seeds which have not germinated should be removed since they are of low vigour and usually produce low-yielding trees. While in the nursery, careful attention should be given to weed and insect control.

When the seedlings are 12−15 months old they are transplanted into the field, preferably at the start of the rains. Though it is best not to injure the roots in the transplanting process, coconuts will quickly regenerate a new root system even if all the roots are removed. Dwarf cultivars are planted closer than tall types and a quincunx layout (see Chapter 2) is normally used. In West Africa, a 9 m × 9 m spacing which gives 143 palms per hectare is recommended for tall varieties.

Cultivars Coconut palms are classified by height as tall or dwarf. The tall types are 20−30 m high. They are usually cross-pollinated, producing seedlings which are highly heterozygous. Thus true cultivars are not generally recognized, though named types with similar characteristics are found. Tall palms require 6−10 years to begin fruiting; however, they are long-lived and produce large crops of medium to large coconuts.

Dwarf coconuts reach an average height of only 8−10 m and produce fruit three years after planting. Fruits are smaller than those from tall types but are abundantly produced. A dwarf cultivar 'Fiji', is widely grown in West Africa as it is early-bearing and produces large, good-quality fruits.

Dwarf coconuts have the advantage of early bearing, ease of harvest, increased resistance to lethal yellowing disease, and increased yields per unit area. They are, however, more susceptible to drought and wind damage due to having a shallow root system and are also more susceptible to damage by insects. Copra from dwarf plants is often of lower quality than that from tall varieties due to a reduced oil content.

Some breeding of coconut palms is now being done in Sri Lanka, Jamaica, Malaya, Ivory Coast, and the Philippines. Crosses between dwarf and tall types have been made, resulting in the production of vigorous promising off-spring. In some cases these crosses, called 'maypans', show some resistance to lethal yellowing, the most serious disease of coconuts.

Environment and soils　Coconuts require a warm climate although they will tolerate a light frost. For this reason, the crop is confined mostly to low elevations in the tropics. Well-distributed rainfall in excess of 1000 mm per year is required unless there is underground water close to the surface or unless irrigation is used. Areas where drainage is poor are not satisfactory unless the water rises and falls frequently, as is the case near a tidal estuary. Coconuts tolerate brackish water and so are an ideal crop for coastal areas where salt water intrusion is a problem. Coconuts will grow on most soil types. The optimum pH is between 5.0 and 8.0.

Fertilizer　Potassium is the most important element for fertilizing coconuts and regular applications will produce maximum yields. Nitrogen is also valuable for coconuts and deficiencies of either of these elements result in sparse yellow crown and low yields. Response of coconut palms to phosphorus has generally been limited.

Green manure crops are considered to be beneficial, especially legumes such as *Crotalaria mucronata* and *Pueraria javanica* which are used in Mozambique. When green manure crops are grown they must be regularly incorporated into the soil or they may actually decrease yields, particularly during periods of drought.

Pests and diseases　A number of insects feed on coconut palms. Termites can attack seedling trees in nurseries and are controlled with soil insecticides. *Pseudotheraptus wayi* is a bug which feeds on the sap of young nuts and at the same time injects a toxin which causes distortion, gummosis, and early drop of fruits. This insect is a problem in Kenya, Tanzania, and Mozambique. Insecticidal control is often uneconomical but interplanting the coconut grove with mangoes, cashews, or citrus decreases losses by encouraging a predator ant which favours these trees.

Aphids, particularly *Cerataphis lataniae*, sometimes damage plants by feeding on leaves and fruit. If high populations occur, control is by insecticidal sprays such as malathion. Related scale insects such as *Aspidiotus destructor* cause serious damage wherever coconuts are grown. Control is by predators (ladybird beetles in Mauritius) or by insecticides.

Caterpillars occasionally reach high populations and cause defoliation of trees. In East Africa, *Parasa lepida* is most common. Control is by insecticide such as carbaryl where populations are high.

Insects belonging to the beetle family are among the most serious worldwide pest of coconut palms. The palm weevil is included in this family. Larvae tunnel in the trunks and slow the growth or even kill the tree. Insecticides are sprayed on the trunk and leaves for control.

The other most serious pest of coconuts is the rhinoceros beetle (*Oryctes* spp.) which feeds on the apical bud and also provides breeding sites for the palm weevil (Figure 4-5). Control is by sanitation and the placing of sawdust or sand mixed with a residual insecticide around the apical bud of the palm. The sand or sawdust forms a natural habitat for the beetle which is then killed by the insecticide. Since the beetles breed in decaying organic matter, dead tree trunks and debris should be removed from the grove and burned.

The most serious disease of coconuts is lethal yellowing (or wilt) which causes fruit drop, chlorosis of leaves, and eventual death. Although the disease is most serious in the West Indies and Florida, similar symptoms have been reported on plants in Ghana (Cape St. Paul Wilt), Nigeria (Awka

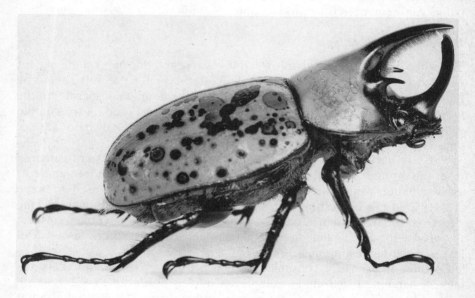

Figure 4-5
Rhinoceros beetle
(USDA Photo)

Wilt), and Togo (Maladie de Kaincope). The disease is believed to be caused by a mycoplasma and can be spread by a plant hopper. Pressure injection of tetracycline or oxytetracycline will cure the disease although this treatment is not usually economical. It is generally recommended that infected trees be removed and replanted with a resistant type such as 'Malayan Dwarf'.

Bud rot caused by *Phytophthora palmivora* causes chlorosis and death of leaves, beginning with the youngest and proceeding to older leaves until only a ring of healthy leaves remains. The disease is most severe in cool areas with low humidity and insect injury makes palms particularly susceptible to the disease. Control is difficult although applications of Bordeaux paste to the bud are of some benefit.

Stem bleeding (*Ceratocystis paradoxa*) is an infrequent disease which causes a brown sap to ooze from the trunk 1.5–2m above the ground. Treatment is by cutting out infected tissue and treatment of the cavity with a fungicidal wound dressing.

The crop cycle

Palms begin to bear between two and eight years after planting. Particular attention should be paid to weed control during the period from planting to first harvest, as palms are particularly susceptible to weed competition when small. Also during this period any plants which die should be replaced and all insects and diseases should be controlled. After bearing has begun the palms can be expected to produce nuts continuously for 30–50 years.

Harvesting and handling

The stage of maturity of the nut when it is harvested depends on the use for which it is intended. For copra and desiccated coconut the nuts should be fully mature, as evidenced by having absorbed the liquid endosperm into the meat. For coconut milk for drinking, immature nuts are used. A tree should yield between 30 and 50 nuts per year although yields of 80 or more are not unusual.

Nuts are harvested in two ways: (a) free fall, in which nuts are gathered from the ground; and (b) picking, where someone climbs the tree and cuts away the nuts which are ready for harvest (monkeys are used in some areas). Free fall harvesting is easier and less expensive. Although picking nuts from the tree is dangerous and expensive, the climber is able to inspect the crown for insect and disease damage and can also prune away dead leaves while harvesting nuts.

Coconut is primarily consumed as a processed product, and the postharvest handling is extremely complex. Copra is produced by splitting the nuts and then drying them in the sun or in flue dryers. Desiccated coconut for culinary use is made by removing the meat from the nut, sterilizing it, and then drying it until the moisture content is around 3 per cent.

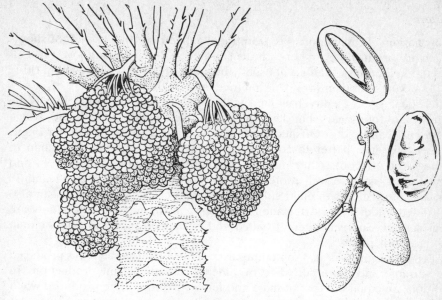

Figure 4-6
Date (*Phoenix dactylifera*)

Date *(Phoenix dactylifera)*

The date (Figure 4-6) is a member of the Aracaceae family which has been cultivated in dry areas of the Middle East since at least 3500 BC. It has many uses including wood for building, fronds for thatching, a palm wine, animal feed, the fruit can be fermented to make an alcoholic drink (arrack), and palm sugar can be extracted from it, but the most important product is the fruit which is eaten either fresh or dried.

The tree has a single trunk often with suckers at the base especially in humid areas. Leaves are pinnately compound, up to 6 m long and blue-green in colour. The tree produces new leaves from the apical bud and old leaves die off below hanging down vertically at first and then eventually dropping off. At any time up to 200 leaves may be present. Trees are dioecious and flowers are borne in a spathe. The fruit is an oblong drupe which may weigh up to 12 g.

Economic importance and distribution

The date is the most important dried tree-fruit in the world and fresh dates are also locally important in producing areas. Dates are produced throughout the Middle East and North Africa and some production also occurs in California and Arizona in the USA and in South Africa. Iraq is the leading producer for export.

Culture

Propagation Propagation has traditionally been by the use of offshoots which are produced from beneath the soil-line in dry areas or above the soil-line in humid areas. Plants will usually begin to produce suckers within three or four years of planting. This method allows the propagation of clones which bear good quality fruit and yield well. Offshoots should ideally be more than four years but less than eight years old, and should weigh between 10 and 25 kg. Offshoots are removed intact from the mother plant. If necessary it is better to cut into the trunk of the mother plant than to damage the offshoot. After removal, offshoots are planted in a nursery and grown until they are well rooted. In a Moroccan study between 31 and 78 per cent of offshoots of eight cultivars rooted. Very small offshoots can also be rooted under mist. After about two years in the nursery, the plants are transplanted. Survival rates of balled and burlapped plants are better than for bare-root plants.

Seed propagation is also sometimes used. However, the seedlings produced are variable and more suited to ornamental use than fruit production. In addition, an equal number of male and female plants will be produced when only a small number of male trees are actually needed. If unsure of whether a nursery plant is a sucker or an undesirable seedling, look for the cut at the base of the plant where it was removed from the parent plant.

Recently, tissue culture has been found to be an inexpensive way of propagating large numbers of dates and the propagation of date plants in this way is becoming common in major date growing countries. This will probably became the most important method of propagating large numbers of trees, though for small scale plantings offshoots remain the best method.

Cultivars

Over 300 named cultivars exist in date growing areas although only a relatively small number are grown on a widespread basis. Cultivars are divided into three categories, soft, semi-dry, and dry. Soft cultivars are grown primarily for export since they are high in fructose but do not keep well. Semi-dry cultivars are usually marketed locally; their keeping quality is better than soft types but not as good as the dry types. Dry types are used as a staple food in date producing areas. Some common cultivars of each type are:

Soft: Halawy, Khadrawy, Sayer, Saidy, Hayany
Semi-dry: Zahidi, Deglet Noor
Dry: Thoory

Fertilizers

Trees benefit from proper fertility and fertilization rates should be according to soil and leaf analysis. Since the fertilizer requirements of dates have not been extensively researched, the following should be regarded as a starting point for further research only. Moroccan research indicates that 160 kg/ha N, and 80 kg/ha each of P and K, plus 2 t/ha of barnyard manure decreased the tendency of biennial bearing when compared with rates approximately half that.

Pests and diseases

Dates are affected by a large number of pests and diseases including insects, mites, nematodes and both fungal and bacterial diseases which are described in Carpenter and Elmer's handbook, *Pests and Diseases of the Date Palm* (see References). The most serious diseases are Phoenia rust also called graphiola leaf spot, caused by (*Graphiola phoenicis*), and Bayoud, caused by *Fusarium oxysporum* f. sp. *albedenis*.

Graphiola leaf spot is serious wherever dates are grown in humid environments and causes black spots on the leaves followed by premature death of the leaf. Damage is due to premature loss of leaf area. Pruning the infected leaves to remove inoculum is the usual method of control, although where the disease is severe sprays with Bordeaux may be used. There are also several resistant cultivars including: 'Barhee', 'Adbad' 'Rahman', 'Gizaz', 'Iteema', and 'Kustawy'.

Bayoud is the most serious date disease but it is not yet widely distributed, being found mostly in North-west Africa. In Morocco, over 10 million date palms have been killed by the disease. Symptoms occur first on newly matured leaves where pinnae (leaflets) on one side of the rachis (main leaf stem) turn white progressing from base to tip and then back down the other side. This progresses from leaf to leaf and when they are cross-sectioned a reddish brown stain is present. Trunks of infected trees when cut will reveal similar staining. The pathogen is soil-borne and can be spread by water, wind, soil movement, or the movement of infected plant parts. Spread from oasis to oasis in North Africa, the disease has, apparently been passed on through palm products such as baskets, ropes, and saddles discarded in moist places after use. The pathogen can persist for several weeks in these articles. Prevention of the spread of Bayoud is dependent on strict quarantine measures. Since there is no control, the long-range solution to the problem is the planting of resistant cultivars, many of which have been identified. Unfortunately the resistant cultivars are of poorer quality than the highly susceptible Deglet Noor cultivar. Breeding programmes are underway to improve the quality of the fruit of resistant cultivars.

Various species of mite can be a problem in dates and are controlled by the use of sulphur, although other acaracides can be used. They are sometimes a problem in stored dates and can be controlled by fumigation.

Parlatoria date scale (*Parlatoria blanchardi*) and several other species of scale can be serious in many date growing areas. There are a number of predators which will provide partial control if they are not disrupted by the use of insecticides. Where additional control is needed, the removal of heavily infected leaves will help, but in large plantings this is not practical. Insecticides will aid in control where biological controls are not adequate.

Various beetles, borers, leaf-hoppers, and moth larvae can also be problems though infestations are often sporadic and biological controls usually keep populations under economically damaging levels.

The crop cycle

Dates are planted at a density of between 6m × 6m and 10m × 10m and will begin to fruit several years after planting, reaching peak production in eight to 10 years and continuing to produce for up to 75 years. It is beneficial to enrich the soil with organic matter prior to planting. Irrigation is required especially for newly planted groves. Furrow or flood irrigation is most commonly used, however, recent trials with drip irrigation have looked good. The frequency of irrigation required varies with climate and soil. However, even where the water table is relatively high occasional irrigations are beneficial. Trees will survive periods of drought but will not fruit. With some soft cultivars irrigation is withheld for two to three months prior to harvest to help the fruit dry without detrimental effects.

Pollination is by wind. However, if left solely to the wind, yields will be low. Commercially, trees are hand-pollinated by placing male flowers among the female inflorescences, or pollen is spread in the air over the grove at rates of about 250cc of pollen per hectare and applied three times per week. Portable dusters are also effective for pollinating trees. One male tree will provide enough pollen for about 50 female trees. Since dates exhibit meta-zenia (i.e. pollen source affects fruit quality) pollen should come only from trees which are known to produce good fruit.

Pruning is usually done at pollination time and consists of cutting away dead or dying fronds. Fruit thinning is also practised to reduce biennial bearing. Branch lowering is a practice which is used to support fruiting stalks to minimize breaking. The stalk is tied to the midrib of a lower leaf or sometimes the fruit stalk is supported by a forked stick.

Harvesting and handling

The stage of harvest varies with cultivar and intended use. Four stages of ripeness are commonly recognized. These are:

(1) *kimri*, where the fruit is growing rapidly and is green,
(2) *khalal*, where the fruit has reached nearly full size and has changed to the colour characteristic of the cultivar,
(3) *rutab*, where the tips have softened and the fruit has changed to brown or black, and
(4) *tamar*, where the fruit has lost most of its water and is fully ripe.

Harvesting is accomplished by climbing the tree and cutting off the ripe bunches of dates. Where equipment is available a tower on wheels or a truck-mounted telescopic boom can be used. After picking, the bunches are spread on mats to dry. Yields vary from 20 to 100kg/tree and 5 to 10t/ha.

Where the climate is favourable, dates can be ripened on the tree to the tamar stage and can then be packed without any additional drying. However, in marginal climates fruit may only reach the rutab stage on the tree and then have to be further dried in the sun. Dates are usually fumigated to kill all pests, cleaned using various washing machines or dry-air blowers, graded by hand, ripened in controlled temperature rooms (if needed), and either dehydrated or hydrated to the correct moisture level for the cultivar, before being packed. Poor quality dates are used to create by-products such as date syrup, date juice, date paste, date candy, date jam, date marmalade, and date butter, or fermented and distilled into alcohol or vinegar. Date stones can be ground and pressed to yield an oil useful in soap making.

Guava (*Psidium guajava*)

The guava (Figure 4-7) is a member of the Myrtacea family and an important fruit throughout the tropics which has become naturalized in many areas. The fruit is very high in vitamin C and is also a rich source of vitamin A. The plant is a shrubby tree which grows up to 10m tall. The bark is smooth and peels off in small flakes. Leaves are opposite with depressed veins and

Figure 4-7
Guava (*Psidium guajava*)
(Photograph by Rex Parry)

are slightly pubescent. Flowers are axillary with from four to five white petals and numerous stamens. The fruit is a 4−12 cm diameter berry with calyx lobes at the blossom end. The exterior of the fruit is fleshy, and the centre consists of a seedy pulp.

Economic importance and distribution

The guava is found throughout the tropics and, like the mango, has become naturalized in many areas from discarded seeds. Guava seeds are readily spread by birds and in some locations the guava is a serious weed.

Guavas are primarily of importance for their contribution to the local diet, although some countries produce processed guava products. Important guava products include preserves, jam, jelly, paste, juice, and nectar. Principal producing countries are India, Mexico, and Brazil.

Culture

Propagation　The majority of guavas are grown from seed, producing seedy fruit of variable quality. To establish a commercial orchard, clonal plants which produce good-quality fruits should be planted. Such plants must be propagated vegetatively.

The guava is more difficult to propagate vegetatively than many other plants and, until recently, vegatatively propagated plants were not available in most areas. The most efficient way to propagate a large number of plants is by chip budding greenwood buds from selected cultivars onto seedling rootstocks. This operation can be done any time during the warm season using seedlings with stems which are at least 5 mm in diameter.

Softwood cuttings will root fairly satisfactory under mist if treated with 200 ppm of IBA. Layering by mound, tip or air techniques is successful, especially if IBA at 500 ppm is applied to the girdled part. Root cuttings can also be used.

If seedlings are being grown because budwood of improved cultivars is not available, seed should be taken only from plants which are known to produce good-quality fruit; the flowers of these seed parents should be self-fertilized to reduce variability. Seeds germinate in two to three weeks. Seedlings are susceptible to damping off so sterilized soil should be used and careful watering practised. Under good conditions, a seedling will grow to 30 cm tall in six months and can be planted into the field at that time. Existing seedling trees can also be top-worked to improved cultivars.

Cultivars　There are both white-fleshed and pink-fleshed cultivars; the latter is generally preferred for processing. The aim of guava improvement programmes is to select cultivars with a high flesh to seed ratio, high yield, and rich vitamin content. Some of the improved pink-fleshed cultivars are: 'Malherbe', 'Red Indian', 'Blitch', 'Patillo', 'Pink Acid', 'Beumont', 'Rolfs',

'Ruby'. 'Van Retief', and 'Keerweeder'. White-fleshed cultivars include 'Parker's White', 'Patnagola', 'Supreme' and 'Miami White'.

Environment and soils Few fruits are as tolerant of a wide range of environments as are guavas; they will grow in most tropical climates and soils. Optimum production occurs in tropical regions below 1300m in elevations where the soil is fertile and rainfall is regular. Guavas will, however, tolerate drought, low-fertility soils, waterlogged soils and pH values ranging from 4.5 to 8.2.

Fertilizer Specific information on the fertility requirements of guavas is limited but they are known to respond well to fertilizer and manure. In Kenya, 180kg of double superphosphate per hectare is recommended at planting, plus 112kg of sulphate of ammonia per hectare applied twice yearly during the rains.

Diseases and pests Diseases are generally not a problem in most areas, although a leaf and fruit spot sometimes occurs. Fungicidal sprays readily control this disease.

Several insects attack guavas although most are not serious. The Caribbean fruitfly (*Anastrepha suspensa*), and the Mediterranean fruitfly (*Ceratitis capitata*) lay eggs in the fruit and may cause serious losses. Control is by the use of malathion/sugar baits and sprays of fenthion (see Appendix). Scales such as *Coccus alpinus* can be a problem and are controlled by spraying the trunks with a persistent insecticide to control the ants which culture them. *Helopeltis* spp. sometimes feed on new shoots thus causing dieback; control is obtained with sprays of carbaryl. Aphids feed on new growth and are controlled with malathion or other insecticides.

The crop cycle

Guava plants are spaced 6m × 5m. They usually begin to bear from one to three years after planting and normally continue to bear for 30 or more years. The plants usually require comparatively little care. Pruning is confined to the removal of low branches which hang on the ground and of diseased or crossed branches. Flowers are insect-pollinated, and fruits ripen about five months after fertilization. Yields of between 12,000 and 15,000kg/ha can be expected from improved trees.

Harvesting and handling

Fruits should be picked when they begin to turn yellow and will ripen off the plant. Fruits allowed to remain longer on the plant may drop and bruise when they strike the ground. The skin of the guava is very delicate, so fruits must be handled with care. Fruits for local markets should be kept as cool as possible until sold. Guavas can be stored for between three and four weeks

Figure 4-8
Macadamia (*Macadamia integrifolia*)

at 7°–10°C and 85–90 per cent relative humidity. Fruits for processing should be moved rapidly to the cannery. Fruits can also be dried and used later as a vitamin C supplement.

Macadamia nut (*Macadamia integrifolia*)

The macadamia (Figure 4-8) is a member of the proteaceae family, a small upright, evergreen tree reaching a height of 15m. Leaves are elongated, leathery and arranged in whorls of three. Flowers are borne in axillary racemes consisting of 150–300 flowers and are self-fertile. Usually, between three and seven fruits develop from each raceme. The fruit is a follicle which opens along one suture and contains a single seed with a hard seed coat or shell. The edible part, called the kernel, consists of the embryo and cotyledons.

Another species, *Macadamia tetraphylla*, is occasionally grown; however, the nuts of this species are of poorer quality and are not as important commercially. *Macadamia tetraphylla* can be distinguished by its thorny, toothed leaf margins. Its primary use is as a rootstock, although there is considerable interest in its use in subtropical areas such as California and New Zealand.

Economic importance and distribution

The macadamia, which is native to eastern Australia, is produced primarily in Hawaii and California in the USA, in Australia, and in southern Africa, including Zimbabwe, Kenya, and Malawi. The nuts are of high quality and are expensive. The level of planting is increasing in all producing areas as demand continues to grow. At present the USA is the largest market for the nuts.

Culture

Propagation Propagation is primarily by grafting scions from desirable cultivars onto seedling rootstocks of either *Macadamia integrifolia* or *Macadamia tetraphylla*. When available, *M. tetraphylla* is preferred as rootstock because the seed germinates more quickly and uniformly and the seedlings are vigorous in growth. In addition, it is somewhat resistant to *Phytophthora cinnamomi* and encourages early fruiting of *M. integrifolia* scions grafted on to it.

Seeds for rootstock production should be gathered as soon as they fall to the ground and the husk should be removed. They should then be planted immediately or stored for no more than six months. Hardly any seeds will germinate after a year of storage at room temperature. The best germination and the most vigorous seedlings will be produced from large, fresh seeds which do not float in water.

Germination of seeds can be hastened by the removal of the hard inner shell provided that the kernel is not damaged. Shell removal can be made easier by heating nuts to 27°C at 15−20 per cent relative humidity for 14 days. Seed should be planted 25−50 mm deep with the micropyle pointing down. Germination will occur from two to 12 weeks later.

After the rootstocks are 60−180 mm tall they are ready for grafting. In Hawaii the wedge or splice graft is preferred while in India and Africa approach grafting is commonly used. Since macadamia wood is very brittle, experience is necessary for wedge or splice grafting to be successful. Large trees can be top worked using bark or veneer grafts. For grafting to be successful, scion wood should be girdled five weeks prior to the grafting operation. Cuttings can also be rooted under mist although the resulting plants are often slow to establish and growth is variable.

Cultivars It is very important that only recommended cultivars of macadamia are grown since untested cultivars or seedlings may produce unsaleable nuts. The most commonly recommended cultivars are: 'Elimbah' (731), 'Ikaika' (333), 'Kakea' (508), 'Keaau' (660), 'Keauhou' (246), 'Beaumont' and 'Nelmak'.

Environment and soils Macadamia are best adapted to a frost-free subtropical climate with at least 125 cm of well-distributed rainfall per year. They grow best at temperatures which do not drop below freezing and do not rise above 38°C. The optimum mean summer temperature is considered to be 25°C. Wind is detrimental to production due to the brittleness of the wood. Therefore where heavy wind occurs windbreaks are necessary. Macadamias will grow on a wide variety of soils if drainage is adequate. A pH of 7.0 is considered ideal.

Fertilizers   Fertilizer is required for good production and growth and should be applied three or four times a year. The amount can be based on soil and leaf analysis or, if this is not available, upon experience. In Hawaii, fertilizer with a ratio of 1:1:1 or 2:4:1 (young trees) is used. Micronutrient deficiencies are a problem in some soils. The most likely deficiencies are iron, magnesium, and zinc. These are corrected with soil applications of iron, magnesium, and zinc sulphates or by the application of chelated trace elements.

Pests and diseases   Insects do not usually cause severe problems in macadamias although a number of insects do occasionally cause economic damage. The green vegetable bug (*Nezara viridula*) feeds on the developing nuts, causing nut drop or kernel staining. In macadamia-growing areas of Hawaii, control is by a parasitic fly *Trichopoda pennipes*. Control is also by insecticidal sprays but these are only used when damage is severe.

The larvae of three moths sometimes feed on nuts. These are the carob moth (*Spectrobates ceratoniae*), the false codling moth (*Cryptophlebia leucotreta*), and the litchi moth (*Argyroploce pettrastica*). Chemical control is generally not required if the recommended cultivars which have some resistance are grown. Bark borer (*Salagena* spp.) is occasionally important in young macadamias. The larvae feed on live bark and may ring branches thus causing dieback. The primary sign of feeding is excrement which is held together on the surface of the twigs by threadlike filaments. The larvae move freely under this covering. Control is by predators and by pushing a piece of wire up the tunnels to crush the larvae. Spraying is not recommended. The broadmite can be a serious flower pest but is easily controlled with sulphur sprays. Aphids, thrips, and scale are also occasional pests but control measures are seldom necessary.

Several diseases may be of economic importance in macadamias. Among these is stick-tight nuts or anthracnose. Anthracnose is a disease which kills the husks of the nuts, causing the nuts to hang on the tree after maturity. The disease is most serious in humid areas and is controlled by planting a cultivar such as 'Keauhou' which is entirely resistant. Flower blight is a fungal infection caused by *Botrytis* spp. or *Phytophthora* spp. affecting flower racemes. Some reduction of nut set can occur, but this is usually not severe so control measures are unnecessary.

Rats and other animals can cause serious losses as nuts ripen. Control is through the use of poison baits.

The crop cycle

Vegetatively propagated trees are normally planted about a year after grafting. Trees grown in plastic sleeves are planted in large holes to which manure and superphosphate have been added. Spacing of 8 × 8m is recommended with a square layout and 10 × 10m with a tree in the centre if the quincunx system is used.

During the first few years before fruiting begins, plants should be watched for termites and other pests and pruned to the central leader system, selecting scaffold branches with wide crotches. Suckers should be removed periodically. Particular attention to weed control around the tree is necessary during the establishment period and mulching is beneficial.

Flowering usually occurs at the end of the cool season and may continue for between one and two months. Nuts require about 212 days to mature and then will drop to the ground. Most trees will begin to bear about five years after planting and a commercial size crop can be expected by the seventh year. At that harvest 3.5−18kg of nuts per tree may be produced. In Hawaii 45kg of nuts can be expected by the eleventh year and 57kg by the twentieth year; mature orchards may yield up to 7800kg/ha of in-shell nuts per year.

Harvesting and handling consists of collecting fallen ripe nuts from the ground. It is therefore important that the orchard floor be smooth, even, and weed-free. After harvesting, nuts are dried to about 1.5 per cent moisture content before being shelled. Shelling is either by machine, using rollers, or by hand, using hammers, if the quantity is small. After shelling, nuts are sorted for quality and are then cooked in oil before being vacuum packed in cans, glass jars, or foil pouches.

Mango (*Mangifera indica*)

The mango (Figure 4-9), a member of the Anacardiaceae family, is a tropical evergreen tree native to the Indian subcontinent. Trees range from 7−10m in height in the case of some of the compact clones to over 40m in the case of many trees propagated from seed. The shapes range from round-headed to dome-shaped and all are densely foliate, producing a heavy shade. The leaves are alternate, simple, and lanceolate with a leathery texture and dark green colour. The root system consists of a tap root which may grow to a depth of 6m, plus a fibrous feeder root system which is located just below the soil surface.

The inflorescence is a terminal panicle which may appear over the entire tree or in only one portion of the tree at any one time. Each panicle branches three to four times and consists of 1000 or more mostly male and occasionally hermaphrodite flowers. Flowers are mostly 5-merous, green or slightly pink in colour and are not showy.

The fruit is a large drupe which varies from 5−30cm in length and often

Figure 4-9
Mango (*Mangifera indica*)

only one forms from each inflorescence. The skin is thick and leathery and may be green, yellow, or red when ripe, depending on the cultivar. The flesh is orange, sweet, and almost free of fibre in selected cultivars but may be resinous and fibrous in seedling trees. The hard, fibrous endocarp encloses a small brown seed and may comprise up to 25 per cent of the volume of the fruit.

Economic importance and distribution

The mango is generally considered to be one of the most important fruits of tropical regions. India is the largest single producer but the fruit is found throughout the tropics. In most tropical regions fruit is grown only for local consumption but it is produced for export in India, Hawaii, Mexico, Brazil and parts of Africa. In Africa, the mango has become naturalized due to germinating discarded seeds and grows wild in most inhabited areas.

Culture

Propagation The majority of mangoes throughout the world are still propagated from seed. Mango seed may be either polyembryonic (usually with one or more apomictic embryos) or monoembryonic. Some clonal cultivars can be reproduced from apomictic seedlings but it is virtually impossible to distinguish between the nucellar seedling and the sexually produced seedling in most instances.

Clonal cultivars are propagated primarily by grafting or budding onto seedling rootstocks. Little work has been done on the identification of particularly desirable clonal rootstocks although some selection for salt tolerance is in progress in India. It is recommended that scions be grafted onto seedlings from a polyembryonic cultivar so that the mature tress will be uniform. In India, the most common vegetative propagation method is approach grafting but this is a laborious method with a low output potential. In the West Indies and the United States budding using the T- or the inverted T-technique as well as chip budding are practised. Air layering is successful, especially if etiolated shoots are treated with IBA at 10,000 ppm. Although cutting propagation is difficult it can be accomplished by using intermittent mist and IBA treatments.

Seed Mango seeds should be obtained from uniform trees and polyembryonic types are preferred. They are depulped and then sown within a week for optimum germination since viability declines rapidly. There is some debate over whether to remove the seedcoat prior to planting. This operation speeds up germination and will often produce a straighter taproot, however it is time-consuming. If it is to be done, the seed can be soaked for a day or two and then the seedcoat and the parchment-like covering of the seed beneath is removed. The seed is immediately planted with the convex side up just below the surface of the soil. Seeds will often germinate within a week and will be ready to transplant from five to six weeks later when the plant has between five and six reddish leaves. The seeds of polyembryonic types are usually planted in seedbeds, 30−45 cm deep, containing a light media such as pine bark since each seed will produce several seedlings. The gametic or sexual seedling will be the most vigorous seedling emerging from each seed and is generally discarded. The relatively shallow seedbed is used because when the taproot strikes the bottom of the seedbed, it will turn sideways thus eliminating apical dominance and causing the root above to branch. This increases the ease of transplanting. When the seedlings have reached the proper stage they are lifted, separated, and the tips of crooked taproots cut off. They are then transplanted into individual containers or planting pockets to be grafted when they reach an appropriate size.

Cultivars Mango cultivars fall into two broad categories, the Indian types which are characterized by fruit which turn yellow, orange, or reddish upon ripening; and the Indo-China or Philippine types, the fruit of which remain green when ripe. The most important types for export are usually the Indian ones. Within these types there are large differences in quality, particularly in the freedom from fibre and the lack of a 'turpentine' flavour. Some cultivars produce polyembryonic seeds which contain apomictic embryos and can thus be propagated reliably from seed, while many cultivars produce monoembryonic seed without apomictic embryos and cannot therefore be propagated from seed. Cultivars differ in their resistance to some diseases,

especially anthracnose and bacterial black spot, and in the size and environmental adaptation of the tree. The Indo-China or Philippine types tend to
be more resistant to anthracnose and are often of good quality, despite their
lack of colouration, so should be tried where anthracnose is a serious
problem.

The following are important cultivars: 'Apple', 'Banglora', 'Batawi',
'Boribo', 'Caraboa', 'Davis', 'Dodo', 'Fascell', 'Haden', 'Harries', 'Julie',
'Keitt', 'Kensington', 'Kent', 'Ngowe', 'Palmer', 'Peach', 'Pico', 'Sabre',
'Sensation', 'Smith', 'Van.Dyke', and 'Zill'.

Environment and soils Mangoes will grow in most soils providing they are
well drained. The optimum pH range is 5.5–7.5. The mango is drought-
tolerant and will survive on as little as 300 mm of rain per year, but for
commercial production a minimum of 635 mm is recommended. In areas
with two wet seasons, such as parts of West Africa, flowering and fruit
production may occur twice each year, but in most areas one crop per year
is produced. The best production occurs at elevations between 0 and 600 m
and in areas where flowering occurs during periods of dry weather since
rainfall during flowering greatly depresses yield. In upland areas above
1200 m production is often poor although the cultivars 'Sabre' and 'Harries'
are reported to yield well at up to 1800 m in Kenya.

Fertilizers Mangoes are not heavy feeders but the periodic use of nitrogenous fertilizer is beneficial. In addition, Indian research has shown that a
foliar application of urea as a 4 per cent solution enhances the yield and fruit
quality of some cultivars. Tip burn of leaves is associated with potassium
deficiencies. As a guideline, mature trees may be given 700 g of N, 400 g of
P_2O_5, and 750 g of K_2O per tree per year. Newly planted trees are given 70 g
of N, 100 g of P_2O_5, and 250 g of K_2O. Amounts should be split into at least
four applications.

Pests and diseases The primary insect pests of mangoes are the mosquito-
back (*Helopeltis* spp.) and the mango stone weevil (*Sternochetus* spp.).
Helopeltis spp. are sucking insects which feed on fruits and on young shoots
which then die back, causing vigorous secondary branching and subsequent
twigginess. The mango stone weevil feeds on the mango seed, often destroying
the fruit in the process. *Helopeltis* spp. is controlled by spraying with carbaryl
while the disposal of fallen fruit in the field is the best method of controlling
the mango stone weevil. Mealybugs, fruitflies, scale, and mites can also be
serious mango pests.

The two most serious fungal diseases of mangoes are powdery mildew
(*Oidium mangiferae*) and anthracnose (*Colletotrichum gloeosporioides*), both
of which are most prevalent in wet weather. Anthracnose causes leaf spots,
flower blights, and fruit discoloration and rot. Powdery mildew responds
well to sulphur sprays and both diseases can be controlled with benomyl
sprays.

In parts of southern Africa, Zaire, Tanzania, and some other countries, bacterial black spot is a serious problem, causing symptoms similar to those of anthracnose. Where the disease is present, resistant cultivars such as 'Fascell', 'Zill', or 'Sensation' should be grown. Inflorescence malformation is a serious problem in some areas which may be caused by *Fusarium moniliforme*, or by the mango bud mite (*Aceria mangiferae*). Remove infected plant parts and spray to control mites.

The crop cycle

Mangoes are usually planted in a square pattern 10.5 m apart, although the spacing will vary somewhat depending on the cultivar. Irrigation during the period prior to the first flowering will encourage vegetative growth and increase the size of the first crop. In most areas irrigation is stopped when the trees reach bearing age.

Mangoes usually begin to bear three to four years after planting in the case of budded or grafted trees, or five to six years after planting in the case of seedlings. Bearing normally continues for 40 or more years. When trees are young and just beginning to fruit or when trees are tending to bear biennially, chlormequat (cycocel) or daminozide (Alar, or B-nine) can be applied to enhance bearing.

Flowering usually begins after a period of dormancy due to cool and/or dry weather. Potassium nitrate can also be used as a foliar spray to enhance flowering. Pollination is by insects, primarily bees and flies, and some cultivars may be partially self-infertile. After pollination, fruit grows rapidly and, in some cultivars, may mature in as little as seven weeks.

Harvesting and handling

As fruits ripen they will naturally abscise and in many areas fruits are commonly gathered from the ground. Fruits which are to be transported to market should be cut from the tree when they have reached full size to prevent bruising by striking the ground. After picking, any latex which has oozed from the pedicel onto the fruit must be carefully wiped off. If this is not done damage to the skin will appear in two or three days.

Fruit for export is graded for size, maturity, and condition and is then commonly packed in 5 kg net weight cardboard cartons or wooden crates for shipment. Packing in sawdust or wood shavings has generally proved detrimental but wrapping with paper is advantageous. In cases of long-distance shipment, fruit may be waxed to decrease moisture loss, and for export it may be necessary to disinfect the fruit to prevent the spread of Mediterranean fruitfly and mango stone weevil. Dips in hot EDB (ethylene dibromide) have been effective as has fumigation with EDB at 20 ml per cubic metre. Research has also shown that a post-harvest rot caused by anthracnose can be eliminated by dipping fruit for one minute in 12 per cent hot benomyl solution. The benomyl treatment also stimulates the formation of yellow and red colouring in the skins of some cultivars, but after treatment waxing is necessary to prevent shrivelling.

Maximum storage life will be obtained if fruits are quickly cooled to between 7°C and 10°C, although there is some variation in optimum storage temperature among different cultivars. Relative humidity should be maintained at 90—95 per cent. Although storage life varies with cultivar, storage for seven weeks or longer is common. It should be noted that mangoes are susceptible to chilling injury. Since chilling injury is not apparent until fruits are returned to warm temperatures it is important that the optimum storage temperature be determined for each cultivar.

Olive (*Olea europaea*)

The olive (Figure 4-10), a member of the Oleaceae family, is an important fruit in areas with warm summers and cool winters. Unlike most of the other fruits it is consumed only after some type of processing such as brining, fermentation, or pressing for oil. In addition to fruit and oil, the olive tree produces a hard, decay-resistant wood and is used in some areas as an ornamental tree.

The olive is a broad-leaved evergreen tree which grows 9—12 m tall. Leaves are opposite, elliptical in shape and dark green on top and silvery beneath. The yellowish flowers are small, fragrant, and borne in panicles from the axils of one-year-old wood. They may be either perfect or staminate with aborted pistils. The fruit is a drupe, black when fully ripe, and may be up to 4 cm long.

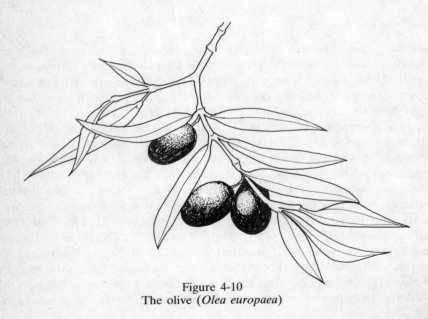

Figure 4-10
The olive (*Olea europaea*)

Economic importance and distribution

World olive production is greater than 8 million tonnes the majority of which is pressed for oil. Production centres around the Mediterranean with Spain, Italy, Greece, Tunisia, and Turkey being major producers. Some production also occurs in California and in southern Africa. In areas where it is climatically adapted it succeeds in dry areas with poor soils where many other crops cannot be grown.

Culture

Propagation Olives are propagated vegetatively, either by cuttings of leafless hardwood 2.5−7.5 cm in diameter, or by leafy semi-hardwood or hardwood cuttings from one or two-year-old growth. The former cuttings are taken in mid-winter, cut to 30 cm lengths, and their bases then soaked for 24 hours in a 13 ppm solution of indole butyric acid. They are then buried in damp sawdust for 30 days at 13°C to 21°C to form root initials. Afterwards they are planted in the nursery with all but 4 cm below ground and kept moist.

Leafy cuttings are taken in the spring and should be 10 to 14 cm long. They are treated with indole butyric acid either by dipping them for five minutes in a 5000 ppm solution of 50 per cent alcohol and 50 per cent water or by using a medium strength rooting powder. Cuttings should then be stuck into a rooting bed with bottom heat; transpiration should be minimized by covering with glass or plastic or by the use of intermittent mist.

Plants can also be propagated by digging up suckers with a root piece attached from around the base of existing trees, by grafting or budding, or by planting knobby-type growth from the trunks of old trees. Where Verticillium is a problem it may be desirable to use the resistant cultivars 'Oblonga' or 'Allegra' as rootstocks. Seedlings should not be used as rootstocks because performance will be variable.

Cultivars There are hundreds of cultivars, many being selections from seedling trees which have been grown for hundreds of years. Cultivars for table use usually produce large fruits while smaller fruited cultivars are used for oil. There are also differences in the chilling requirement with those cultivars originating from southern Greece apparently requiring less cold. Cultivars differ in oil content with some suitable only for table use due to their low oil content, while others may be used for both table and oil. In high rainfall areas cultivars with some resistance to peacock spot such as Manzanillo should be considered.

Environment and soils The olive grows well only in areas with hot summers and considerable chilling during the winter. Unlike deciduous fruits, where chilling is required for breaking rest, chilling in the olive is required to stimulate the initiation of flowers. The amount of cold required varies from cultivar to cultivar. Areas where the average of the mean high temperature and the mean low temperature during winter are around 12.5°C are likely to have sufficient chilling for olive production. There is considerable disagreement among researchers as to the optimum temperatures and number of chilling days required, however there is no doubt that in areas where cold weather does not occur during winter, there will be no flower initiation. Olives will also be injured by winter temperatures below −10°C and late spring frosts or early fall frosts are undesirable. Fruit set will be impaired by hot, dry winds or cool, wet weather during flowering.

Olives are quite tolerant of a wide variety of soils as long as they are well drained. In some cases they will yield better on soils of low to moderate fertility than on deep, rich soils which tend to encourage too much vegetative growth. They are tolerant of high boron and salt levels as well as alkaline soils and can thus be grown in soils and utilize irrigation water which would be damaging to most other crops. Soils which remain wet for long periods or those with a pH greater than 8.5 will, however, impair growth.

Fertilizer   Where soil and leaf analysis data are not available, nitrogen can be applied at the rate of 0.5 to 1 kg of N per tree per year. The nitrogen status of the trees can be judged as follows. If leaves are dark green and new growth grows 25 to 50 cm per year, no additional nitrogen is needed. If trees are light green and growth is less than 25 cm per year, nitrogen fertilization will probably be of benefit. In soils low in potassium applications of 0.5 kg of K_2O will be beneficial. In some cases boron deficiency will occur and can be corrected by the application of borax at 0.25 to 0.5 kg per tree. Boron deficiency is exhibited by the formation of short branched twigs instead of the long shoots normally formed.

Pests and diseases   The most serious pests on olives are various scales particularly in areas where temperatures are cool in the summer. Nematodes can also be a problem so only nematode-free plants should be grown and soils should be checked for nematode infestation prior to planting.

Peacock spot (*Cycloconium oleaginum*) disease is a serious problem in high rainfall areas or where trees have been planted too densely. The symptoms are dark green or black circular spots on the upper leaf surface which gradually expands causing the leaf to yellow and drop. Control is by the application of copper sprays beginning soon after the harvest of the previous crop.

Verticillium alboatrum causes trees or sometimes only single branches of trees to wilt, due to the clogging of their vascular tissue. Leaves and flowers hang on after the plant is dead. The disease is soil-borne and its incidence can be decreased by not planting where cotton, tomatoes, potatoes, or other susceptible crops have grown before. Do not intercrop susceptible plants in orchards. Resistant rootstocks can also be used. In some cases, the use of grass cover crops will decrease the inoculum in the soil.

Olive-knot (*Bacterium savastanoi*) can be a serious problem which causes galls to form all over the tree. It is transmitted primarily by pruning tools and by propagation from infected trees. Control is difficult and involves the removal of galls where only a few are present and the removal of heavily infected trees. The sterilization of pruning equipment between trees is essential in an infected orchard.

The crop cycle

Nursery-raised trees are planted either bareroot or from containers at a spacing of from 7m × 7m to 13m × 13m. When closely spaced, every other tree should be removed when the trees begin to crowd. Since fruit is borne on one-year-old wood on the outer surface of the tree canopy, it is essential that sunlight penetrates into the tree. Although trees are drought tolerant when established, they should be irrigated regularly during establishment and regular irrigation thereafter will greatly increase yields. Young trees should be pruned to three scaffold branches using the vase system (see Chapter 6) and thereafter annual pruning should be used to remove old wood in the bearing area of the tree and to encourage new growth which will fruit the following year. Biennial bearing is a serious problem with many olive cultivars which can be alleviated by fruit thinning which will also produce larger and better quality fruit. Thinning can be done by hand, which is laborious, or by the application of napthylene acetic acid from four to 18 days after full flowering. In many cases olives are self-fertile and a pollinator is not required. However, in many environments cross-pollination is required and even when not required will increase the quantity and quality of fruit. While bees are the primary pollinators, the wind dissemination of pollen is also important for trees within 30m of each other. Trees will normally begin to bear from three to four years after planting and yields will increase up to about 15 years of age. Trees are long lived and may remain productive for 100 years or more.

Harvesting and post-harvest handling

The stage of harvest is dependent on the way that the olive will be processed. Fermented olives are usually harvested when they have reached full size but are still green. These are allowed to undergo a lactic acid fermentation, followed by a treatment with sodium hydroxide, and are finally stored in a strong brine solution. Black 'ripe' olives are treated in the same way, except that they are allowed to oxidize during the process which creates the black colour. Truly ripe olives are picked when they are black and firm on the tree and then usually packed in dry salt. After the water has been drawn out by the salt they are stored in oil to which spices have been added. Olives for oil are allowed to remain on the tree until early or mid-winter (when they have reached their highest oil content) by which time they may have shrivelled.

They are usually picked by shaking the tree or by beating the branches with beating sticks. The latter method is not ideal since it tends to remove fruiting twigs and can damage the bark permitting olive-knot bacteria to enter. In good growing areas olives can yield 750 to 900 kg/ha per year.

Papaya *(Carica papaya)*

The papaya (Figure 4-11) is a member of the Caricaceae family and is also known as paw paw. It is a small tropical tree native to South America which normally grows with a single unbranched trunk up to 10 m in height, but is more commonly 4−5 m tall. The crown is covered by large, palmately-lobed leaves up to 75 cm across on long, hollow petioles up to 100 cm in length.

Trees are primarily dioecious although hermaphrodite forms which bear perfect flowers are common in some cultivars. Flowers are tubular, fragrant, white to orange, and 2−5 cm across. Male plants bear a long, pendulous inflorescence on which small fruits occasionally form. Female flowers are nearly sessile on the trunk and are larger than male flowers. Hermaphrodite flowers are most commonly borne in the manner of female flowers but are

(a)   (b)

Figure 4-11
Papaya *(Carica papaya)*:
(a) female plant; (b) male plant
(Photographs by Rex Parry)

occasionally found on the tips of male panicles.

The fruit is a berry with a green rind which changes to yellow in most cultivars when ripe. The inside flesh is yellow or occasionally pink, and seeds are black located in a central cavity in the fruit and surrounded by a gelatinous material.

Small latex vessels extend throughout the tree and are particularly abundant in fruit which has reached full size but has not yet begun to ripen. The latex contains a high percentage of papain which is a proteolytic enzyme used in meat tenderizing, brewing, tanning, and textile manufacture.

Economic importance and distribution

Papayas are locally important throughout the tropics where they thrive in frost-free areas below 1500 m in elevation. Little fresh fruit is exported since it is highly perishable and requires careful handling to reach distant markets in a saleable condition. Hawaii is the major producer of papaya for fresh fruit export while Sri Lanka is the primary producer of papain.

Culture

Propagation Propagation is primarily from seed which germinates from two to four weeks after planting if the soil is warm. To obtain plants of the best quality, seed should be obtained from superior plants. In the cultivar 'Solo', which is the most important commercial cultivar, seed should be obtained only from hermaphrodite plants which have been self-pollinated or crossed with another hermaphrodite. If this procedure is followed, seeds will produce 67 per cent hermaphrodite offspring and 33 per cent female offspring. Dioecious cultivars normally produce a 1:1 ratio of male to female plants which will result in the production of four or five times as many male plants as are needed for pollination.

Propagation by cuttings is possible but only entire branches including the basal swelling should be used. These should be taken early in the growth cycle. Due to the non-branching growth habit of the papaya, trees produce few cuttings and propagation by cuttings is therefore impractical in commercial operations. Grafting of scions from desirable plants onto seedlings is also possible although this is not practised on a commercial scale. Patch budding in which a rectangle of bark containing a bud is placed on the stock in an identical size rectangle from which bark has been removed is the most successful technique used.

Cultivars In most areas, papayas are grown from open-pollinated seed of no known identity. In many situations, this is preferable to growing selected cultivars since the papaya is sensitive to environmental changes and naturalized plants will probably be well-suited to the local environment due to generations of natural selection pressure. There are, however, a number of cultivars which breed fairly true and which are worthy of trial. The most common is

'Solo', a hermaphrodite cultivar which produces small (0.5 kg), pear-shaped fruits on hermaphrodite trees and larger, rounded fruits on female trees. There are now a number of solo selections available from Hawaii. Other cultivars of merit are 'Bluestem' (hemaphrodite), 'Graham' (dioecious), 'Fairchild' (dioecious), 'Hortus Gold' (dioecious), and 'Honey Gold' (female, vegetatively propagated).

Environment and soils   The papaya succeeds on a wide range of soil types providing that drainage is good. A windbreak is advisable if the area is subject to high winds since the plants are susceptible to breakage. Papayas are tolerant of drought once established but in areas with a pronounced dry season little fruit will be set except during the wet season. Irrigation will increase yields in low rainfall areas but has a disadvantage in that, if irrigation is excessive, the flavour of the fruits may be poor. Flood and furrow irrigation are the most common methods used.

Fertilizers   Papayas are very responsive to fertilizer and yields can be significantly improved by proper fertilization. Since the fruit is formed in the axils of the leaves, plants must be kept growing continually for maximum yield. In Australia, 0.9–1.4kg of 8–12–6 fertilizer or equivalent per year per tree divided into three applications is recommended. Fertilizing of papaya seldom causes too much vegetative growth and sparse fruiting as is the case with many other species.

Pests and diseases   Few insect pests seriously damage papayas but damage by diseases can be severe and often accounts for the short life of the tree. Seedlings are very susceptible to damping off and older plants are susceptible to root and collar rots. All of these diseases are most serious in waterlogged soils, therefore the importance of good drainage cannot be over-emphasized. Perhaps the most potentially serious disease is bunchy top caused by a virus transmitted by homopterous insects. Bunchy top-infected plants yield little or no fruit and should be removed to prevent spread of the virus to healthy plants.

Root-knot nematodes can also be a serious problem, especially in sandy soils. If moisture and fertility levels are high, plants will produce fairly well despite nematode infection. However, under dry, hot conditions severe yield reductions can be expected. The application of mulch around the base of the tree is often beneficial in reducing nematode populations.

Anthracnose (*Colletotrichum gloeosporioides*) causes spotting of the fruits which renders them unusable. Powdery mildew on the leaves may also be a problem. Fixed copper or benomyl sprays will control both of these diseases.

The crop cycle

Papayas are usually spaced 2.7m apart with up to four seedlings at each planting station. These are thinned to one in each location after flowering has begun and the sex of the plants has been determined. One male tree for every 12 female trees is necessary for pollination in dioecious cultivars. In hermaphrodite cultivars male trees are unnecessary.

Papaya trees begin to bear very quickly after planting and will generally produce ripe fruit within a year, with the exception of 'Solo' which may take up to 18 months. Trees should be maintained with only one main stem and will generally continue to produce profitable crops for between two and three years. After three years yield usually begins to decline and the fruits are borne too high on the tree for easy picking. At this stage it is best to remove the trees and replant. A nematode-resistant crop should be planted for a season or two before rotating back into papayas. Yields vary tremendously, but 30 fruits per tree year is considered to be a minimum acceptable yield in many areas.

Harvesting and handling

Fruit is harvested at the first signs of yellowing if it is to be sent to distant markets; it may remain on the tree a day or two longer if intended for local markets. Papayas should be stored at temperatures between 10° and 13° for maximum storage life. Lower temperatures will cause chilling injury and fruits will fail to ripen properly. Since the skin is extremely delicate, very careful handling is necessary. A post-harvest thiobendazole dip will aid in anthracnose control.

Pineapple (*Ananas comosus*)

The pineapple (Figure 4-12) is a member of the Bromeliaceae family and native to tropical America. It is a perennial monocotyledonous herb with a terminal inflorescence and fruit. Typical of the family Bromeliaceae, each shoot flowers only once and is replaced by off-shoots which arise from the base of the stem. Leaves are leathery, strap-shaped to lanceolate, and slightly to fully serrate. Leaves are borne at closely-spaced nodes forming a rosette and are curved and positioned so as to channel water which falls on them into the funnel-shaped centre of the plant and then downward to the roots. A layer of colourless, water-storage cells is located just below the epidermis. These cells shrink or expand according to the moisture status of the plant and allow the pineapple to survive lengthy periods of drought.

The fruit is a rounded to cylindrical berry composed of 100−200 segments called fruitlets which are seedless in cultivated varieties. The fruit is capped by a short, shoot-like growth called a crown. The crown can be used for asexual reproduction and shades the developing fruit, thus decreasing the risk of sunscald.

The root system consists mainly of secondary roots. These roots are often not developed adequately enough to anchor large plants and, since they are shallow, tend to inhabit the drier surface layers of the soil.

Figure 4-12
Pineapple (*Ananas comosus*)

Economic importance and distribution

The pineapple is an important luxury fruit of tropical regions. The primary areas of world production are Hawaii, Thailand, Taiwan, the Philippines, Brazil, Mexico, Puerto Rico, Australia, and parts of Africa.

Culture

Propagation   Pineapples are propagated vegetatively by planting crowns, slips (shoots growing from the stem beneath the fruit), or off-shoots (suckers from the lower part of the plant). These propagules are extremely resistant to desiccation and root readily when planted in the nursery or in the field. By planting various types of propagules the period of harvest can be extended since off-shoots fruit in about 17 months, slips in 20 months, and crowns in 22–24 months.

Planting generally takes place at the start of or during the rainy season. In areas with regular rainfall or where irrigation is practised, planting can take place at any time. Leaves should be removed from the lower 3 cm of the propagule stem and propagules should be allowed to air dry for a week or two prior to planting so that a callus layer forms over the damaged tissue. Drying can be done easily by laying the cut propagules on top of the plant from which they were harvested. Research has shown that the best-quality fruits are generally produced by the largest propagules.

Cultivars 'Smooth Cayenne' is by far the most important pineapple cultivar throughout the tropics; the leaves are almost spineless and large fruits of good quality are produced. 'Queen' is still grown in some areas. This cultivar produces smaller but sweeter fruits than 'Smooth Cayenne' but the leaves are spiny and unpleasant to work with. 'Red Spanish' is a semi-spineless cultivar grown primarily in West Africa and South Africa. Its fruits are intermediate in size between those of 'Queen' and 'Smooth Cayenne' but are of better quality than 'Smooth Cayenne'. It has some resistance to mealybug wilt disease. 'Hilo' and 'Baron de Rothschild' are 'Smooth Cayenne' types which are occasionally grown.

Environment and soils Pineapple production is generally limited to low or middle elevation tropical areas and to frost-free microclimates in subtropical areas. Pineapples will grow in most soils providing that they are well drained and not highly alkaline, although sandy soils with a pH range of 5.0−6.5 are preferred.

Fertilizers Nitrogen is the nutrient most used by pineapples. A general recommendation is 50 kg of nitrogen per hectare applied as a top dressing about a month after planting, followed by another application of the same amount two months later. Additional nitrogen can be applied at between three and six monthly intervals until a total of 450−670 kg per hectare has been applied. In areas with very heavy rainfall and porous soils it is advisable to make smaller applications at intervals of between two and three weeks. An additional 200 kg per hectare of nitrogen can be applied to each ratoon crop (see crop-cycle section).

Phosphorus is required in very small amounts by pineapples and is not needed where soil analysis by the resin-extraction method shows a content greater than 6 mg per kg of soil. Where phosphorus is lacking, plants have narrow, brittle leaves with a dark red colour spreading over the entire leaf.

Potassium is an important nutrient and when it is deficient fruit quality will be poor and few suckers or slips will be produced. Soil or leaf analysis techniques can be used to determine the exact amount of potassium fertilizer required, but where these are unavailable, approximately 300 kg/ha of potassium should be applied to sandy soils or 200 kg/ha to clay soils. The best method of application is a pre-plant broadcast application mixed with the top 15 cm of soil.

Micronutrients are sometimes deficient, particularly iron and zinc. Iron deficiency is identified by a general chlorosis similar to that typical of nitrogen deficiency. Zinc deficiency causes a mottled yellowing of the leaves. These deficiencies can be corrected by foliar sprays of 2 per cent iron or zinc sulphate.

Pests and diseases Weeds are the primary pest affecting pineapple production. To obtain maximum production, perennial weeds should be dug out prior to planting and all newly germinating weeds should be controlled while in the seedling stage. A mulch of leaves, grass, or other organic material such as chopped debris from previous pineapple crops is beneficial in restricting weed growth, decreasing erosion, adding humus to the soil, and conserving moisture. In soils which are poorly drained a mulch may, however, increase disease problems. Selective herbicides are available which provide effective weed control (see Appendix).

Nematodes, particularly the root-knot nematode (*Meloidogyne* spp.) may become a problem where pineapples have been grown for several cycles on the same land. Although they have not yet become serious in most areas of smallholder production, rotations with nematode-resistant crops must be followed or pineapple yields will decline. Where nematodes are present initially, the soil should be fumigated prior to planting.

In some areas white grubs of various kinds and the adult black maize beetle (*Heteronychus orator*) are a problem. Control is obtained primarily by the use of soil insecticides. Pineapple scale (*Diaspis bromeliae*) occurs on pineapples wherever they are grown but is rarely severe enough to warrant spraying. These insects generally infest individual plants, causing chlorosis and wilting. Populations decline after periods of rain.

The pineapple mealybug attacks leaves, roots, and fruits of pineapples and is a serious pest in many pineapple-growing regions. Infestation by mealybugs results in yellow spots appearing on the leaves but, more importantly, the insects spread black spot (*Penicillium funiculosum* and *Fusarium moniliforme*) and are also responsible for mealybug wilt, which is thought to be due to a toxin produced by the mealybug. Mealybugs, and consequently the diseases associated with them, can be controlled with insecticides.

Mites and thrips are also pests of pineapple but they are not usually of economic importance. Thrips may be important as vectors of yellow spot virus which causes deformed fruits or the production of fruits with several dead fruitlets. Hosts of the virus other than pineapple are weeds such as *Bidens pilosa*, *Datura stramonium*, *Emila sonchifolia* and vegetable crops such as pepper, tomato, egg plant, potato, broadbean, spinach, and peas. Yellow spot of pineapples is usually more severe when these plants are grown nearby.

Heart and root rots caused by the fungus *Phytophthora cinnamomi* and *P. parasitica* are sometimes serious. Pink disease, caused by a bacterium transmitted to ripe fruits by insects, causes a browning of canned fruits but can be prevented by early harvesting. Yeast fermentation, caused by *Saccharomyces* spp., is primarily a problem on ripe-harvested fruits although it may also be a problem on damaged fruits in the field. This disease enters fruits via wounds and causes the fruit to ferment, producing a yeasty, alcoholic smell and an undesirable off-flavour. Control is by careful handling to prevent injury and by sorting so that injured fruits are consumed or processed rapidly and only uninjured fruits are sent to market.

The crop cycle

'Smooth Cayenne' pineapples are planted on 90 cm wide beds, two rows per bed with the rows 60 cm apart. Beds are separated by 60 cm pathways for maintenance. When plants are grown on a large scale, plant populations of 45,000/ha for canning and 55,000/ha for fresh fruit are used. The different densities determine the size and weight of the fruits produced. Propagules should be planted sufficiently deep to ensure that they are firmly supported and should be selected to be of uniform size and type within a given field.

A crop cycle usually lasts from three to four years. Normally, after the first fruit has been harvested (plant crop) two or sometimes more new shoots will be produced at the base of the plant. These will produce fruit in about a year and are called the first ratoon crop. They complete a three year cycle for a crop propagated from slips or crowns. A second ratoon crop will follow in the same fashion if the plant is vigorous and this will complete a four year cycle. However, in most instances, the plant will either not be vigorous enough to produce an economic second ratoon crop or will be so poorly anchored by its small root system that it will not be able to support the weight of the by now very large plant. Whether or not to allow the plants to produce this second ratoon is a decision which should be based on the condition of the plants. It should also be realized that fruit size in the first ratoon corp is smaller than that of the plant crop and will become smaller yet in the second ratoon crop.

In Hawaii and other areas where pineapples are grown on a large scale, plant growth regulators are commonly used to promote earliness and uniformity of fruit set. The primary growth regulators involved are the naturally occurring hormone ethylene, and the artificial plant growth regulators acetylene and NAA (alpha-naphthalene acetic acid). Ethylene is applied either dissolved in water or in the form of Ethrel or Ethephon which are organophosphate compounds which produce ethylene upon exposure to the atmosphere. Acetylene is generated by dropping calcium carbide crystals into the centre of the plant. Acetylene will be formed when the crystals contact water. NAA is dissolved in water and sprayed on the plant. At the present time, the use of plant growth regulators in pineapple is not generally thought to be practical for small holder production due to the cost of the chemicals, the small size of the farms, and the precision timing and application methods required.

Harvesting and handling

The degree of ripeness at which fruit is harvested depends on whether it is to be canned or sold fresh. Since ripening begins at the bottom of the pineapple fruit and continues upward to the top, it is impossible to harvest the fruit when all the fruitlets are at the peak of ripeness. When fruit is to be taken immediately to the cannery, a pineapple with 50 per cent of its fruitlets ripe is preferred. For fresh markets, the degree of ripeness at harvesting will depend on the distance to market and the mode of transportation. As a minimum guide, the bottom tier of fruitlets should have begun to yellow before the fruits are harvested. Pineapples which are picked before this stage will not ripen to their proper sweetness.

During transport, fruits should be kept cool and care should be taken to prevent bruising since this will hasten their decline. Fruits which are to be stored or exported should be graded according to size and condition and all damaged fruits should be discarded. Pineapples which are to be stored should be cooled to 8°C as quickly as possible after harvest. They can then be stored for between two and six weeks if the relative humidity is maintained at 90 per cent.

Questions for study and review

1. Based on how different races of avocados are adapted to different climates, which race would you choose for your area? Explain your answer.
2. List and explain the factors which are limiting expansion of the export market for tropical fruits produced in your country.
3. What are the advantages of growing grafted rather than seedling avocados? What are the disadvantages?
4. How can you determine when an avocado is ready to harvest?
5. Why is it advisable to cut many types of fruits rather than simply picking them from the tree?
6. Every banana shoot has the potential to produce a stalk of bananas. Why then should banana clumps be thinned to three shoots?
7. You are starting a banana plantation but the only propagating material available is infested with root nematodes. What will you do about this problem?
8. What makes the cashew nut an especially appropriate crop for developing nations?
9. Explain how cashew nuts are processed for eating, starting at harvest.
10. Would you plant dwarf or tall coconut palm varieties in your area and why?
11. In your coconut palm orchard the trees have been turning yellow and dropping nuts early and a few trees have died. What is possibly wrong with the trees? What can you do about it?
12. Explain how you would propagate guavas for a commercial guava orchard in your country.
13. What is a macadamia? Describe three activities associated with maintaining a macadamia orchard.
14. Using the information in this chapter and your own knowledge, explain several ways in which mango production could be improved in your country.

15. What is a hermaphrodite papaya? Why is a hermaphrodite papaya cultivar valuable?
16. In addition to fruit, papayas yield another valuable product. What is it and for what is it used?
17. Why should papayas be replanted every 3 years?
18. Draw a diagram of a pineapple plant with a fruit on it and show the location of the following propagules: off-shoots, slips, crown.
19. What are some of the desirable fruit characteristics for a mango cultivar?
20. Briefly explain the crop cycle of a pineapple plant from planting of the propagule through the third ratoon crop.
21. What are the climatic requirements for the economical production of olives?
22. Bayoud disease is potentially the most serious disease of dates. What steps should be taken to halt the spread of this disease?

CHAPTER 5

Minor Fruits

The fruits grouped in this chapter have no botanical similarities. They are referred to as minor fruits because they are not grown on a large scale. Some, such as litchi, passion fruit, and strawberry are, however, potentially important fruits for commercial production. Others, such as carambola, are primarily 'dooryard' or home garden fruits which, without a great deal of investment in breeding and research, are not likely to be grown on a commercial scale in the immediate future.

Acerola (*Malpighia glabra*)

The acerola (Figure 5-1), also known as Barbados cherry, is a member of the Malpighiaceae family. It may actually be a hybrid between *M. glabra* and *M. punicifolia* and is native to the Caribbean and Central and South America. It is an evergreen shrub which grows to a height of 10 m with simple, entire, ovate-lanceolate leaves from 3 to 8 cm long. Flowers are red, 2 cm in diameter and borne in umbels of between three and five flowers. The fruit is globular, red when mature, 1–3 cm in diameter, and has yellow flesh with a sub-acid to acid flavour. It is very high in vitamin C, ranging from 1000 to 2000 mg per 100 g of fully ripe fruit and up to 4500 mg per 100 g of partially ripe fruit. The most acidic fruits have the highest levels of vitamin C. Therefore, a single fruit could supply the daily adult requirement of vitamin C. Various cultivars have been selected to provide better fruit and higher vitamin C concentrations, of which 'Florida Sweet', 'B-15' and 'B-17' are notable. Propagation is by air layering during spring and summer and requires six to eight weeks. Leafy hardwood cuttings with an IBA treatment will root in two months, and intermittent mist is helpful. There is considerable variation in ease of rooting between clones. In several tests of rooting hormones, a five to 30 second dip into a 2500 ppm IBA solution was the best method, although there was considerable variation in results. Rooted cuttings or layers develop rapidly and are ready to transplant to the field between six and 12 months later. Seeds for rootstocks should be removed from the fruits

Figure 5-1
Acerola (*Malpighia punicifolia*)

and planted, although germination seldom exceeds 50 per cent due to a high percentage of non-viable embryos. *M. suberosa* seedlings which are somewhat nematode-resistant have been successfully used as rootstocks. Side-veneer grafting, cleft grafting, or shield budding is then used.

Planting sites should be well drained and free of nematodes. Young plants will be injured at −1°C, while mature plants will tolerate −3°C for short periods. Plantations should be planted with a 5m × 4m spacing; or if a hedgerow planting is used, a 5m × 2−3m spacing. In some environments fruit set is a problem which can be partially overcome by cross-pollination between several clones. Parthenocarpy resulting in seedless fruits is common. In vegetatively propagated plants, fruit production will begin in the second year after planting and harvesting must occur frequently since fruit does not store on the plant. Partially ripe fruit will store for several days under refrigeration at 8°C.

Akee (*Blighia sapida*)

The akee (Figure 5-2), a member of the Sapindaceae family, is a broad-spreading tree native to West Africa. Its fruit, which is a three-celled capsule, is gathered from the wild or occasionally from cultivated trees. The edible part of the fruit is the white pulp which surrounds the bottom half of each seed, and this is eaten raw or cooked. The pulp is edible only when

Figure 5-2
Akee (*Blighia sapida*)

ripe but not overripe. If eaten at the wrong time, poisoning may occur due to the presence of hypoglycin A and B. Care must be taken to remove the pink connective tissue to the seed as it also is poisonous. Plants are propagated from seed and will begin to fruit after about five years.

Ambarella (*Spondias cytherea*)

The ambarella (Figure 5-3), a member of the Anacardiaceae family, is also known as wi tree, golden apple, and otaheite apple. It is an upright tree, which grows to a height of 20m with smooth, grey bark and compound leaves with 11 to 23 elliptical leaflets each up to 8cm long. Trees are deciduous in areas with a pronounced dry season. The flowers are small, white, perfect, and borne in terminal panicles. They are self-fertile. The fruit is ovoid, up to 7.5cm long, and consists of a yellowish, firm, juicy flesh containing up to five seeds surrounded by a tough, orange-yellow skin. Ambarellas are adapted to the hot tropics but will grow in the frost-free subtropics or in areas where only light frosts occur. They require well-drained soils. Propagation is by seed, hardwood cuttings, veneer grafting, or shield budding. Large limbs can be stuck in the ground during the rains, and will root — this method is often used to create 'living fences'. There is considerable variation in fruit quality among seedlings so the vegetative propagation of superior individuals will produce a considerable improvement. Trees should be spaced 15–18m apart and irrigated during dry weather

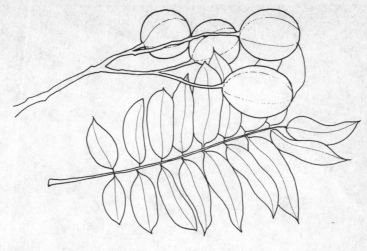

Figure 5-3
Ambarella (*Spondias cytherea*)

when young. After their establishment, trees are somewhat drought-tolerant and will fruit better where there is a distinct dry season. Fruit is harvested when it begins to soften on the tree and changes from green to yellow. It can be eaten fresh, dried, pickled, or made into preserves.

There are several related species, including *S. mombin* and *S. purpurea*, called the yellow mombin and red mombin respectively. The culture of these trees is similar to *S. cytherea*, although they are slightly smaller trees and can be planted at higher densities. The red mombin does not produce fertile seed so must be propagated vegetatively.

Annona (*Annona* spp.)

Plants of the genus *Annona* (Figure 5-4), members of the Annonaceae family, are native to South and Central America. They are mostly small trees which produce compound fruits consisting of many fused ovaries each containing only one seed. Although the flowers have both male and female parts, the stigmas are generally not receptive at the time the pollen is shed. Beetles of several species are important in carrying out natural pollination but complete pollination seldom occurs and misshapen fruit frequently results. Hand pollination often improves both the yield and the quality of the fruit.

The three species most commonly grown are the soursop (*A. muricata*), the sweetsop (*A. squamosa*), and the cherimoya (*A. cherimola*). However, the fruit of the greatest potential is the atemoya (*A. cherimola* × *A. squamosa*) of which several cultivars (i.e. 'Page' and 'Bradley') are available. The soursop is semi-evergreen, upright and should be planted 5 m apart. Fruits are white-fleshed, 15–25 cm long and weight about 2 kg. Their flavour

(a)

(b)

Figure 5-4
(a) Sweetsop (*Annona squamosa*) and (b) Soursop (*A. muricata*)

is slightly acid and they are used in the preparation of cold drinks in their native habitat. They will survive in most areas of the tropics & subtropics.

The sweetsop, also called sugar apple or custard apple, bears heartshaped fruit which is mostly produced on new growth. Fruit set is poor during hot dry weather even with hand pollination. The carpels of the fruit are not tightly joined as in other annonas and may even separate sufficiently for the underlying white flesh to be visible at ripening. The flavour is sweet but the fruits are less acid than the soursop.

The cherimoya grows well in the tropics at elevations above 700 m. It is a bushy tree which is planted at intervals of 6−7 m. Growth of young trees is rapid but slows quickly with age. Fruits are 0.25−1 kg in weight with the carpels strongly joined. Yield is light unless flowers are hand-pollinated. The flavour is variable and can be extremely good in some selections but may be poor in others. Trees require little care other than hand pollination. Pruning generally reduces yield. Fruits should be harvested when nearly ripe and can be stored at 2−4°C. Ripening fruits develop the best flavour at 12−14°C.

The atemoya is grown commercially in Australia and the Middle East. Propagation is by seed, or preferably by grafting onto *A. reticulata*, *A. squamosa* or *A. glabra* rootstocks. A 9−12 m spacing is used in the field. Frequent topping is performed to encourage branching. Hand pollination may be necessary. Fruit cracking is common. The primary disease affecting annonas is anthracnose (*Colletotrichum gloeosporioides*). It occurs in humid climates and infection induces both dark spots on flowers (causing them to drop) and the mummification of young fruits. Leaf symptoms are the occurrence of small, light green spots and premature leaf drop. Control is by sprays of fermate, phygon, and copper fungicides.

Breadfruit (*Artocarpus altilis*)

The breadfruit (Figure 5-5), a member of the Moraceae family, is one of the staple foods of some Pacific islands but is of minor commercial importance. The fruits are borne on tall trees up to 20 m in height, with large, deeply lobed leaves. The trees are attractive, monoecious, and self-fertile. The fruits are borne near the ends of the branches and in the axils of the leaves. A mature fruit is commonly 15−25 cm in diameter and may weigh up to 4 kg. The shape varies from round to oblong and the skin is rough but thin. Depending on preference, the fruit can be harvested when it is fairly ripe or earlier while it is still starchy. The breadfruit is a good source of carbohydrate and calcium and contains some vitamin A and B. Fruits may be either seeded or seedless. The seeded varieties are called breadnuts and the seeds may be roasted and eaten as nuts.

Propagation is by seed, in the case of breadnuts, or by root cuttings in the case of breadfruit. Roots can be dug during the rains, cut into 20-cm-long pieces and then buried diagonally in sand with the end which was near the

(a)

(b)

(Photograph by Tropical Products Institute)
Figure 5-5
Breadfruit (*Artocarpus altilis*):
(a) tree and (b) fruit

trunk protruding slightly from the sand. Stem cuttings 3–4mm in diameter with 3 or 4 nodes can also be rooted using IBA. Shield budding and approach grafting is sometimes used.

Breadfruits require warm, humid conditions to thrive and generally grow poorly at elevations above 2000m. Trees are planted 10–15m apart and will bear fruit when only three or four years old. The fruit should be picked while still firm, about 2–3 months after fruit set. The fruits are quite perishable and should be consumed within a few days of harvest. Yields of 700–3500kg per tree can be expected from a mature tree.

Canefruits (*Rubus* spp.)

The canefruits, also called brambles (Figure 5-6), are members of the Rosaceae (rose) family. Most require a cold climate but a few will grow in the tropics and wild species are found in some high elevation areas. Two species of brambles grown in the tropics are *Rubus niveus*, the hill or Mysore raspberry, and *Rubus rosifolius*, the Mauritius raspberry, both of which have red fruit. Propagation is primarily by tip layering or root cuttings. Many low-chilling blackberries will also succeed well in upland areas.

Canefruits require careful pruning to yield well. Fruit is borne only on one-year-old wood, so to maximize yield the amount of one-year-old wood also has to be maximized. This is done by removing all old wood as soon as the

Figure 5-6
Canefruits (*Rubus* spp.)
(Photograph by Alan Thomas)

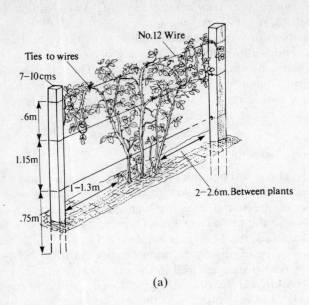

(a)

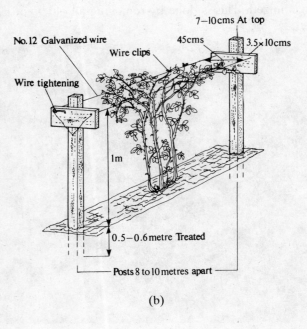

(b)

Figure 5-7
Training systems for (a) vining and (b) upright canefruit

fruit has been picked each year and by keeping new wood separate from one-year-old wood. Some blackberries have long, weak canes and are treated like vines. Others have stiff, upright canes and do not require support. Training systems for both types of canefruits are shown in Figure 5-7. Berries are picked when they reach full colour and are kept as cool as possible until sold. Since the berries are very soft they should be picked directly into the shallow containers in which they will be sold.

Carambola (*Averrhoa carambola*)

The carambola (Figure 5-8), a member of the Oxalidaceae family, is a small tree reaching 5–10 m in height which is grown as a dooryard tree in parts of Africa, Asia and South America. It produces deeply ribbed fruits 8–16 cm in length and yellow to yellow-brown in colour. The fruit has a pleasant flavour when eaten fresh or preserved as jam or jelly. Because it is very perishable it must be consumed locally.

Carambola trees grow best in the hot, humid tropics but will tolerate some cool weather. Propagation is by seed, budding and grafting onto seedling rootstocks, or air layering. Plants are set 7–8 m apart and will begin to bear between two and three years after planting. Flowering continues throughout the year and fruit is available for most of the year. In some areas, fruit-piercing moths may damage ripe fruit. Otherwise trees are not seriously affected by pests.

Durian (*Durio zibethinus*)

The durian (Figure 5-9), a member of the Bombacaceae family, is an evergreen tree which is native to South-east Asia, particularly Malaysia, where it is found in lowland forests of up to 300 m altitude. It is a fruit of the humid tropics which prefers a deep well-drained soil with a pH in the 4.5–6 range. Though the durian grows best at low elevations, successful production has been reported at 600 m in Sri Lanka and 760 m in India. The fruit is actively gathered from the wild and after the selection and propagation of cultivars is cultivated on a small scale. The fruit has a short shelf life and must be consumed within a short time after ripening if kept at room temperature, although it is possible to store it for several weeks at 4–5°C. Where it is known, the fruit is extremely popular and often quite expensive. It is a large semi-deciduous tree up to 35 m tall, although grafted plants seldom exceed 20 m. Leaves are alternate and simple, elliptical in shape, and up to 17 cm long. Flowers are borne on major stems in cymes with between three and 30 perfect flowers. Petals are white or greenish white, up to 6 cm long, and leathery in texture. Flowers usually open in the evening and fall by morning. Pollination is by bats or insects. Fruits are elliptical to globose, up to 30 cm long, weigh up to 3 kg, and have a spiny covering. Seeds are 4 cm long and surrounded by a sweet, custard-like flesh which is white to pale

(a)

(b)

Figure 5-8
Carambola (*Averrhoa carambola*)
(Reproduced by permission of Dr Henry Nakosone)

Figure 5-9
Durian (*Durio zibethinus*)

green. Fruits have a strong disagreeable odour.

The durian has been propagated primarily by seed which must be removed from the fruit, washed and planted immediately, as the period of viability is short. Storage for more than a week may result in poor germination. If fresh seed is used, seed should germinate within three to five days with nearly 100 per cent germination.

Vegetative propagation is recommended to take advantage of clonal material. Durians have been successfully propagated by cuttings rooted with bottom heat, by air layering, and more commonly by patch budding or approach grafting onto seedling rootstocks. Some work is now underway to identify clonal rootstocks and rootstocks of graft-compatible species such as *Cullenia excelsa*.

Patch budding is done when plants are actively growing. The bud is removed from a mature twig from which the leaf petiole subtending the bud has been previously removed. The budpatch should be 0.5 x 2.5 cm in size. Buds should heal in about three weeks after which the wrapping material can be removed. The rootstock is then cut 30 cm above the bud to promote bud sprouting. After growth of the bud has occurred the stock can be removed just above the bud. The bud healing process should occur in a humid, partially shaded environment.

Approach grafting is also widely used, though it is an arduous process which produces relatively few trees. There is some indication that time of year is an important determinant of success. Cuttings and layering can also be successful methods of propagation though they are not widely used.

A number of clones have been developed in Malaysia and Thailand. These differ in fruit quality and yield. In Malaysia the clone D-24 is recommended although, since plants require cross-pollination, other clones such as D-16, D-10, D-8, D-7, and D-2 should also be planted.

Orchards are planted at a 10–15 m tree spacing, but spacings closer than 12 m will require tree thinning later on. Intercropping is commonly practiced while trees are small and there is some evidence that young trees may benefit from shade. Lower limbs are removed to produce 1–2 m of clear trunk, although little work has been done on tree training techniques. Little is known about the fertility requirements of the durian. Vegetatively propagated trees will bear five to seven years later, while seedlings often require 12 years. The economic life of a planting is from 20 to 30 years. Up to 50 fruits per tree per year can be expected, weighing from 50 to 150 kg in total. Fruits are either collected from the ground or cut from the tree just before they drop. Diseases and pests are described in the Appendix, Tables 7 and 8. Diuran, a pre-emergence herbicide, is used for the control of weeds in Malaysia.

Eugenia and Syzygium

The genus *Eugenia* of the Myrtaceae family (which has been recently divided so that many of the South-east Asian species are now included in the genus *Syzygium*) contains several fruits which are primarily of localized importance. They are seldom commercially grown.

Eugenia uniflora is known as the Surinam cherry, Brazil cherry, Barbados cherry, cayenne cherry, and the pitanga cherry (Figure 5-10). It is a shrub or small tree with solitary, white flowers 1 cm across which appear to be a cluster of stamens lacking petals. Fruits, which may be up to 3 cm in diameter, are yellow to red when ripe with eight grooves. Though native to South America, *E. uniflora* is widely grown in the tropics and subtropics, both as an ornamental and for its fruit. Fruiting is better in the humid tropics than in dryer areas, and a mature tree will tolerate short periods when temperatures fall to as low as −3°C. It is tolerant of high pH and a variety of soil types although sandy well-drained soils are preferred. Irrigation is necessary for high yields where regular rainfall does not occur, though established plants will survive droughts. Propagation has been primarily by seed which is sown as soon as it is removed from the fruit and will germinate three to five weeks later. Seed viability is lost between two and three weeks after removal from the fruit. However, for commercial production, selections from trees with superior characteristics should be made and propagated vegetatively. Cleft grafting has been successfully used, as has the rooting of semi-hardwood cuttings in sand with bottom heat under high humidity or under mist.

Syzygium aqueum, known as the water rose-apple, is a small tree 3–10 m

Figure 5-10
Pitanga cherry (*Syzygium uniflora*)
(Reproduced by permission of Milwant Singh Sandhu)

tall native to Malaysia and Borneo. It bears ovate leaves 21 cm long and white, purple or red flowers in terminal or axillary clusters between three and seven in number. The fruit is red or white, top shaped, up to 3 cm across, and has several seeds (although seedless cultivars are available). It is propagated by seed, producing variable offspring, or can be budded onto seedlings or *S. pycnanthum* or *S. samarangense* rootstocks. Propagation by air and mound layering has also been reported. Trees are planted 6−8 m apart in the field.

Syzygium cumini is commonly known as the jambolan, jambool, jambu or black plum. It is native to India, Burma and Sri Lanka where it is found in moist regions up to elevations of 1300 m. It is a tall tree which can reach 30 m, though 12 to 18 m is more common. Leaves are elliptic to 13 cm long. Flowers are borne in small panicles and are followed by purplish red, oval berries up to 2 cm in diameter. A period of dry weather before flowering increases flowering and fruit set, although abundant moisture is needed after the fruit has set.

Trees are mainly propagated by seed which loses viability within a month; seedlings grow slowly in their first year but growth increases dramatically thereafter. Seedling trees often bear fruit of poor quality, however, trees with good flavour and seedless fruit do exist and should be propagated vegetatively. Grafting can be accomplished using T-budding, modified Forkert budding and approach grafting. Air layering is highly successful and cuttings

Figure 5-11
Syzygium spp.
(Bruce Coleman photograph by Gerald Cubitt)

will root with high concentrations of IBA.

Syzygium jambos, the rose apple or Malabar plum is a 6 m to 10 m tall tree native to South-east Asia where it is cultivated for its rose-scented fruits which are eaten raw or made into preserves. Leaves are lanceolate and grow to 20 cm in length. Flowers are greenish-white, 5–8 cm in diameter, and borne in small terminal clusters. Fruits are creamy yellow to rose coloured, up to 4 cm in diameter, and have one or two brown seeds which are often polyembryonic. Propagation is by seed which produces variable offspring since the degree of polyembryony is not known. Grafting by shield and modified Forkert methods onto seedling rootstocks of *S. jambos, S. pycnanthum*, or *S samarangense* is possible. Layering and rooting under mist are also successful. Trees are planted 6–8 m apart and are reported to require heavy fertilization for good cropping.

Syzygium malaccense, commonly known as the Malay rose-apple, the Malay apple, and the large-fruited rose-apple is native to Malaysia but is widely planted throughout the tropics. The tree may reach a height of 15 m, has ovate leaves 15–30 cm long and showy red flowers in cymes on old wood. The fruit is pear-shaped, up to 5 cm long, purple, and contains one large seed. Propagation by seed, cuttings, budding, or layering is possible. Well-drained soil which retains moisture is recommended, and plants are spaced 8 m–10 m apart in the orchard. Fruits ripen 2–3 months after flowering. Termites are reported to be a problem in some areas.

Feijoa (*Feijoa sellowiana*)

The feijoa (Figure 5-12), a member of the Myrtaceae family, is sometimes known as the pineapple guava, and is a shrub or small tree which thrives in cool tropical or subtropical areas. The flowers are purple and the fleshy petals are sweet and edible. The fruit is green, oblong, 4–10cm in length and the inside pulp is consumed. Plants are highly ornamental and can be used in the landscape as well as for fruit production.

Propagation is mainly by seed which, however, produces variable offspring. Leafy softwood cuttings treated with IBA will root in small percentages and is the preferred method of propagating the named cultivars, 'Triumph', 'Mammoth', and 'Coolidge'. Grafting is also used although the success rate is low. Plants can also be grown in hedgerows or as individual trees spaced 3–3.5m apart. Pollinators are needed for the cultivars, 'Triumph' and 'Mammoth'.

Jackfruit (*Artocarpus heterophyllus*)

The jackfruit (Figure 5–13), a member of the Moraceae family, is an important fruit in India and is grown sporadically in other parts of the tropics, mostly as a dooryard tree. The tree commonly reaches 20m or more in height and has simple, entire leaves. The fruit may weigh up to 50kg, but is more commonly in the range of 12–15kg. The pulp can be eaten fresh or dried

Figure 5-12
Feijoa (*Feijoa sellowiana*)

Figure 5-13
Jackfruit (*Artocarpus heterophyllus*).
(Reproduced by permission of Dr Henry Nakosone)

and is fairly sweet; the seeds are roasted and eaten as nuts. The rind is prickly and so gloves are normally worn when picking. In addition to their being valuable for fruit, jackfruit latex is extracted and used for repairing pottery, and the yellow heartwood is used for timber. Trees are also used to shade coffee and as living trellises for black pepper vines (*Piper nigrum*).

Plants are grown from seeds planted either where they are to grow or in plastic plant bags. Selected cultivars can also be grafted onto seedling rootstocks. The seedling quickly produces a delicate taproot and it must therefore be transplanted carefully. Although it is regarded as a lowland tropical tree, the jackfruit will tolerate a cooler climate than the breadfruit. Trees should be set 13–15 m apart in any well-drained soil. Pests are generally not a serious problem. The first fruit is borne about eight years after planting with the exception of such precocious cultivars as 'Singapore Jack'. A yield of 500–700 kg per tree annually is considered good.

Langsat (*Lansium domesticum*)

The langsat (Figure 5-14), a member of the Meliaceae family, is also called the duku (although this is frequently used to describe the types with rounder, thicker-skinned fruit which are free of latex). It is native to Malaysia, the Philippines, and Java and is cultivated primarily in that area, though plantings have been made in Sri Lanka and India. It is a tree of the humid, lowland (below 600 m) tropics and does not fruit well where there is a pronounced dry season. A well-drained soil is required, and filtered shade when it is

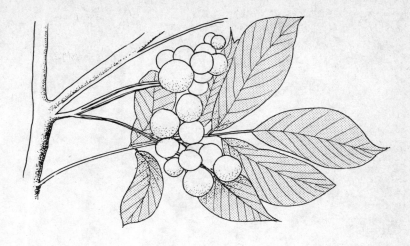

Figure 5-14
Langsat (*Lansium domesticum*)

young is beneficial. The tree typically grows 15−20 m tall and bears large, simple, elliptical leaves with depressed veins. Flowers are small, cream coloured, perfect and borne in racemes on the trunk or large branches. The fruits, 2−5 cm in diameter are surrounded by a leathery skin which exudes latex when injured and are borne in clusters of between five and 30. The fruits are divided into five citrus-like sections which are easily separated and may be seeded or seedless, as the fruits are frequently parthenocarpic. The langsat is most frequently propagated by seed which is removed from the fruit, cleaned and dried, and planted while fresh. Germination occurs from one to three weeks after planting, but seedlings grow slowly. Grafting can be accomplished using cleft grafting, side grafting, or the modified Forkert method. Air layering is successful with rooting often occurring within two months, and layered plants beginning to bear as soon as two years later. Hardwood and semi-hardwood cuttings can also be rooted. Trees are planted at a spacing of 8−10 m and seedling trees begin to fruit from 5 to 8 years later.

Litchi (*Litchi chinensis*)

The litchi (Figure 5-15), a member of the Sapindaceae family, is a semi-tropical evergreen tree which succeeds well in areas with cool or dry winters and warm, humid summers. It has pinnately compound leaves and produces a round-headed tree up to 12 m tall. The flowers, which are perfect, are produced in racemes. The fruit is 2.3−3 cm in diameter and consists of a large seed surrounded by sweet flesh and enclosed in a leathery covering

Figure 5-15
Litchi (*Litchi chinensis*)

called a pericarp. The pericarp is red or occasionally yellow when mature.

Litchis thrive in upland regions of the tropics and will survive the dry season in most areas without irrigation. In lowland areas, trees often fruit poorly due to lack of a cold period, which triggers flower initiation. Trees are fairly tolerant of wet soils and prefer a slightly acid pH. They are relatively sensitive to nitrogen and over-fertilization will slow growth and may be fatal to newly planted trees.

Propagation is mainly by air layering. Large branches (1.5 cm or more in diameter) root better than small ones and best results are obtained during periods of active growth although plants will root any time of the year. Tip cuttings from a flush of growth early in the wet season will root under intermittent mist treatment. Seeds will germinate within two or three weeks if planted immediately after removal from the fruit, but plants raised from seed may take 15 years to fruit and then the fruit is often of poor quality. Grafting or budding has generally not been successful; however, a fair success rate may be obtained if the scion wood is ringed between three and four weeks before grafting. Approach grafting is also successful although it is a low-yielding technique.

After trees have been well established in containers in the nursery they are planted in the orchard at a spacing of 8−18 m. Because of their sensitivity to nitrogen, the amount of manure in the planting holes should be minimal. Only after plants are well established and in active growth should a fertilizer programme begin. Fertilizer should be applied in a split application before flowering and after harvesting by sprinkling it on the soil surface at a distance of 20 cm from the trunk to 50 cm outside the drip line of the tree. Fertilization should be followed by a thorough irrigation. Table 5-1 gives recommended fertilizer application rates.

The litchi fruit is harvested when the pericarp turns colour. The procedure is to cut the entire cluster, usually with at least one leaf attached. Since the point where the pedicel joins the fruit is quite brittle, fruit should be cut from the tree with the pedicel attached. Failure to do this will commonly result in breakage of the pericarp, allowing decay organisms to enter and causing rotting. Fruit should not be picked before it is fully ripe since unripe fruit will not ripen adequately off the tree. Ripening of the fruit is not enhanced by ethylene treatment. Storage is best at 0−1°C, and fruits will store for between two and three weeks at these temperatures although there will be some deterioration in appearance. Fruits may also be canned, frozen or sun-dried.

During the harvest period, litchi orchards often suffer from vandalism caused by thieves attempting to bend branches down to reach the fruit. Since the branches are very brittle, large branches are commonly broken this way. A fruit as valuable as litchi may warrant the posting of guards to watch the trees.

A number of cultivars are grown in China, but relatively few are grown elsewhere. In Africa the cultivar 'Mauritius' is most common. 'Brewster', an important Florida cultivar, is also worthy of trial.

The litchi is usually regarded as a tree with few pests. Despite this, several insects can be a serious problem. The most serious are fruitflies which can result in almost total crop loss. Baiting, as described under Citrus, can be used. Since the fruits are borne in clusters, they are sometimes bagged in waterproof paper or muslin bags immediately after fruit set to prevent fly attack.

The litchi moth (*Argyroploce peltastica*) lays eggs on the fruit, causing damage similar to that caused by fruitflies. The only control is bagging of the

Table 5−1
Amount of fertilizer per year for litchis (kg per tree)

Tree age	Manure	Nitrogen (elemental)	Superphosphate	Potassium Chloride
1−2 yrs	3−5	0.06	0.20	—
3−4 yrs	6−8	0.11	0.20	0.05
5−6 yrs	9−12	0.17	0.3	0.20
7−8 yrs	15−23	0.22	0.45	0.30
9−10 yrs	30−45	0.28	0.90	0.60
11−12 yrs	60−75	0.56	1.50	1.25
13−14 yrs	90−105	0.84	2.50	1.75
15 yrs +	120	1.12	3.50	2.50

fruits. Fruitbats, which feed on ripening fruits, are serious pests in some areas. Unfortunately, the only successful control is floodlighting of the trees which is not economical. Bagging of fruit will, however, reduce damage.

The litchi is resistant to most common nematodes, but two nematode species may cause a problem called litchi decline. These are *Hemicricone-moides mangiferae* and *Xiphinema brevicolle*. These two species are found singly or together and cause considerable damage to roots. For this reason, unsterilized orchard soil should not be used for potting or air-layering plants.

Longan (*Euphoria longan*)

The longan or lungan (Figure 5-16) is a member of the Sapindaceae family and native to India. South China and the Indo-Chinese peninsula. It has become an important fruit in Indo-China and in northern India. The tree is evergreen and grows to a height of 13 m, with compound leaves consisting of between four and 10 elliptic to lanceolate leaflets, each up to 16 cm long. Flowers are small, yellowish brown, and borne in large panicles at the tips of the branches. Fruits are round, from 2 to 4 cm in diameter, and consist of a leathery skin surrounding translucent, white flesh which encases a brown seed. In appearance the flesh is similar to that of the litchi, although the flavour is different. Trees propagated by seed require at least six years to fruit so air layering, which allows the selection of superior types, has been the preferred propagation technique and offspring bear fruit within two or three years. Layers will root in 10 to 12 weeks. Veneer grafting onto seedlings is also used and trees can be topworked by veneer grafting suckers from other

Figure 5-16
Longan (*Euphoria longan*)

trees, which have been severely cut back. There are several named cultivars, one of which is 'Kohala', a Hawaiian introduction.

The climatic requirements for longan are similar to those for litchi — they will be injured at temperatures below −1°C, however, mature trees will survive −5°C, but with substantial injury. A short dry period stimulates flowering. Soils should be moist but well drained. The longan is more tolerant of calcareous soils than the litchi. Plants should be planted 8−10m apart. Alternate bearing is frequently a problem which is overcome by fruit thinning and by ringbarking. Since trees get quite large, occasional pruning to keep them to a manageable size is desirable. Yields are variable although 100kg per tree is considered to be an average yield, with yields from large trees sometimes exceeding 300kg.

Longan fruits are consumed fresh or can be dried, frozen, or canned. In areas where litchis are popular, the longan (since it ripens later) is useful for extending the season with a similar fruit.

Loquat (*Eriobotrya japonica*)

The loquat (Figure 5-17), a member of the Rosaceae family, is native to China where it is an important fruit tree. It is now grown extensively in India and Japan and is of some importance around the Mediterranean. In other areas it is of minor importance, being cultivated mainly as a dooryard or ornamental tree.

Figure 5-17
Loquat (*Eriobotrya japonica*)

The tree is a dense evergreen which grows to about 8m in height. Leaves are elongated, serrate, dark green, and when young are covered with brown fuzz. The white flowers are borne in dense racemes and are fragrant. Pollination is by bees and flies, and cross-pollination may increase yield although trees are normally self-fertile.

Loquats are propagated by seed, air layering, T-budding, and side grafting. Since seedling trees may produce poor quality fruit, vegetative propagation is preferable. Quince, hawthorn (*Crataegus*), and loquat seedlings can be used as rootstocks. Trees are planted 4–5m apart and will grow fairly rapidly once established. Some thinning of branches is desirable when trees are young. Flowering usually occurs during the cool season of the year and thinning of fruits is often necessary to produce large, good quality loquats. The fruits are 3–8cm long and have a tough skin covering a soft flesh with large seeds in the centre. They are harvested when they turn yellow to orange and begin to soften.

Black spot or scab (*Spilorea eriobotryae*) may be serious in some locations. The disease causes rounded dead spots on the leaves and fruit and may cause leaf fall and death of young fruits. Control is through sanitation and twice-weekly sprays of captan, dodine or maneb.

Cultivars which can be tried include: 'Golden Yellow' (requires a pollinator), 'Oliver', 'Fletcher', 'Red Royal', and 'Tanaka'.

Mammey apple (*Mammea americana*)

The mammey apple (Figure 5-18), a member of the Guttiferae family, is a large (18m) evergreen tree native to the Caribbean region. It is cultivated there and in Central and South America for its 20cm diameter fruit which has russetted yellow skin and firm juicy flesh enclosing between one and four large seeds. Leaves are opposite, simple, up to 22cm long, obovate in shape, thick and glossy. Flowers are white, 2.5cm in diameter, and fragrant. The species is propagated by seed, budding, and approach grafting. Fruit is consumed fresh, stewed, and as a preserve. A liqueur is made from the flowers in the French Antilles.

Mamey sapote (*Pouteria sapota*)

The mamey sapote (Figure 5-19), a member of the Sapotaceae family, is a tree of the lowland tropics. Young trees are injured at 0°C while mature trees can withstand short periods of −2°C. It will tolerate a wide variety of soil types as long as drainage is good. Trees should be planted from 6 to 8m apart and provided with regular irrigation. For commercial use, fruit should be harvested when it begins to redden, and for home use when it is completely red. Fruit may be eaten fresh or made into preserves, and the seed when ground has a bitter chocolate flavour.

Taxonomically there is a great deal of confusion over this species which is

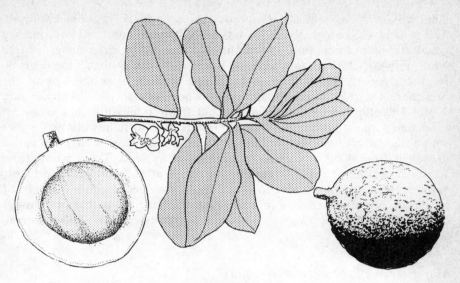

Figure 5-18
Mammey apple (*Mammea americana*)

Figure 5-19
Mamey sapote (*Pouteria sapota*)

commonly misclassified as *Calocarpum sapota* or *Calocarpum mammosum*. The fruit is also commonly called sapote, mammee sapote, sapota, mamey colorado, marmalade plum, and marmalade fruit. It is a large evergreen tree which grows to 25 m (though size and shape vary with cultivar) and bears 30 by 15 cm leaves, which tend to be borne in clusters at the tips of the branches. The 1 cm diameter flowers, borne along smaller branches, are white. The ellipsoid fruit is between 8 and 15 cm in length with a persistent calyx and a russetted brown surface. Flesh is pink to reddish brown, sweet with a slightly granular texture, and encloses a single large seed. Fruit varies in weight from 400 g to 2400 g. Propagation is by seed. Trees grown from seed will require at least 7 years to start to bear and will produce highly variable trees. It is preferable to veneer graft, chip bud which is difficult, or to approach graft which is a more reliable method. A number of cultivars are available of which, 'Copan', 'Magana', 'Mayapan', 'Pantin', and 'Tazumal' are notable.

Mangosteen (*Garcinia mangostana*)

The mangosteen (Figure 5-20), a member of the Guttiferae family, is a native of Malaysia and thrives in hot, humid climates. The tree is an evergreen reaching 15 m in height and is very slow-growing. A mangosteen takes between eight and 12 years to produce its first fruit.

Propagation is by apomictic seed. The seed is, however, highly perishable

Figure 5-20
Mangosteen (*Garcinia mangostana*)

Figure 5-21
Mulberry (*Morus* spp.).
(Bruce Coleman photograph by Michael Viard)

and must be planted within a week of being removed from the fruit. The seedlings require from two to three years of growth in containers to reach a size large enough to transplant and are normally planted 8−10 m apart. Trees produce a tap root with few laterals and a great deal of care must therefore be taken in transplanting to prevent the root ball from falling apart and damaging the roots. Cuttings will also root under mist and approach grafting is possible though difficult.

The fruit, which has the reputation of being the most delicious of all tropical fruits, is a 4−7 cm diameter berry with a tough purple skin covering translucent white segments. Fruits should be allowed to ripen on the tree and will keep only a few days after picking. Because of the perishable nature of the fruit, it is normally consumed near the growing site although storage for up to 49 days is possible at 4−6°C and 85−90 per cent relative humidity.

Mulberry (*Morus* spp.)

The mulberry (Figure 5-21), a member of the Moraceae family, is a temperate fruit but is extensively grown in the middle to high-elevation tropics as a dooryard tree. Two species are commonly grown: the white mulberry (*Morus alba*) which produces white, pink, or purple fruits and grows to a height of 24 m, and the black mulberry (*M. nigra*) which grows only about 9 m tall and produces black fruits. Both species have a spreading form and grow rapidly.

The white mulberry grows at altitude above 300m. The black mulberry, however, requires colder temperatures so should be grown at elevations above 1000m.

The fruits are berries 1−2cm in diameter which are very sweet but low in acid. They can be eaten fresh or made into preserves with the addition of lemon juice or other acid. The leaves can be used for raising silkworms and have also been used as a feed for domestic rabbits. Trees are self-fertile and commonly produce two crops per year. Propagation is by 1-year-old hardwood cuttings which root easily.

Naranjilla (*Solanum quitoense*)

The naranjilla or lulo (Figure 5-22), a member of the Solanaceae family, is a shrub to 2.5m tall which is native to the Andes of Colombia and Ecuador. Leaves are alternate, pubescent, from 20 to 45cm long, and pointed with between 11 and 14 lobes. Veins are bright purple when leaves are young; stems and leaves may have spines. Flowers are white or light purple, perfect, 5mm in diameter, and are borne in axillary cymes. The fruit is a round berry, 3−5cm in diameter, which is yellow to orange at maturity and covered with easily removed hairs. The flesh is greenish and encloses many small seeds. Propagation is by seed, cuttings or by grafting onto the more nematode-resistant rootstocks, *S. macranthum*, *S. verbascifolium* or *S. torvum*. The naranjilla grows best in humid upland areas between 1300 and 1800m in the tropics with rainfall of 1500−3800mm per year. It requires well-drained soil and since it is very nematode-sensitive should be grown where nematode populations are low unless grafted onto resistant rootstocks.

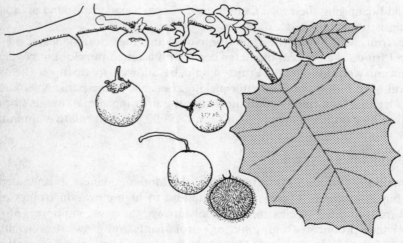

Figure 5-22
Naranjilla (*Solanum quitoense*)

Plants are planted 2−2.5 m apart and will begin to fruit after six to 12 months, and in favorable environments will fruit continuously over a period of two to three years.

Naranjilla is susceptible to a number of fungal and viral diseases which have not been well studied. Yields can be expected to be from 1500 to 3000 kg/ha per year with fruits averaging 40 to 70 g each. The primary use is for juice, jam, and jelly, and as flavoring for ice cream. In Guatemala the juice is made into a frozen concentrate.

Passionfruit (*Passiflora* spp.) — Passifloraceae family

The passionfruit (Figure 5-23), known in Africa as the granadilla, is a member of the Passifloraceae family. It is a vine of potential importance as a vitamin source and as an export in the form of juice, pulp, and syrup for flavouring. Four species are cultivated: the purple passionfruit (*Passiflora edulis*), the golden passionfruit (*P. edulis* f. *flavicarpa*), the giant passionfruit (*P. quadrangularis*), and the sweet passionfruit (*P. ligularis*). Of these only the purple and golden varieties are of commercial importance.

The purple passionfruit produces two crops per year of purple, egg-shaped fruits containing hard seeds surrounded by a fleshy pulp. It grows best at middle to high altitudes and will tolerate some frost. The golden passionfruit produces one crop per year of yellow fruit with a harder rind and a more acid flavour. It requires warmer temperatures and will not tolerate frost.

Plants can be propagated from seed or from cuttings, although seed is the most common method. The purple passionfruit is also sometimes grafted onto other rootstocks to provide resistance to fusarium wilt disease. Vines are planted 3−7 m apart, although the purple variety is less vigorous and can be planted 3 m apart. The vines are trained on one- or two-wire trellises similar to those used for grapes. On one-wire trellises, only one stem is permitted to develop and this is trained in one direction on the trellis. Side branches will grow and hang downward. If a two-wire trellis is used, two stems are allowed to develop and both grow in the same direction with one on each wire. Fruit is borne on the side branches about a year after establishment. Pruning consists mainly of cutting back the side branches before they reach the ground and the removal of branches which develop close to the ground from the main stem.

The passionfruit is cross-pollinated by insects and some clones may be self-sterile. In addition, pollen grains are killed by contact with water so that fruit set may be poor during rainy periods. In Kenya, yields of 16,800 kg per year per hectare are obtained. Plants continue to give economic yields for between five and six years.

Passionfruit respond well to fertilizer, and nitrogen applications once or twice a year are beneficial. In Hawaii, a 10−5−20 NPK mixture at about 1 kg per plant per year is recommended. Under high rainfall conditions this should be split into three or four applications.

(a)

(b) (Bruce Coleman photograph by Eric Chrichton)

Figure 5-23
Passionfruit:
(a) Golden (*Passiflora edulis* f. *flavicarpa*);
(b) Sweet (*Passiflora ligularis*)

Figure 5-23
(c) Purple (*Passiflora edulis*)
(Reproduced by permission of Dr Henry Nakosone)

Three diseases are serious: fusarium wilt, leaf spot, and woodiness. The fungal fusarium wilt is soil-borne and causes a complete collapse of the plant within 48 hours of the appearance of the first symptoms. The disease may be prevented by grafting on to resistant rootstocks such as *P. aurantia, P. herbertiana, P. coerulea, P. incarnata*, and a selected strain of *P. edulis f. flavicarpa*. The latter is most commonly used.

Woodiness is caused by a group of viruses transmitted by vegetative propagation and by aphids. The disease causes stunting, leaf discolouration and, most seriously, thick-walled, misshapen fruit which may shrivel and dry before ripening. All isolated plants should be destroyed. Where symptoms are not severe, fruit can sometimes be picked before ripening for local use, although the sugar content will be low.

Leaf spot (*Alternaria* or *Septoria*) is serious during very wet weather. Benomyl and mancozeb sprays are effective as a control measure.

Pomegranate (*Punica granatum*)

The pomegranate (Figure 5-24), a member of the Punicaceae family, is native to the Middle East but has spread to most tropical and subtropical areas of the world. The plant is an ornamental bush or small tree with small,

Figure 5-24
Pomegranate (*Punica granatum*)
(USDA Photo)

dark green leaves and bright orange flowers. The fruit is a leathery-skinned berry with many seeds, each surrounded by a pink juicy pulp.

The pomegranate grows best at altitudes below 1000 m where the summers are hot and dry. Fruit production is poor in humid areas. The plant will tolerate cool conditions and temperatures as low as −6°C do not harm it. Though tolerant of any well-drained soil, the pomegranate grows best in heavy loam with a neutral pH.

Propagation is by seeds, cuttings, or layers. One-year-old hardwood cuttings root readily and softwood cuttings taken during active growth periods will also root if kept under high humidity. Seed propagation will produce variable plants and therefore is not recommended unless the plants are to be used solely as ornamentals.

Plants are established 5−7 m apart. Early training involves the selection of between three and four scaffold branches which are kept free from suckers. Fruit production will begin between one and three years after planting and 100−200 kg of fruit may be produced per plant per year from fully developed trees.

Fruit is picked before it is fully ripe to prevent splitting and can be stored at cold temperatures for six months. Flavour generally improves in storage. Improved cultivars include 'Wonderful', 'Papershell', 'Spanish', 'Ruby', and 'Purple'.

Rambutan (*Nephelium lappaceum*)

The rambutan (Figure 5-25), a member of the Sapindaceae family, is a tree 12−25 m in height which is native to Malaysia. It is extremely popular there and is also cultivated in many other areas in South-east Asia, as well as in Zaire. The rambutan requires tropical conditions and seldom succeeds in subtropical areas. Regular rainfall is required, though dry weather two or three weeks before flowering increases flower production. Soils should be well drained and preferably high in organic matter with a pH of 4.5 to 5.5. Leaves are pinnately compound, with between two and four pairs of leaflets, and 7−30 cm in length. Trees are dioecious but sometimes bear separate male and female flowers in the same axillary or terminal erect racemes. Flowers are covered with a greenish or reddish pubescence and lack petals. The fruit is yellowish to red, elliptical in shape, and covered with a leathery pericarp with short, soft spines. Inside, a single large seed is surrounded by a white juicy flesh. Fruits are borne in clusters of between ten and twelve and ripen from four to five months after flowering.

Rambutans are propagated by seed which must be planted immediately after removal from the fruit. This, however, produces variable offspring with a high percentage of male plants. For commercial production, patch budding, or modified Forkert budding is most commonly used. Rootstocks are either seedlings or are propagated by layering. Air layering of 12−18-month-old wood is also successful and will root within six to twelve weeks.

Figure 5-25
Rambutan (*Nephelium lappaceum*)

There are a number of clones which have been selected, primarily in Malaysia, of which R3, R4 and R7 are recommended for good yields and canning quality. R137, R139 and R156 are also recommended. In Indonesia, 'Seematjan', 'Seejonja', 'Seetangkoowah', 'Seelengkeng' and 'Seekonto' are recommended. In the Philippines, 'Zamora' is a high-yielding cultivar which produced over 11,000 fruits weighing 234 kg nine years after planting.

Trees are spaced 10−12 m apart in the orchard and vegetatively propagated plants will begin to bear between three and four years later, while seedlings will take a year or two longer. Plants should be shaded and staked when first transplanted to the field and pruned to produce an open centre. Sprays for powdery mildew on inflorescences may be necessary and where fruit bats are abundant bagging of immature fruit may be necessary. Fruit is harvested by cutting the entire raceme from the tree. Yields of 150 to 250 kg per tree can be expected at peak bearing which is 8−10 years after planting.

Sapodilla (*Manilkara achras*)

The sapodilla (Figure 5-26), a member of the Sapotaceae family, is native to tropical America is grown both for its milky sap which is tapped every two or three years as a source of chicle (the main ingredient in chewing gum, although artificial ingredients are commonly used instead today) and for its fruit. The wood is also valuable for timber.

Figure 5-26
Sapodilla (*Manilkara achras*)

The sapodilla is a 20 m high evergreen tree. The leaves are elliptic and are commonly clustered near the tips of the branches. The small flowers are produced in the leaf axil and are usually borne singly. The fruit is rounded and up to 10 cm in diameter with yellow-brown flesh and a brown skin.

Propagation is by seed, which produces variable seedlings, by air layering, or by budding or grafting. For budding or grafting, seedling rootstocks 1 cm in diameter are used. Just prior to grafting, the bark of the stock should be cut just above the grafting site to allow the latex to drain out. The scion should also be conditioned by girdling the branch from six to 12 weeks before grafting. Even with this conditioning, the number of plants which are successfully grafted is only about one-third.

Trees are difficult to transplant and so are best grown in containers in the nursery. Trees are spaced 7−12 m apart, depending on the growth habit of the cultivar. Trees require irrigation during the dry season for the first three or four years, after which they are able to withstand drought. Response to fertilizer is good and from three to four small applications of a nitrogen-containing fertilizer per year are beneficial. Serious diseases of the sapodilla have not been reported. The fruit may, however, be damaged by fruitflies and several scale insects are occasionally damaging.

The first fruits are produced on vegetatively propagated trees four years after planting and 2500–3000 fruits per tree may be borne over a long harvest period each year. Fruits are picked when they begin to drop and then require two weeks to soften sufficiently for eating.

Strawberry (*Fragaria* spp.)

The strawberry (Figure 5-27), a member of the Rosaceae family, is a popular fruit wherever it is grown and, if suitable short-day, or day-neutral, cultivars are chosen, will produce well in the tropics at altitudes above 1000 m. The plant is a low-growing herb and reproduces naturally by runners or stolons. The production of flowers and stolons is day-length controlled. Flowers are produced in short days and stolons in long days. Since days are uniformly short in the equatorial tropics, many cultivars produce fruit over a long period but produce fewer runners than when grown in temperate regions. Storing plants at 0°C for three weeks prior to planting, and supplying long days through artificial lighting will often increase runner production when it is desirable for propagation purposes.

Most strawberries are propagated by removing rooted stolons but the European strawberry (*Fragaria vesca*), which produces small berries, can be propagated by seed as can the large-fruited cultivar 'Sweetheart'. Plants are set on 90 cm wide beds with three or four rows on each bed. Strawberries must be planted carefully so that the crowns are level with the soil surface because deep or shallow planting is harmful to the plants. Roots should be spread out and not bent. After planting, irrigation must be frequent since the root system is shallow.

After the plants are established they can either be grown in the matted row system (Figure 5-28), where all runners are allowed to remain and a solid cover of the bed is obtained, or the hill or mother plant system where runners are removed and large single plants are formed. The matted row system has the advantage of smothering weeds but yields may be reduced if the plants become too crowded. The mother plant system which is generally preferred in warm climates, is high-yielding and allows for better air penetration, thus reducing disease. Whichever system is chosen, plants should be mulched with straw, plastic sheeting, or other materials. Beds should be replanted each year or, at the most, every two years since older plants do not yield well.

The strawberry is susceptible to many diseases and to nematodes and some spraying is almost always required. In addition, strawberries are commonly infected with several virus diseases. These diseases often produce no symptoms except reduced yield. Nematode problems are prevented either

Figure 5-27
Strawberry (*Fragaria* spp.)
(Photograph by Alan Thomas)

Figure 5-28
Matted row system of strawberry production

by growing plants in soil which has not grown fruit or vegetable crops previously or by sterilizing the soil either chemically or by the use of clear plastic, as previously described in Chapter 1. Plants used for propagation should also be free from nematodes. Viruses can be prevented only by obtaining plants from nurseries producing virus-free plants. In many areas this is impossible so the best procedure is to use only those plants which are free of obvious virus symptoms such as discoloured, crinkled, or misshapen leaves.

Leaf spot (*Mycosphaerella fragariae*) and grey mould (*Botrytis cinerea*) are the most serious fungus diseases. Leaf spot causes small black spots which develop on the leaves in the rainy season and is controlled with captan or copper oxychloride sprays. Grey mould is a problem on fruits during wet weather. The grey-coloured mycelium infects the fruits which rapidly soften and are not edible. Control is by mulching to keep fruits off the ground, adequate plant spacing to allow air circulation and periodic sprays with captan, carbendazim, dicloran, or vinclozolin.

As with most horticultural crops, screening trials are needed to identify the cultivars best suited for different areas of the tropics. Some suggested cultivars are 'Brightmore', 'Daybreak', 'Douglas', 'Florida 90', 'Fresno', 'Klondike', 'Klonmore', 'Lassen', 'Rolinda', 'Solano', 'Texas Ranger, and 'Torrey'. In Malawi, 'Cambridge Favourite' and 'Red Gauntlet' are recommended. In Zimbabwe recommended cultivars are 'Earlibelle', 'Parfaite', 'Robyn', 'Selekta', 'Tioga', and 'Torrey'. The day-neutral cultivars, 'Fern' 'Soquel' and 'Selva' are also worthy of trial.

The yield and thus the potential income from stawberries is high so the use of fertilizers can be justified. Prior to planting, 500−750 kg per hectare of superphosphate and 150−250 kg per hectare of potassium sulphate should be applied. The fertilizers should be broadcast on the beds and mixed to depth of 30 cm. After plants have become established, 75−100 kg per hectare of ammonium nitrate should be applied as a top dressing. A similar application should be repeated a month later. When periods of active growth commence, 100 kg of potassium sulphate and 75−100 kg of ammonium nitrate should be applied per hectare. This is repeated monthly throughout the cropping season.

Strawberries can be expected to produce yields well in excess of 6t per hectare. The fruit is picked when it turns fully red for the local market or canned preserves, or when it is slightly pink for more distant markets. Fruit picked before it is fully ripe will develop full flavour if held in the dark at 25°C. Strawberry fruit is highly perishable and so should be picked carefully with the stem attached and placed directly into the containers in which the fruits will be sold. These containers should be no more than 10 cm deep.

Tree tomato (*Cyphomandra betacea*)

The tree tomato or tomatillo (Figure 5-29), a member of the Solanaceae

(b)

(a)

Figure 5-29
Tree tomato (*Cyphomandra betacea*)

family, is native to the Andes mountains of Peru. It is easily grown from seed and will produce a small tree 3–6m tall. Fruit production will begin as soon as a year from planting. The tree is easily damaged by wind and does not tolerate drought. The red, yellow, or orange fruit is the size and shape of a hen's egg, slightly acid in flavour and, after peeling, can be eaten raw, cooked, or made into preserves.

Propagation is by seed which is sown in a container or nursery bed. At 15–25cm tall they are transplanted to the field at spacings of 3m x 3m. When plants reach 1m in height, the tip should be cut out to encourage branching. Plants respond well to nitrogen fertilizer, particularly after the first fruit has set.

Aphids are the most common pest and transmit cucumber mosaic virus and potato virus 'Y'. Cucumber mosaic causes stunting and mottling of leaves. Potato virus 'Y' causes mottling of leaves and dark spots on the fruit. Yield is also reduced. *Phytophthora* blights (*P. palmivora* and *P. infestans*) also affect plants, particularly during the rainy season. These blights can be prevented by applying copper, daconil, or maneb sprays.

Questions for study and review

1. Describe the difference between a breadfruit and a breadnut.

2. Which of the minor fruits are useful not only for their fruit but also as ornamentals?
3. List three minor fruits which you think would be suitable for export from your country and explain why you selected each.
4. You have been given an orchard site in a humid lowland area near a major city. Select three minor fruit crops which you would plant in the orchard and explain why you chose each one.
5. Describe the ideal orchard location and characteristics for litchi production.
6. You have been assigned to start an orchard and your boss wants it to be producing fruit within the next year and a half. Select four minor fruits which will produce fruit on time.
7. Explain how day length affects strawberry plants.
8. The government is interested in opening a tinned fruit and preserve factory in a rural area. Which of the minor fruits would you recommend growing in order to supply the factory with suitable fruit for processing.
9. Devise an intercropping plan for an orchard using strawberries and another one or two minor fruits.
10. Which of the minor fruits do you think has the greatest potential in your country? Explain your answer.

CHAPTER 6

Deciduous Fruits in the Tropics

Introduction

Deciduous fruits are native to temperate regions where they drop their leaves when winter comes and resume growth when temperatures rise in the spring. The most important deciduous fruits are the pome fruits (including apples, pears, and quinces), the stone fruits (including peaches, plums, nectarines, apricots, almonds, and cherries) and miscellaneous fruits such as grape, persimmon, fig, and kiwi.

The deciduous fruits are important in all the temperate regions of the world. Apples are the most important deciduous fruit and the largest single producing country is the United States, although the bulk of apple production is centred in Europe. Pears are less important than apples and quinces are of only minor importance as deciduous fruits. Of the stone fruits, peaches and plums are the most important. The majority of the world plum production is in Europe, particularly Yugoslavia, Romania, and Germany. The fruits are either consumed fresh or dried as prunes. The most important peach-producing country is the United States. Nectarines, which are smooth-skinned or 'fuzzless' peaches, are produced primarily in California. Cherries are most important in Europe where sweet cultivars for fresh eating are grown (*Prunus avium*) while the United States is the primary producer of sour cherries (*Prunus cerasus*) for cooking purposes. Apricots are a relatively minor fruit, produced mainly in California and in Syria. Due to their perishability most of the crop is either canned or dried. The tropics do not produce appreciable quantities of deciduous fruit, with the exception of Brazil, Indonesia, and Zimbabwe, where these fruits are important. Where adapted to the climatic conditions, these fruits are potentially important as export crops, for processing as jams and juices, for drying and, in the case of grapes, for wine production. In addition, where temperate fruits are currently imported (either fresh or preserved) foreign exchange can be conserved by local production.

147

Dormancy and rest requirements

To ensure that plants remain dormant during the winter and do not start to grow during temporary warm periods, deciduous fruits have evolved a biological cycle involving a minimum yearly 'rest' period before they will begin active growth. This ensures that the plants will remain dormant until spring arrives.

The rest requirement (also called the chilling requirement) of a tree is a requirement for a certain number of hours of cold before the buds are capable of growing when placed in a suitable environment. Each species, and even different cultivars of the same species, have different rest requirements both in number of hours of cold required and the actual temperatures required. Generally, 7°C is regarded as the highest temperature at which chilling can occur, and the rest requirement of a specific fruit is the number of hours of temperatures below 7°C which must pass before bud dormancy can be broken. The effect of chilling is cumulative and, in broad terms, the plant is able physiologically to 'remember' how much cold it has received. When its chilling requirement has been satisfied, it will be ready to begin growing again when placed in a suitable environment.

Just as chilling hours can be accumulated by the plant, they can also be lost or cancelled-out by warm temperatures. At temperatures above 16–18°C, chilling is reversed at the approximate rate of one hour of reversal for every hour that temperatures remain above these critical temperatures. Temperatures between 7 and 16°C do not reverse or contribute to chilling. The temperatures stated here are for general reference purposes as they vary between fruits and cultivars.

In the tropics, the limiting factor in the production of deciduous fruits is likely to be cold. Where cold is lacking the trees will suffer from a physiological disorder called delayed foliation. This can be recognized by the failure of trees to produce leaves properly in the spring. Normally, trees of a certain cultivar in the same orchard will all flower at the same time. Flowering will end within two weeks and the trees will form leaves. However, when there has not been enough cold, trees will flower over an abnormally long period and leaf production will be sporadic. Usually the terminal buds will grow first (Figure 6-1). In some instances, buds will die. The result of this is a poor yield and possibly the death of the tree.

In other cases, flowering will be delayed so that fruit ripening occurs later in the season. This, in turn, means that fruit will be more likely to rot due to the beginning of the rains (particularly in East Africa) or be more susceptible to attack from late season pests. The problem of insufficient rest or chilling is overcome in the tropics by a combination of techniques including: (a) cultivar selection; (b) careful site selection; (c) the use of rest-breaking treatments; or (d) rest avoidance.

The selection of appropriate cultivars is the most important factor in preventing problems arising from insufficient rest period. The chilling require-

Figure 6-1
Delayed foliation of a peach tree due to insufficient chilling.
Note that only the buds at the tips of the branches have broken dormancy

ments of apple cultivars, for example, range from as low as 250 hours to as high as 1600 hours and similar variations occur in other deciduous fruits. It is apparent that while an apple cultivar, requiring only 250 hours of cold could succeed in carefully selected sites in the tropics, one requiring 1600 hours would never have its rest requirements satisfied. Only by selecting cultivars with low chilling requirements can deciduous fruits be raised successfully in the tropics unless special rest-breaking treatments are applied.

The next important factor is site selection. In general, high altitudes have colder temperatures. Therefore sites should be chosen where the altitude is high enough for sufficiently cold temperatures to occur over a long enough period in the winter to meet chilling requirements. Temperature records for a proposed deciduous fruit orchard location should be studied carefully before deciding on the site.

Where temperatures are too warm and insufficient rest occurs, the chilling requirement can be partially overcome through the use of rest-breaking chemicals, the most common of which is DNOC (dinitro-ortho-cresol). DNOC, when sprayed uniformly on the plant at the first signs of budswell in the spring, will improve budbreak and encourage higher yields. It is, however, highly poisonous and must be applied carefully, otherwise the spray operator or the tree may be injured. For peaches, 3 per cent DNOC is applied in an 80 per cent summer oil emulsion at the rate of 10 litres per 500 litres of water. For plums, 3 per cent DNOC is applied in an 80 per cent winter oil at the rate of 6−12 litres per 500 litres of water. Application to apples and pears is the same as for plums except that 25 litres of emulsion per 500 litres of water are used. In addition to the use of DNOC sprays, sprays of vegetable oils, surfactants, and even of household detergents may be somewhat effective as substitutes for DNOC where this chemical is not available. DNOC substitutes are applied in the same manner as DNOC.

Though DNOC sprays are the most widely used rest-breaking treatment, other techniques might also be effective. The training of branches so that they grow nearly horizontally is often effective in reducing the harmful effects of delayed foliation by causing more of the axillary buds to grow. In Java a system of growing apples is used in which plants are grown in areas where little or no chilling occurs. Instead the orchards are manipulated by a system of branch bending to induce many lateral buds to grow. About a month after the terminal buds form on the resulting vertical shoots, the plants are defoliated by hand which induces flowering to occur. This technique is termed 'rest avoidance'. By repeating this process, trees are induced to bear twice yearly and a yield of 30 kg/ha is produced. It is possible that the technique would permit the production of apples in areas which do not receive chilling temperatures.

Maintenance

The planting and maintenance operations carried out in deciduous fruit orchards are similar to those followed for subtropical and tropical fruits except that more attention is given to pruning and fruit thinning.

Pruning techniques Pruning is begun at planting and continues annually throughout the life of the tree. The main objective of pruning is to produce a tree with branches that are strong and well placed so that they can support a heavy weight of fruit without breaking. In addition, pruning is used to eliminate dead or diseased branches and to keep plants from growing larger than is practical.

Though pruning does achieve some important goals it can be detrimental to the tree if too much is done. This is because pruning removes photosynthetic areas and stored carbohydrates from the tree. This results in slowing of the growth of both roots and shoots, leading to reduced early yield. For

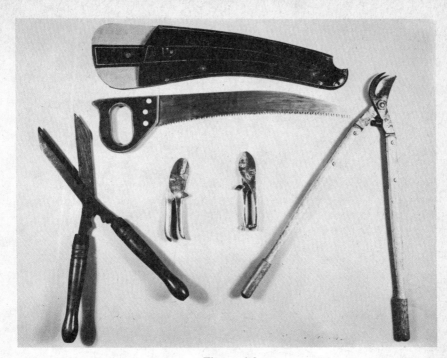

Figure 6-2
Pruning tools: pruning saw and sheath (top), hand pruners or secateurs (centre), and
loppers (right)
(Reproduced by permission of Saunders College Publishing)

this reason branches should be pruned when small so that the reserves of
carbohydrates stored in the wood are not severely reduced. In addition, the
removal of large branches will stimulate the undesirable growth of many
vigorous non-productive shoots, called water-sprouts, which often develop
near the pruning cut.

For pruning to be carried out correctly, well-maintained tools of the
proper type should be used (Figure 6-2). Hand pruners or secateurs are used
for the removal of branches up to 2 cm in diameter and are the most
commonly used tool. If larger branches must be removed, lopping shears are
useful since their long handles provide additional leverage. For branches
greater than 5 cm in diameter a pruning saw is used. All pruning tools
should be kept sharp, and when pruning is in progress they should be dipped
in a 10 per cent sodium hypochlorite solution between trees to prevent the
spread of disease.

There are two important types of pruning cuts used on fruit trees, the
heading cut and the thinning cut (Figure 6-3). The heading cut removes part
of a branch just above a bud, causing side buds just below the cut to grow.
The result is an increase in the number of branches and consequently in the

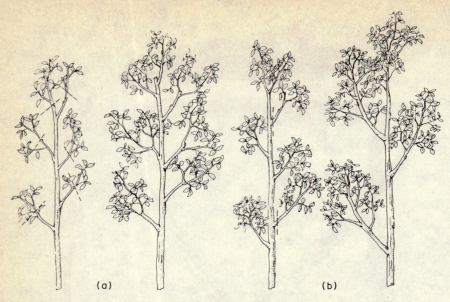

Figure 6-3
Response of identical trees to: (a) heading cuts and (b) thinning

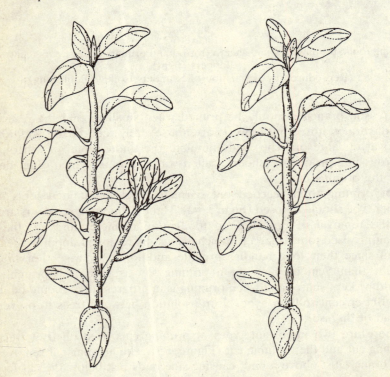

Figure 6-4
Correct procedure in performing a thinning cut. Note that the branch is removed
flush with the parent branch

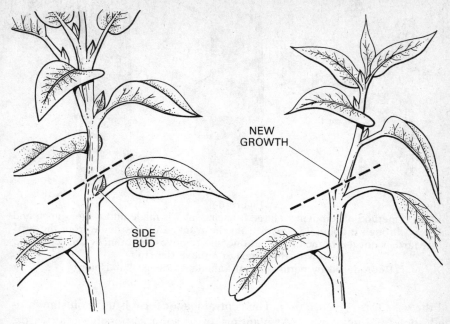

Figure 6-5
Correct procedure for a heading cut showing a slanting cut just above a bud and the
probable response
(Reproduced by permission of Vocational Education Productions)

bushiness or density of the plant. The thinning cut removes an entire branch
at its origin, resulting in a decrease in density of the plant.

With either type of pruning cut, care should be taken to ensure that the
cut is made flush with the main branch or trunk in the case of thinning cuts
(Figure 6-4), or just above a bud in the case of heading cuts (Figure 6-5). If
this is not done, stubs are left which will not heal and are therefore a
potential source of infection.

When it is necessary to remove large branches, as is the case with orchard
renovation, a special pruning technique is used, called the 'three-cut pruning
method' (Figure 6-6). This cutting method is used to prevent possible injury
to the tree caused by a large branch ripping the bark down the trunk as it is
being removed. The first cut is made on the underside of the branch about
30 cm from the trunk and extends about halfway through the branch. This
cut is the insurance that prevents bark damage to the trunk when the main
branch is removed. The second cut is made slightly farther out on the
branch than the first and cuts all the way through so that the branch is
removed. The third cut is made flush with the trunk so that a stub is not left.

Tree training forms   Three main training forms are used for deciduous fruit
trees. These are the central leader form, the modified central leader form,

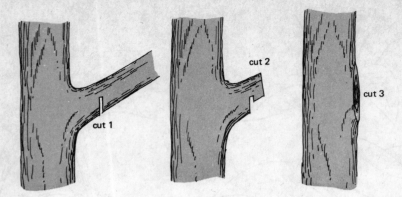

Figure 6-6
'Three-cut' method of removing a large branch. Cut 1 is made under the branch and half way through it about 15−20cm from the trunk. Cut 2 is made from the top slightly farther out the branch from cut one and removes the branch. Cut 3 is made flush with the trunk to remove the stub
(Reproduced by permission of Saunders College Publishing)

and the vase form (Figure 6-7). The central leader form is used on standard-sized apple and pear trees. At planting, or as soon as possible afterwards, the central leader or main upright shoot is selected. Then, as side branches develop, several main branches called scaffold branches are selected from among these side branches and the rest are pruned off. The scaffold branches should, if possible, be evenly spaced around the leader and should be vertically separated by at least 15−29cm. These branches should also pre-ferably be growing parallel to the ground since a wide angle between trunk and branch will make the branch able to support heavy fruit yields. As the tree grows, scaffold branches continue to be selected and interfering branches are removed.

The modified central leader form produces a tree which is shorter than the central leader form. It is used for apples, pears, and sometimes plums. For the first three or four years after planting, pruning is the same as for the central leader form. However, after four or five scaffold branches have been selected, the central leader is cut back to a strong scaffold branch. The result is that the ultimate height of the tree is reduced and the scaffold branches grow more vigorously.

The vase form is used for peaches, plums, apricots, nectarines, and almonds. In training young trees for the vase form, three to five scaffold branches are selected about 45−60cm above the ground. All the branches originate near the same point on the trunk and the central leader is removed flush with a side branch. The resulting tree has a open centre to admit sunlight and encourage fruiting, and a low-growing, vase-like form.

In addition to pruning, various other training techniques are used to create strong branches. The most common is the use of branch spreaders

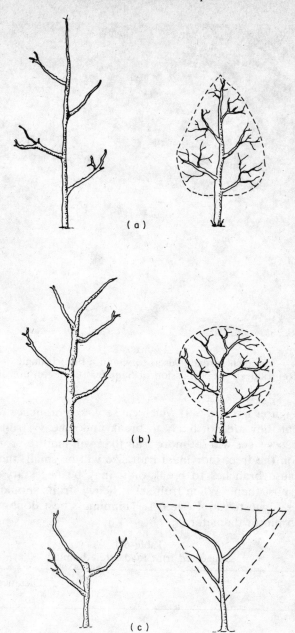

Figure 6-7
Fruit tree-training forms:
(a) central leader,
(b) modified central leader,
(c) vase

Fruit and Vegetable Production

Figure 6-8
Branch spreader used to widen a branch angle
(Reproduced by permission of Saunders College Publishing)

and wires (Figures 6-8 and 6-9) which cause young branches to grow at wide angles so that they are not likely to break under heavy fruit loads.

In many cases trees will set more fruit than will mature properly. If all the fruit is left on the tree, individual fruit size will be small, the weight of the fruit may cause branches to break or it may be necessary to use branch props to support them. When fruit set is heavy, fruits should be thinned to the distances suggested in Table 6-1. Thinning is best done when the fruits are about 15 mm in diameter.

Table 6-1
Suggested fruit load after thinning

Species	Distance between fruits (cm)
Apples	6−8
Apricots	4−5
Nectarines	10−13
Peaches	15−20
Pears, persimmons, quinces	Thinning not necessary
Plums (Japanese)	5−10

Figure 6-9
Wire used to widen a branch angle
(Reproduced by permission of Saunders College Publishing)

Stone fruits

Introduction

Of the stone fruits, all but cherries are potential crops in the upland tropics. Cherries have high chilling requirements and therefore are generally unsuitable unless low-chilling cultivars are developed. Peaches have the lowest chilling requirements followed by nectarines, plums, and apricots.

Peaches and nectarines (*Prunus persica*)   The peach (Figure 6-10) is native

Figure 6-10
Peach (*Prunus persica*)

Figure 6-11
Nectarine (*Prunus persica*)

to Iran but is widely grown throughout the temperate regions of the world. The tree is relatively small and short-lived. Most cultivars are self-fertile. The fruit is hairy except in nectarines (Figure 6-11) which are a seed mutation with smooth skin. Peaches bear fruit quickly and are therefore termed 'precocious'.

Cultivars are divided into two classes: *freestone* which are used for eating fresh and have a soft flesh which separates easily from the stone, and *clingstone* which have a firm, rubbery flesh which adheres to the stone and are used primarily for canning. There are four distinct races of peaches of which the South China race and its hybrids are most likely to be adapted to tropical conditions. The characteristics of this race are small fruit, a beaked blossom end, sweet flesh, and a low chilling requirement. Peaches are generally grown on seedling rootstocks; however, other rootstocks are listed in Table 6-2. Table 1 in the Appendix lists some peach cultivars which are appropriate for the tropics, together with their chilling requirements.

Plums (*Prunus* spp.) There are many species of plums (Figure 6-12) of which three groups are important: the North American plum (over 100 species), the European plum (*Prunus domestica*), and the Oriental plum (*Prunus salicina*). Of these, only the Oriental plum is likely to succeed in the tropics, although both the Oriental and the European plum are grown in some parts of southern Africa. Oriental plums bear conical or heart-shaped

Figure 6-12
Plum (*Prunus* spp.)
(Photograph by Alan Thomas)

Table 6-2

Some rootstocks for peach, plum, and apricot trees

Rootstocks	Scion species	Resistance	Remarks
Peach:			
Lovell seedlings	Peach, plum, apricot		Well-drained soil; do not replant peach after peach
Nemaguard seedlings	Peach, plum, apricot	Resistant to *Meloidogyne incognita* and *M. javanica*[a]	Well-drained soil
Rancho resistant seedlings	Peach, plum, apricot	Immune to *M. incognita*	Well-drained soil
S-37 seedlings	Peach, plum, apricot	Immune to *M. incognita*	Well-drained soil
Okinawa seedlings	Peach. No information on others	Immune to *M. incognita* Resistant to *M. javanica*	Well-drained, not alkaline soil
Plum:			
Myrobalan (*P. cerasifera*) seedlings	Plum	Some resistance to crown gall, verticillium	For heavy, wet soil
Apricot seedlings	Best for apricot. Satisfactory for plum in nematode-infested soil	Some resistance to peach borer, crown gall. Mostly immune to *M. incognita* and *javanica*. Resistant to *Pratylenchus vulnus*	Well-drained soil; somewhat tolerant to alkaline soil

[a]*Meloidogyne incognita* is a root-knot nematode and *M.javanica* is root lesion nematode.

Source: Adapted from:
1. Norton, R. A., C. J. Hansen, H. J. O'Reilly, W. H. Hart (1963), *Rootstocks for Peaches and Nectarines in California,* Calif. Agric. Extension Service.
2. Ibid. (1963), *Rootstocks for Plums and Prunes in California,* Calif. Agric. Extension Service.
3. Ibid. (1963), *Rootstocks for Apricots in California,* Calif. Agric. Extension Service.

Figure 6-13
Apricot (*Prunus armeniaca*)

fruits with pointed tips. Most Oriental cultivars are self-sterile and require other cultivars for pollination. Table 1 in the Appendix lists some plum cultivars together with their pollination and chilling requirements.

Apricots (*Prunus armeniaca*)   The apricot (Figure 6-13) is a highly perishable fruit resembling the peach, but is smaller with a deep orange colour and a somewhat dry flesh. It is not as widely grown as other stone fruits due to its poor keeping quality. The crop is labour-intensive due to the need for hand-thinning and hand-picking. Since the fruit does not all ripen at the same time, several passes through the orchard are required to pick the full crop. Apricots prefer high summer temperatures and some cultivars have relatively low chilling requirements. Most apricot cultivars are self-fertile and can be grown on apricot, peach, or plum rootstocks. Apricot rootstocks are naturally nematode-resistant. Table 1 in the Appendix lists pollination and chilling requirements of apricot cultivars. Table 6-2 lists apricot rootstocks.

Almonds (*Prunus amygdalus*)   Almonds (Figure 6-14) are similar to peaches except that it is the seed inside the stone which is eaten. The almond, like the apricot, prefers high temperatures, and some cultivars have low chilling

Figure 6-14
Almond (*Prunus amygdalus*)
(Bruce Coleman photograph by Michel Viard)

requirements. Most almonds are self-sterile, so pollinators must be provided. Almonds are grown on seedling almond rootstocks or, where nematodes are a problem, on peach rootstocks. Table 1 in the Appendix lists almond cultivars together with their pollination and chilling requirements.

Diseases and pests

Stone fruits are attacked by several potentially serious insects and diseases. At the present time, diseases are not severe over much of the tropics because of the isolated nature of deciduous fruit plantings. However, since many of the organisms are present, their spread can be expected. In the meantime, only certified disease-free trees should be planted.

Leaf curl (*Taphrina deformans*) is perhaps the most widespread peach disease. Young leaves are attacked, causing thickening, curling, and defoliation. Control is by fungicidal spray such as Bordeaux, or copper and oil applied once during the dormant period and again when buds are swelling.

Insect pests of stone fruits include three species of scale: pernicious scale (*Quadraspidiotus perniciosus*), red scale (*Aonidiella aurantii*) and grey scale (*Diaspidiotus africanus*). All of these pests can cause serious tree damage by their feeding and, in addition, the pernicious and grey scale are quarantined in most countries so that infected fruits or plants cannot be sold. Control is primarily through the use of dormant sprays. Fruitflies and fruit-piercing moths may also be serious pests. Control is as described under citrus pests.

Pome Fruits

The pome fruits, with the exception of the pear, are fairly well adapted to upland areas in the tropics although there are even some cultivars of pear which can succeed in the colder areas. Several apple cultivars have been bred for low chilling requirements and are suitable for the tropics. The quince has a particularly low chilling requirement and can be grown in many areas where other pome fruits fail.

Apple (*Malus domestica*)   Apples (Figure 6-15) are adapted to altitudes above 1300m in many tropical areas although the low-chilling cultivars are often not as good quality as many cultivars requiring colder weather. Trees are propagated by grafting or budding onto either seedling or vegetatively-reproduced rootstocks. Though most trees in the tropics are grown on seedling rootstocks, vegetative rootstocks such as the East Malling series (see Table 6-3) are worthy of trial for increasing pest resistance, decreasing chilling requirement,and to create dwarf trees for easier maintenance and picking. Most apples are partially self-fertile but yield is improved with a pollinator. Suggested low-chilling cultivars and their pollinators are shown in Table 1 in the Appendix.

Figure 6-15
Apple (*Malus domestica*)
(Photograph by Alan Thomas)

Table 6-3
Apple rootstocks

Rootstock	Resistance	Comments
Seedling rootstocks:		
French Crabapple (*Malus sylvestris*)	Crown gall, hairy root, root knot nematode, root lesion nematode. Some resistance to *Armillaria*, high resistance to verticillium	Grade in nursery to remove off types
Delicious, Golden Delicious, McIntosh, Winesap, Yellow Newton, Rome Beauty	Same as French crab but no resistance to crown gall or hairy root. McIntosh susceptible to powdery mildew	
Clonal rootstocks:		
Northern Spy	Resistant to wooly aphids but not S. African strain	Not for infertile soil or replanting
Malling 27	Susceptible to S. African wooly aphid	Extremely dwarfing. Mature trees 1.3 m useful for high-density planting
Malling 9	Resistant to crown rot. Susceptible to S. African wooly aphid	Dwarfing; mature trees 3 m. For high-density planting. Precocious. Prefers cool soil. Staking or trellising required
Malling 26	Same as Malling 27	Propagate by softwood cuttings under mist
Malling 7	Same as Malling 27	Produces trees half normal size. Staking required first few years. Suckers badly. Propagate by softwood cutting under mist

Table 6-3
Apple rootstocks

Rootstock	Resistance	Comments
Malling-Merton 106	Susceptible to crown rot. Resistant to wooly aphid	Produces trees half normal size. Staking not required. More productive than Malling 7. Propagated by hardwood or softwood cuttings.
Malling 2	Resistant to crown rot. Some resistance to crown gall	Slightly dwarfing. Bud 15–20cm above soil and plant deeply for anchorage. Difficult to propagate
Malling-Merton 111	Resistant to wooly aphid	Slightly dwarfing. Bears earlier and more drought-tolerant than Malling 2. Propagate from hardwood/softwood cuttings
Malling-Merton 104	Susceptible to collar rot	Requires well-drained soil. Difficult to propagate. Drought-tolerant
Malling 25	Tolerant of high soil temperatures	Trees earlier bearing and higher yielding than seedlings
Malling-Merton 109	Tolerant of high soil temperatures	Earlier bearing than seedlings. Not for wet soil

Adapted from Hartmann, H. T. and D. E. Kester (1975), *Plant Propagation Principles and Practices*, Englewood Cliffs, New Jersey, USA.

Pear (*Pyrus communis*)   The pear (Figure 6-16) has a higher chilling requirement than the apple and can be expected to succeed only in the coolest
areas. Like the apple, low-chilling pear cultivars often have poorer flavour
and texture than temperate climate selections. Pears are grafted onto several
different seedling pear rootstocks as well as seedling quince. The most
common cultivars grown in the tropics are 'Kieffer', 'Le Conte', 'Pineapple',
and 'Orient'. Most pears are self-sterile so another cultivar should be planted
close by for pollination.

Quince (*Cydonia oblonga*)   The quince (Figure 6-17) is similar to the apple
except that it is harder and has a peach-like fuzz covering its skin. It is not
popular for fresh eating but is good for preserves. Of the pome fruits it is
the easiest to grow in the tropics, surviving wherever peaches will grow.
Propagation is by cuttings or layers. Quinces are self-fertile.

Diseases and pests

The two most serious diseases of pome fruits are scab and powdery mildew.
Scab (*Venturia inaequalis*) is a fungus which causes brown spots on fruits
and leaves. The spores germinate only when leaf and fruit surfaces are moist
for 12 or more hours (depending on the temperature). Thus when conditions
are favourable for spore germination, fungicidal sprays are used.

    Powdery mildew (*Podosphaera leucotricha*) is a fungus which is recogniz-
able by the greyish-white mycelia which appear on the leaves. Control is by
applying fungicidal sprays, beginning at budbreak. Various mites can also be
a serious problem. Integrated pest management using predator mites is the
most effective control.

Grapes (*Vitis* spp.)

There are three important species of grapes: (Figure 6-18) *Vitis vinifera*
called the French grape, *Vitis labrusca* called the American grape, and *Vitis
rotundifolia* called the muscadine grape. *V. vinifera* is grown extensively in
many parts of the world both for winemaking and as a table grape. *V.
labrusca* is grown primarily for juice, preserves, and eating fresh and, to
some extent, for wine. The wine produced from *V. labrusca* grapes is,
however, inferior to that from *V. vinifera*. *V. rotundifolia* is used for eating
fresh and for preserves. In the tropics, both the French and the American
grapes are grown; the muscadine is probably suited though it is not currently
grown.

    Grapes are potentially one of the most important deciduous fruits in the
tropics for eating fresh, drying into raisins, and making wine. Within tropical
Africa, grapes are grown fairly extensively throughout southern Africa,
Zimbabwe, and the Malagasy Republic. Smaller plantings exist in Ethiopia,
Kenya, Tanzania, Mozambique, and Malawi. The most serious problems

Figure 6-17
Quince (*Cydonia oblonga*)
(Photographed by Alan Thomas)

Figure 6-16
Pear (*Pyrus communis*)
(Photograph by Alan Thomas)

Figure 6-18
Grape (*Vitis* spp.)
(Photograph by Alan Thomas)

experienced with grape-growing in the tropics have concerned disease control and lack of research to identify suitable cultivars.

The grape is a vining crop which requires support and careful pruning and training to yield well. There are many training techniques used, depending on the type of grape being grown and the climate. One common type of training is the four-cane Kniffen system which is used for vigorous cultivars, generally *V. labrusca* selections. The head-trained, spur-pruned system is more commonly used with less vigorous cultivars, generally *V. vinifera* selections.

The four-cane Kniffen system utilizes a trellis with two horizontal wires, one 0.75 m above the ground and the other 1.5 m above the ground (Figure 6-19). One strong cane is allowed to grow after planting until it reaches the bottom wire after which it is tied to the wire and is then cut just above it. Side branches then develop and one is tied along each wire. An upright-growing shoot is then selected to grow to the top wire and this is then tied and cut to develop two additional side branches.

Yearly pruning consists of removing all but four canes, but also leaving a short section of stem with between four and six nodes (called the renewal spur) near the origin of each cane (Figure 6-20). Since grapes bear fruit only on the one-year-old wood, the four canes to remain should be selected from the previous year's growth.

In the head-trained spur-pruned system, vines are staked at planting but are not trellised. As the vine grows it is tied to the stake until it is 1–1.3 m

(a)

(b)

Figure 6-19
Standard two-wire trellis:
(a) centre section and (b) end supports
(Photographs by Alan Thomas)

(a)

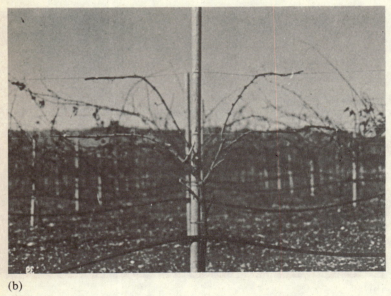

(b)

Figure 6-20
Four-cane Kniffen system of training grapes showing grapes (a) before and (b) after pruning.
(Reproduced by permission of Vocational Education Productions)

tall, at which time the tip is cut to promote branching. The result is a miniature tree which, in four or five years, will develop a strong trunk so that staking is no longer required. Yearly pruning consists of cutting canes which have already fruited back to one or two buds and cutting canes which have grown that year back to four or five buds. Fruit will be borne on the one-year-old canes (Figure 6-21).

The head-trained, spur-pruned system is inexpensive to start since trellising is not required. The system cannot, however, be used on cultivars in which the first two to four buds on each cane are sterile, nor can it be used on very vigorous cultivars such as 'Sultanina ('Thompson Seedless').

In some tropical areas, grapes do not become dormant naturally due to insufficient cold. In these areas, dormancy is induced by pruning and leaf removal twice each year. The first pruning is carried out just before start of the rainy season. The following growth will produce a few fruits; however, this is usually not a large commercial crop. At the end of the rains the process is repeated and fruit is then borne during the dry season under irrigation. Fruit ripened during dry weather is less likely to develop fruit rot than that ripening during rainy weather.

Harvesting

Grapes are ready for harvesting when they have reached the best combination of taste, appearance and keeping quality. One of the bottom grapes in the cluster can be tasted when ripeness is suspected and, if the flavour is good, the remaining berries will also be ripe. The appearance of the trial cluster can then be used as a guide for picking the remainder of the crop. Clusters should be handled only by the stem (peduncle) and should be cut from the vine. Damaged berries are then cut from the cluster before packing. The waxy coating on the berries is called bloom and should not be disturbed. If fruits are to be transported any distance they should be tightly packed in shallow boxes and kept as cool as possible. Grapes which are to be used for wine need not be handled as carefully as table grapes, but they should be taken to the winery as rapidly as possible after picking to prevent off-flavours from developing.

Diseases and pests

The most serious grape disease is powdery mildew (*Oidium tuckeri*). The presence of white, powdery mycelia on the surface of the leaves or any green part of the vine signals the disease, which is spread by wind-borne spores which germinate when leaves are dry. During warm, dry weather leaves should be dusted with sulphur to prevent germination of the spores. Once the disease has begun, benomyl or wettable sulphur sprays are used for control.

Downy mildew (*Peronospora viticola*) is another serious disease which can be recognized by the appearance of translucent yellow spots on the upper

(a)

(b)

Figure 6-21
Head-trained grapes (a) before and (b) after pruning
(Reproduced by permission of Vocational Education Production)

surface of the leaves. On the lower surface a white downy mass will be visible. As the disease progresses, the spots will turn brown and the leaves may drop. Prevention is by sprays of antracol, benomyl, mancozeb, pomuran, basfungin, folpet, or polygram-combi. After the active growing season is over copper sprays may be used. Unlike powdery mildew, downy mildew is a problem during moist weather.

Insect pests include the *Phylloxera* or root louse which is effectively controlled only through the use of resistant rootstocks. Mites, fruitflies and fruit-piercing moths also attack grapes and control is as previously described (see Citrus pests).

Kiwi or Chinese gooseberry (*Actinidia chinensis*)

The kiwi (Figure 6-22) is an unusual fruit with excellent potential in the tropics where at least 400 hours of chilling occurs. The fruit, which is shaped like a hen's egg and is covered with soft brown hair, is borne on vigorous vines which require trellising. Kiwi fruits are higher in vitamin C than oranges.

The kiwi is propagated by grafting or budding onto seedling rootstocks or by rooting firm, semi-hardwood, 1.5 cm diameter cuttings. Cuttings should be treated with IBA and rooted under mist or in a very humid environment.

Vines are planted 5 m apart in the row with rows 4−6 m apart and supported on 2 m tall trellises resembling grape trellises. Pruning is required

Figure 6-22
Kiwi fruit or Chinese gooseberry (*Actinidia chinensis*)

to remove wood which has already fruited, since fruit is borne on one-year-old wood only. The plants are dioecious and one male plant should be planted for every nine females. With the exception of nematodes, pests are not a serious problem. However, well-drained soils are required to prevent root rot.

Fruit is borne three or four years after planting on new growth from one-year-old wood and is picked when it begins to soften slightly. The fruit is snapped from the vine with the calyx attached and should ideally be cooled after picking. If fruit can be refrigerated and wrapped in plastic film after picking it will store for six months at 0°C. The fruit can be peeled and eaten fresh, made into juice, canned, or frozen.

Fig (*Ficus carica*)

The fig (Figure 6-23) is a subtropical fruit which is particularly important in the Mediterranean region. The fruit is used for eating fresh, canning, and drying. It is tolerant of a wide range of environmental conditions, has a low chilling requirement, will withstand some frost and is tolerant of drought although it grows most vigorously with abundant water.

There are two primary types of figs, the Adriatic fig which produces fruit without pollination and the Symrna fig which requires pollination by a fig wasp (*Blastophaga* spp.). To avoid pollination problems, the Adriatic fig

Figure 6-23
Fig (*Ficus carica*)

should be grown unless the fig wasp is known to occur naturally on wild or Capri fig.

Propagation is by 30 cm long hardwood cuttings rooted during the dormant season. Cuttings root easily and can be planted in the orchard during the rainy season within a year of rooting. Alternatively, cuttings can be inserted directly where they are to grow, with a high success rate. Trees are planted 4 m apart with 6 m between rows. Trees respond to nitrogen and regular fertilization will increase yields without reducing fruit quality.

Figs may bear a small crop the first year after planting, although full production is not reached for another four or six years. Two crops per year are normal. Fruits are picked from the tree when they begin to soften and colour changes indicate maturity. When picking, gloves should be worn to prevent damaging fruit and to prevent skin irritation caused by the white sap exuded from the fruit when the stem is broken at picking. Since fruit ripens irregularly, picking should be done daily during the harvest period. After harvest, fruit can be stored for a few days at 0−1°C or dried in the sun.

Although a number of cultivars will fruit well in the tropics, the choice of cultivar will depend on the market (white-fleshed fruits are preferred for export) and the climate. Many cultivars have an opening called an eye at the bottom end. In climates where rain falls at the same time as the fruit is ripening, diseases tend to enter the eye, causing fruit rot. In these climates, cultivars with a closed eye such as 'Black Mission' and 'Kadota' are preferred.

Table 6-4
Fig cultivars

Cultivars	Description
Adam	Large, dark purple fruit; vigorous, heavy-bearing trees
Black Mission[a]	Purple fruit of excellent quality; heavy-bearing trees
Cape Brown (Eva)	Sweet, medium-sized fruit, brown to green-purple; heavy producing but not vigorous trees
Cape White	Small to medium green fruit; firm, sweet, suitable for drying
White Genoa	Like Cape white but larger fruit; vigorous tree but not for hot, dry conditions
Kadota[a] (Pacific, Dottato-white, Endich-white)	Large, yellow-green fruits; 3 crops/year in hot, dry climates
New Brunswick[a]	Excellent quality, large brown fruit; vigorous trees not as high yielding as Cape brown
White Adriatic (Grosso, VerteNebian)	Green-yellow fruit; first crop small, second crop fruit large; not for sandy soil

[a]Indicates closed eye fruits

Where rainfall during ripening is not a problem other selections may be preferred (Table 6-4).

Figs are not usually seriously affected by pests except in high rainfall areas. In these areas and during the rainy season in other areas, fungicidal sprays may be necessary to control leaf spot (*Phyllachora pseudes*) and rust (*Cerotelium fici*). Aphids, fruitflies, and scale insects are occasionally a problem. Figs are highly susceptible to nematodes and should not be planted in infested soils. Weevils and bark-eating beetles have been reported in East Africa.

Persimmon (*Diospyros kaki*)

The Japanese persimmon (Figure 6-24) is adaptable to tropical regions above 1000 m where temperatures do not drop below 10°C. The tree is slow-growing, eventually reaching a maximum height of 15 m. The fruit is shaped like a tomato and may be yellow, orange, or red when ripe. Both seeded and seedless cultivars are available although the latter are most commonly grown. The fruit is sweet but most cultivars have an astringent taste until they are fully ripe. 'Fuyu' is a non-astringent cultivar. Table 1 in the Appendix lists pollination and chilling requirements of selected cultivars.

Propagation is by root suckers or budding and grafting onto seedling rootstocks. Trees are planted 8 x 8 m apart and will begin to bear four years

Figure 6-24
Persimmon (*Diospyros kaki*)
(Bruce Coleman photograph by Michel Viard)

after planting. A well-drained, fertile soil is preferred as is abundant rainfall. Established trees are, however, highly drought-resistant. Diseases and insects are usually not serious. The fruit may, however, be attacked by fruitflies and fruit-piercing moths.

Fruit is of best quality when harvested fully ripe and will store for about a week at 4°C. Astringency of fruits can be reduced by freezing just prior to eating or by soaking for 24 hours in a dilute sodium hydroxide solution. Fruit is eaten fresh, pureed, or steamed to soften and may also be sun-dried.

Questions for study and review

1. Why are deciduous fruits, which are native to and mainly eaten in the temperate zones, of interest in the tropics?
2. What is the main problem associated with growing deciduous fruits in the tropics?
3. Explain how chilling hours can be accumulated and lost according to weather conditions.
4. What symptoms would indicate that a deciduous fruit tree might not have received its rest requirement?
5. What can be done to ensure that deciduous fruit trees growing in the tropics have their chilling requirements met?
6. Physiologically, why is it not generally recommended to remove too much wood and foliage during pruning?
7. Explain the difference in technique and purpose between a heading pruning cut and a thinning pruning cut.
8. Diagram and explain how you would remove a large branch from a fruit tree?
9. What are the advantages of using the vase-training method for a fruit tree?
10. Diagram and explain how scaffold branches should be located on a tree and trained to make the tree strong and able to bear a heavy load of fruit?
11. Why should fruit be thinned when obviously this decreases the number of fruits one will harvest from the tree?
12. What is the botanical relationship between peaches and nectarines?
13. How can the problem of insufficient chilling in grapes be handled?
14. Explain how the timing and duration of the rains in your particular climate area would affect the fig cultivar you would select. Name a cultivar which would be likely to succeed in your area.
15. What region of your country (if any) would you recommend for growing deciduous fruits? Why?
16. If you had an area which only got slightly cold in the winter but you wanted to grow some deciduous fruits, which ones would you try?

Section II — Vegetables

CHAPTER 7

Vegetables

Introduction

Vegetables are those plants which are consumed in relatively small quantities as a side-dish or a relish with the staple food. Most vegetables are the leaves, roots or stems of herbaceous plants, although flowers, calyces, immature seeds or fruits may also be consumed as vegetables. Nutritionally, vegetables are good sources of vitamins, proteins, minerals, and fibre and the composition of some of the important tropical vegetables is given in the Appendix, Table 10.

Three major classes of vegetables are consumed: (a) those that are gathered from the wild such as the leaves of purslane (*Portulaca*); (b) indigenous vegetables which are often gathered but are also cultivated, such as *Amaranthus*; (c) imported vegetable species which are cultivated. Only those vegetables included in the latter two categories have been described in this text since little is known about the collection and use of the wild or naturalized local vegetables. This selection should not, however, be construed as diminishing the importance of gathered vegetables. In many cases, they are far more nutritious than those species which are commonly cultivated and their continued consumption should be encouraged. In some instances, there is the possibility of selecting and improving wild forms of vegetables to make them more suitable for cultivation.

Planning the vegetable garden

Vegetables, like fruits, require intensive cultural practices and the financial and labour inputs involved are therefore greater than those required for most staple crops such as rice or maize. Due to these factors and also to the perishable nature of most vegetables, the vegetable garden must be carefully planned.

Climatic conditions

Although many vegetable producers have little choice in selecting land for cropping, where there are alternatives some of the following factors which affect crop growth should be considered:

(1) *Rainfall*, both the total annual amount and the monthly distribution should be considered carefully in selecting the site. Many crops are sensitive to excessive water in the root zone and the rainfall pattern is therefore likely to determine the range of crops which can be grown. Some crops are severely damaged by heavy rainfall and a site which is exposed to periods of prolonged wet weather may not be suitable for these crops. Where the wet season is limited and there are either one or two distinct dry periods, irrigation may be required to provide the soil water necessary for normal crop growth.

(2) *Temperature* is another major factor affecting growth of many vegetables but one which cannot easily be modified in any practical manner. Those crops which are either sensitive or tolerant to high temperatures should be selected for the local prevailing conditions; with some crops there are cultivars which show more tolerance to high temperatures than others. Where seasonal differences in temperature occur, crops should be chosen accordingly.

(3) *Topography* of an area is important in relation to temperature and temperature fluctuation. The diurnal or day-night temperature variation which occurs at elevations over 800−1000 m is beneficial to many crops, particularly onions and leaf crops such as Brassicas. Reference to the effect on crop growth of the topography of both upland and lowland areas has been made in Chapter 2. Where land is sloping or uneven, soil conservation measures will be required.

(4) *Daylength* is also an important factor which has to be considered in the selection of some introduced or exotic types of vegetables, particularly those which originated in subtropical or temperate regions. Many of these are adapted to daylengths longer than the 12−13 hours which is typical of most tropical areas and it is therefore necessary to ensure that only cultivars which respond to relatively short days, or are daylength neutral, are grown in most tropical regions.

Soils

Vegetables will grow on a wide range of soil types, the main requirement being that they are fertile and well drained. Heavy clays are generally difficult to work particularly during the wet season, but can often be improved by the addition of organic material. Sandy soils will also become more fertile with the addition of organic material which improves the humus content and therefore the water-holding capacity.

The ideal soils for most vegetables are the medium clay loams, but these have to be properly used and well supplied with both nutrients and organic reserves in order to retain their fertility. The acidity or alkalinity of soils can be determined using relatively simple soil-testing techniques. In general, suitable pH values for soils carrying intensively cropped vegetables are in the range 5.5–7.5, although some crops such as tomatoes and peppers can tolerate slightly more acid soil conditions.

A soil analysis, used to determine the reserve of chemical nutrients in soils, is more difficult and has normally to be undertaken by a government soil analytical laboratory. Where an excess or deficiency of soil elements is suspected, an analysis of this type is strongly recommended.

Nematodes are becoming a serious problem in many vegetable-growing areas, and since they are difficult and expensive to control, sites which are nematode-free are preferred. This situation is most likely to occur where vegetable crops have not previously been grown on an intensive scale. Nematodes are very small and cannot normally be seen without a microscope, therefore samples should be sent for examination to the nearest plant pathology laboratory if a serious level of nematode infestation is suspected. Further reference to nematode infestation of soils will be found in Chapter 2.

Water availability and quality

During the dry season, most vegetable crops will require irrigation and it is therefore essential to select a site which has the possibility of providing a year-round water supply. The source of water may be from wells, streams, or rivers and where the possibility exists, the damming of streams or rivers at a higher point will give a suitable rate of flow for furrow irrigation to be used.

The quality of irrigation water is also important, and, if saline or brackish water is used, many vegetable crops will either die or be retarded in growth. An analysis for water quality, particularly the presence of potentially toxic salts, should therefore be considered if crops grown previously have shown symptoms of leaf scorch or reduced growth.

Market outlet and transport

Reference has already been made in Chapter 2 to the importance of selecting a production site which has access to a market, if the crops are either to be sold locally or exported to more distant markets. Most vegetables are highly perishable and are often harvested and sold on the same day. Where they have to be transported for some distance, however, it is essential that they be kept cool and moist; the use of containers which will protect them from damage and overheating is therefore essential.

A reliable form of transport to market is necessary if regular market supplies are to be maintained but this if often difficult to achieve unless the grower also has the means of transporting his produce directly to market. In this situation, he is able to control all the essential operations from harvesting, washing, grading or trimming and packing in containers to the final presentation in the market.

Preparation and layout of site

Before the area selected for cropping is cleared, a plan should be prepared to show the location and extent of the various operations proposed. This plan should mark the boundary lines to be fenced and the area to be planted with windbreaks, the siting of the compost heap and the nursery, the exact position and size of the beds and pathways, and the proposed areas to be planted with each crop. Plans for the rotation system to be adopted should also be included in the main plan.

Fencing and windbreaks

In most locations, fencing will be required to prevent cattle, goats, chickens and other animals from destroying the crop. Fences may be made from local materials such as posts from nearby trees intertwined with branches, or a more permanent barrier may be established by planting thorny hedge plants.

If the site is exposed to strong winds, a windbreak may also be necessary. A windbreak can consist of natural vegetation which is allowed to grow up around the holding or rapidly growing trees may be planted at right-angles to the prevailing wind. A grass fence may also be used as a windbreak for small plots. Due to their shading effect and root competition, living windbreaks should be established at least 6 m away from the nearest crops. Grass windbreaks may be placed closer. Bananas are frequently used as windbreaks since they are quick growing and also produce a bonus of fruit.

Site clearance

All unwanted vegetation should be removed from the area, including trees, bushes, and weeds; the woody remains should be burned and the remainder placed on the compost heap. Stones should also be removed and the soil roughly levelled. Minor obstructions such as termite mounds should be broken down.

Compost area

The compost pit or heap provides a place for the disposal of organic debris and is a source of organic material for soil improvement. It is generally located close to the nursery and may be sited in an area which is unsuitable for cultivation. Whether a compost heap or pit is utilized depends on the climate of the area. In low rainfall areas, a pit will hold moisture well whereas, where rainfall is abundant, a pit is likely to become waterlogged and therefore an above-ground compost heap is preferred. To construct the compost pit or heap, organic debris such as weeds and crop residues, provided they are not affected by diseases or pests, are laid in layers of about 10–20 cm thick. Each layer is followed by a thin layer of soil and a small amount of animal manure, if available. These layers are alternated

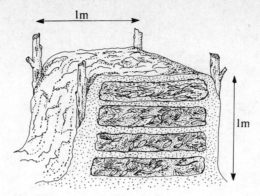

Figure 7-1
Compost heap construction

until the pit is full or a reasonable height for the heap has been reached (Figure 7-1). The compost is then kept moist but not waterlogged while decomposition occurs. This is due to the bacteria present in both the animal manure and the soil. During the early stages, this decomposition results in the evolution of considerable heat. A stick pushed into the heap and removed after about 10 minutes should be warm to the touch, indicating that decomposition is proceeding satisfactorily.

It is desirable to have the nursery area and the compost heap or pit fairly close together since the raising of young seedlings involves the regular use of decayed organic material.

In many parts of South-east Asia, the use of cement-lined slurry pits is preferred; all organic wastes, including livestock residues, are allowed to ferment in these pits. The resulting slurry, when diluted with water, produces an excellent source of liquid manure.

Nursery siting

A lightly shaded area is preferable for at least a section of the nursery and if possible, a plant house should be constructed as indicated in Figure 7-2. This can be built from local materials such as poles from local trees, preferably hardwoods which will withstand termite attack; the roof may be of thatch. The nursery area should cover approximately 2 per cent of the total area of the holding, including beds for raising and transplanting seedlings which will later be transferred to their permanent beds. A regular supply of water throughout the year is essential for the nursery. Reference has been made in Chapter 2 to the requirements for a fruit propagation nursery and many of these are similar to those needed for vegetable seedling propagation.

Irrigation and distribution methods

The irrigation of vegetables is a necessity in most tropical and subtropical

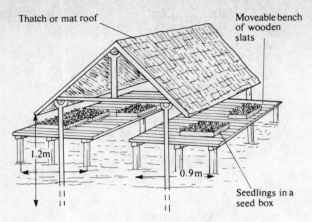

Figure 7-2
Plant house for seedling propagation

areas and plans for a system to provide adequate water, particularly in the dry season, should be made well in advance of planting. Many irrigation systems are used and the system chosen will depend largely on cost and the availability of materials.

The simplest system is the use of watering cans, with water being carried from a stream or water hole. This system is dependent on the availability of family or hired labour but works well for the small farmer and requires little capital input. Since vegetables produce large yields from small areas, hand watering can be economically feasible on a limited scale of operation. Another basic form of irrigation is furrow irrigation which is operated by diverting all or part of a stream through the vegetable garden. This system depends only on the use of locally available materials and labour and is therefore inexpensive. The primary disadvantage is that the site of the holding must be carefully chosen so that it is slightly downhill from the source of the water or, alternatively, the water source must be dammed at a higher point. For efficient operation, the furrows have to be planned to run along a slope and therefore the location of the beds and any terraces required will have to be determined by the demands of the irrigation system.

Sprinkler irrigation is used for large-scale operations but the investment in a pump, distribution pipes and sprinklers is large. In addition, due to the cost of fuel, these systems are relatively expensive to operate and will probably continue to increase in cost as fuel costs rise.

The drip system of irrigation, in which water from a raised storage tank is fed to beds via plastic tubing and metered nozzles, is expensive to install but the maintenance costs are minimal. It has potential usefulness with high-value vegetable crops, particularly those grown under plastic covers or tunnels, but is more applicable to fruit crops.

Whichever system is chosen, it should be reliable and economical. If a mechanical system is chosen, spare parts should be available at short notice since lengthy periods without irrigation can result in substantial decreases in crop yield or may even mean a total crop loss.

Cropping pattern

The area which is to be allocated to each crop has to be planned well in advance so that seeds can be obtained and the land adequately prepared. The type and size of bed to be used for specific crops has also to be decided in advance although it is customary for most vegetables to be grown on beds of equal dimensions, varying the planting or sowing distances in accordance with the size of the crop and its nutrient requirements.

The main choice is between the use of ridges or beds and if the latter, between sunken or raised beds. Climatic conditions generally determine the traditional use in any given area. In low rainfall areas sunken beds are used, particularly for dry season cultivation, so that all water supplied is directed around the root system. Flat beds are sometimes used for the dry season cultivation of more robust crops such as tomatoes or peppers but the use of slightly raised mounds or low ridges is generally preferred.

Rotational cropping

Vegetables, like all crops, should be rotated. The aim of these rotations is to decrease disease and nematode problems and to increase the efficient use of soil nutrients. Whenever possible, vegetables from the same plant family should not be grown in the same location during the same year and, in some cases, a period of at least two years should be allowed to elapse between crops of the same family. In addition, shallow-rooted vegetables should be followed by deep-rooted ones in order to improve the utilization of nutrients. The inclusion of legumes in a rotation has the added advantage of improving soil fertility by adding nitrogen to the soil through the activity of the nitrogen-fixing bacteria which are associated with the roots of legumes and which inhabit the nodules formed on their roots. If the residues of legumes are returned to the soil, some of the nitrogenous material remaining in the roots and other plant parts will contribute to soil fertility. Certain plants, particularly grasses, aid in the suppression of nematodes and are useful in a rotation scheme. Table 7-1 lists groups of vegetables which are commonly used in rotations.

When the rotation is being planned it is necessary to check whether the crop selected for planting will grow during the season of the year for which it is scheduled. Most vegetables are grouped into either warm-season vegetables which require hot weather or cool-season vegetables which require lower temperature conditions (Table 7-1). In medium or high elevation areas, distinct seasonal and sometimes diurnal (day-night variation) differences in temperature occur. In other areas, variations in humidity and rainfall may

Table 7-1
Rotational cropping vegetable groups for use in rotations: their depth of rooting
(D = deep, M = moderate, S = shallow),
and temperature preferences

1. *Brassica or cole crops* (cool season)
Brussels sprouts (S)
Cabbage (S)
Cauliflower (S)
Chinese cabbage (S)
Collard (M)
Kale (S)
Kohl rabi (S)
Leaf mustard or rape (S)
Turnip (M)

2. *Curcurbits* (warm season)
Bitter Gourd (M)
Chayote or cho-cho (D)
Cucumber (M)
Melon (M) to (D)
Pumpkin (D)
Snake gourd (M)
Sponge gourd (M)
Squash (D)
Vegetable marrow (M)
Wax gourd (M)

3. *Solanaceae* (warm season, except potato)
Egg–plant or aubergine (M)
Pepper (M)
Potato (S)
Tomato (D)

4. *Root, tuber, and bulb crops*
Beetroot (M) (cool season)
Bunching onions (S) (warm season)
Carrots (M) (cool season)
Chinese yam (M) (warm season)
Chives (S) (warm season)
Cocoyam (D) (warm season)
Garlic (S) (warm season)
Leek (S) (cool season)
Onion (S) (cool season)
Radish (S) (cool season)

Shallot (S) (warm season)
Sweet potato (M) (warm season)
Tannia (M) (warm season)

5. *Legumes*
Asparagus bean (M) (warm season)
Chick pea (M) (warm season)
Dolichos bean (M) (warm season)
French or snap bean (M) (warm season)
Jack bean (M) (warm season)
Lima bean (M) (warm season)
Mung bean (M) (warm season)
Peas (M) (cool season)
Sword bean (M) (warm season)
Winged bean (M) (warm season)
Yam bean (M) (warm season)

6. *Leaf and miscellaneous crops*
Amaranth (S) (warm season)
Asparagus (M) (cool season)
Cassava (D) (warm season)
Celosia (S) (warm season)
Endive (S) (cool season)
Fennel (M) (cool season)
Indian spinach (M) (warm season)
Lettuce (S) (cool season)
Long–fruited jute (M) (warm season)
New Zealand spinach (S) (cool season)
Okra (D) (warm season)
Roselle (M) (warm season)
Spinach or Swiss chard (M) (cool season)
Sweet corn or green maize (S) (warm season)

affect the growth of a particular crop. In such locations, these factors must be considered in planning the rotation.

Multiple cropping

A more effective use of land is achieved when different crops are grown in association and since their nutritional requirements vary, the rate of depletion of specific soil nutrients is reduced. This is a traditional practice in many warm climates and has the main advantage of reducing the proliferation of pests and diseases, compared with monoculture systems.

Successional cropping

The practice of planting the same vegetable on several dates two or three weeks apart, or planting early, mid-season, and late cultivars at the same time, is called successional cropping. Successional cropping is used because many vegetables have a very short but relatively uniform growth period and therefore tend to be ready for harvest at one time. For a continuous supply, therefore, a particular vegetable should be sown at one to three-week intervals; alternatively several cultivars which have different maturity dates may be sown more or less at the same time.

Intercropping

Intercropping, which is sometimes also called catch-cropping, is the growing of two crops in the same area at the same time. This practice makes the maximum efficient use of the land. In general, rapidly maturing crops such as amaranth, okra, radish, lettuce, or spinach are grown between the rows of a slower growing crop such as tomatoes or pumpkins. By the time the space is required by the longer duration crop, the early maturing crop will have been harvested. When intercropping, the plants which are being grown together must be compatible in their cultural requirements so that operations such as spraying and irrigating can be carried out on both crops at the same time.

Relay intercropping

This practice is a variation on the intercropping described in the previous paragraph in that the second crop is sown or transplanted after the first crop has become well established and may be approaching the flowering stage.

Cultural practices

Tools and equipment required

The basic tools required for small-scale vegetable production may be limited to a cutlass, panga, or matchete for clearing debris or general cutting operations; hoes or jembe for preparing beds or ridges and burying organic residues; a rake for levelling beds prior to planting; a trowel for transplanting seedlings; and a watering can for watering newly planted seedlings and growing crops in beds.

However, for full efficiency and labour-saving, a wider range of tools and equipment is desirable and these are illustrated in Figure 7-3. Some of these can be made locally but, in general, reliable and long-lasting tools should be purchased from local suppliers.

A portion of the nursery plant house can be used for housing these tools to protect them from rusting during wet weather; if tools with metal parts are greased or oiled after use, their period of usefulness will be considerably extendend.

Propagation and crop establishment

Vegetable seeds may be sown in boxes or other suitable containers, in nursery beds for transplanting as seedlings, or they may be direct-seeded in their permanent positions. The method chosen mainly depend on the type of vegetable since some are better transplanted as seedlings about 10–15 cm in height, while other types of vegetables are better sown directly in drills in beds where they are to mature (see Appendix, Table 11). Other factors affecting the method of sowing are the type of cultural system used, the prevailing climatic conditions, the facilities available and the preference of the grower.

Containers may be of various sizes but the main requirement is that they should be provided with adequate drainage holes in the bottom since many seedlings require watering. Rapid root development is prevented if the soil is deficient in oxygen or if excess water accumulates in the container. A well-prepared finely sieved soil, containing both well-decayed organic material and a small amount of fertilizer is considered suitable; heavy soils should have coarse sand or compost added to improve drainage. The seeds are scattered on the firmed and level surface of the container and should then be covered with fine soil to a depth two or three times greater than the diameter of the seed. The seeds of some vegetables are better sown in rows to reduce the risk of 'damping off' disease.

After sowing, the container should be well watered but watering should be limited to maintaining the soil in a moist condition until the seedlings are well established in order to avoid damping off, which is caused by soil-borne fungi. Soil should be sterilized by heating or by applying chemicals as recommended in Chapter 1.

The containers should be kept in a shaded position, preferably in a plant house, until germination begins. Seedlings should then be exposed to increasing amounts of sunlight until they are ready for transplanting, by which time they should be adapted to full light conditions by gradually transferring them to a location where they are exposed to the sun.

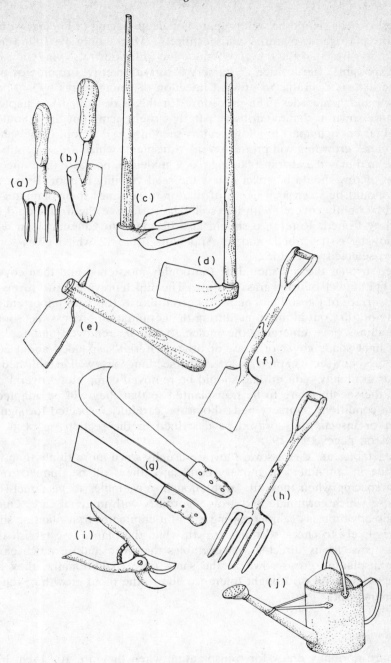

Figure 7-3
Hand tools in general use: (a) hand fork, (b) trowel, (c) forked or khodali hoe,
(d) long handled or West Indian hoe, (e) short handled hoe, (f) spade, (g) cutlass or
machete, (h) fork, (i) secateurs, (j) watering can

Nursery beds should be well prepared by deep digging or hoeing, with the addition of organic material and fertilizers. This should be followed by raking the surface of the bed to obtain a fine tilth prior to sowing the seeds. Where possible, sterilization of nursery beds with methyl bromide or other soil sterilant is desirable to prevent infection of young plants by soil-borne diseases and nematodes. The procedure for this is described in Chapter 2. Seeds are sown in shallow drills at a fairly close spacing but they should be spaced far enough apart to allow healthy seedlings to develop. If the spacing is too close, crowding will produce weak transplants which are susceptible to disease and may die at transplanting. As a guideline, no more than one seed per cm of row should be sown unless the seed viability is poor. In general, seeds should be sown at a depth of two or three times their diameter and should be lightly covered with fine soil. The surface of the drill should then be lightly firmed. Reference should also be made to Chapter 2, in which seed sowing is described, and to Appendix Table 12 which gives sowing rates of selected vegetables.

After sowing, the soil should be thoroughly moistened and then covered with a light mulch of dried grass or straw. The mulch prevents crust formation on the surface of the soil and helps to keep it moist. It is very important that beds do not dry out at any time during the germination process. As soon as the seedlings have emerged, the mulch should be removed and replaced with a light shade cover of straw or other material suspended about 30 cm above the bed (see Chapter 2). After the seedlings are well established the shade is gradually reduced and should be removed completely several days before the seedlings are to be transplanted so that they will be adapted to full sun conditions. Nursery beds should be carefully inspected for signs of disease or insects and sprayed as described in the section on pests and diseases on pages 195–199.

If vegetables are direct-sown they are usually sown more thinly than they would be in containers or nursery beds since they will be thinned to the correct spacing when they are large enough. Vegetables to be direct-sown are those which germinate and grow vigorously with minimal care. One of the main advantages of direct sowing is that it eliminates transplanting shock which is likely to check seedling growth while the plants re-establish their root systems. Many direct-sown vegetables therefore mature much earlier than transplanted crops sown at the same time. Additionally, they may develop some level of drought tolerance due to the rapid growth of the tap or primary root.

Transplanting

Plants are normally ready for transplanting when they are 10–15 cm high. For about a week prior to transplanting, the seedlings should be 'hardened' by reducing watering and giving them full exposure to the sun. This process toughens the plant so that it can withstand transplanting.

The actual process of transplanting begins only after the permanent beds have been completely prepared and holes dug at the planting stations. The plants should be thoroughly watered three or four hours before transplanting is to take place. If possible, transplanting should take place in the evening or on a cloudy day. The transplants are lifted with a spade or trowel, care being taken to damage them as little as possible. Loose soil is removed from the roots and the plants are placed in a basket or similar container with moist sacking or banana leaves in which they are transported to the planting area.

During the transplanting operation, only a small number of plants should be removed from the shaded container at a time. The seedlings can be inspected as they are being prepared for planting and any diseased or very small plants should be discarded. Plants are set in the holes at the same level or slightly deeper than they were growing in the nursery and a transplanting or 'starter' solution, consisting of about three tablespoons of a 5−15−15 fertilizer or similar mixture dissolved in 4 litres of water, is used to water the plants immediately after transplanting. If the weather is sunny, banana leaves, palm leaves or leafy twigs are useful for shading the transplants for the first few days. The transplants should be regularly watered until they become established.

Mulching

One of the simplest but most beneficial practices in vegetable production is mulching. Mulching reduces weed growth by smothering the seedlings, decreases the spread of disease by diminishing the splashing of pathogen-containing soil particles on to the plant, conserves soil moisture by decreasing evaporation of water from the soil surface, decreases erosion by protecting the soil surface from contact by rain, lowers soil temperature, adds organic material to the soil on decomposition, and decreases nematode populations. However, when mulches are used regularly there may be an increase in the numbers of some kinds of soil insects, termites, and snails. In most cases, however, the advantages far outweigh any disadvantages.

The practice of mulching consists of applying an 8−12 cm layer of dried straw, leaves or other plant refuse, excluding weeds, on to the surface of the vegetable beds. Plastic sheeting and other materials can also be used as a mulch but they are more costly.

Staking

Some tall-growing vegetables such as tomatoes, twining plants such as various climbing forms of bean, and cho-cho or chayote benefit from staking. Staking improves production and quality by keeping the fruit off the ground, thus minimizing disease infection and rotting. In addition, harvesting is made much easier and photosynthesis may be increased through improved light penetration into the canopy of the crop.

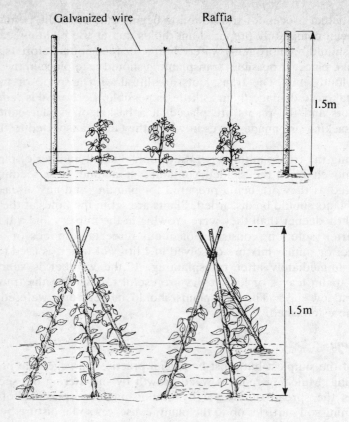

Figure 7-4
Two methods of staking

Staking can be carried out in many different ways, two of which are shown in Figure 7-4. Whatever method is chosen, it is essential that the stakes used are strong enough to support the mature, fruit-laden crop. A heavy crop such as tomatoes or egg-plant (garden egg or aubergine) may all too often be ruined when stakes break due to the weight of a heavy crop.

Fertilizers and manures

Most vegetables respond to high levels of fertility. Although it is impossible to prescribe accurately for all vegetables and soil types in warm climates, due to local soil type variation, the following generalizations are useful where specific data are lacking:

(1) The use of well-rotted manure prior to planting is beneficial to most vegetables although heavy applications may cause cracking in some root crops. Fresh manure should be avoided since it can decompose in

the soil, causing damage to the root system. Useful forms of organic matter, in addition to compost, include sheep and goat manure, poultry manure, blood meal, dried blood and wood ash.

(2) Fertilizers with a 1−3−3 or 1−4−4 ratio or similar composition are useful as pre-plant applications on most vegetables. These should be followed by side dressings with fertilizers with a good nitrogen content such as sulphate of ammonia, calcium ammonium nitate, or nitrate of potash.

(3) Brassica or cole crops and other vegetables of which the leaf is the edible portion respond well to applications of nitrogen.

(4) Nitrogen should be applied frequently, but in small quantities to prevent leaching.

(5) Root crops, in general, respond to high levels of phosphorus and potassium.

Where crops are grown in rows, either on beds or ridges, the fertilizer is usually applied as a dressing in bands alongside the base of the plants, a few centimetres from the stem. For larger plants such as fairly mature tomatoes, peppers, and garden egg, the fertilizer can be placed in a circle around the base of the plant. To avoid loss by surface wash due to rain, it is advisable to fork in the fertilizer lightly so that it is well mixed with the surface soil. Appendix Table 13 contains some suggestions for fertilizer rates where local recommendations are lacking.

Healthy crops which are growing in a relatively fertile soil adequately supplied with organic material and to which surface dressings of fertilizer are added at regular intervals will normally develop to full maturity and give high yields. However, some soils are deficient in some of the minor or 'trace' elements which, although required in minute quantities, are nevertheless essential for normal plant growth. See Table 5 in the Appendix for general comments on the nutrient deficiencies of vegetables.

Some of these trace elements are boron, molybdenum, iron, and manganese and, if these are lacking in soils, poor crop growth will result. Since the exact identification of these deficiency diseases can only be done in an analytical laboratory, the grower can only suspect that his crops may be suffering from a deficiency disease if the leaves develop an abnormal shape and colour although the plants have been given normal fertilizer treatments. Treatment consists of applying small amounts of the missing elements in solution, as directed by the local Extension Service.

Insects and diseases

Insects and diseases can both cause serious reductions in vegetable yields, the extent of which is affected by factors such as the vegetable species and cultivar, the environment, the location and the cultural practices used in the production of a particular crop. In order to minimize the harmful effects of insects and diseases, the above factors are manipulated so that insect and disease damage is reduced.

Vegetables species and cultivar　The choice of species and cultivar can affect the seriousness of pest problems since some vegetables are more likely to be attacked than others due either to the nature of their leaf or stem structure or to the attraction of their taste or appearance to some insects. In addition, cultivars differ in their tolerance or resistance to diseases and insects, and therefore cultivars which are known to be resistant should be grown whenever possible.

Environment　The environment has a major influence on the recurrence and severity of pest problems. Unfortunately, it is difficult to manipulate the environment on any large scale although the micro-environment (the environment immediately surrounding the plant) can be manipulated and modified through the application of cultural practices such as shading, mulching, and irrigation. The timing of sowing and transplanting and choosing of vegetable cultivars which are well adapted to the local environment, also allowing for seasonal variation, will both contribute to the production of vigorous crops which have some degree of natural resistance to pests.

Location　The location of the vegetable garden will affect pest problems on both a large and a small scale since small changes in location can affect the micro-climate and the presence or absence of pests. The location must therefore be carefully selected to minimize potential disease and insect problems.

Cultural practices　To some extent, insects and diseases can be controlled through cultural practices such as staking and mulching to minimize the splashing of pathogen-laden soil on to leaves. Correct plant spacing, allowing air circulation between plants so that leaves dry quickly thus limiting spore germination and the use of rotations so that a crop which is susceptible to a particular soil-borne pathogen does not follow a similarly susceptible crop, are basic precautions which can be taken to reduce insect and disease damage.

In many cases, insects are best controlled by hand picking and destroying, especially when the pest is sedentary and is not present in large numbers. Many caterpillars fall into this category. Garden hygiene or sanitation is another cultural practice which can greatly decrease pest problems. All plant debris in and around the vegetable garden, including weeds, should be removed to the compost heap or pit, unless it is infected with disease or infested with insects, in which case it should be burned to prevent the pest from spreading. When mature fruits, leaves, and other plant residues are allowed to remain on the ground where they fall they are likely to attract and harbour insects and diseases which can spread to developing vegetable plants and they should therefore be regularly removed and composted.

Where other methods of control fail, the practice of spraying and dusting with pesticides is employed. Although spraying is effective, it should be remembered that beneficial insects or predators may also be killed, leading to more severe pest outbreaks later. In addition, pesticides are expensive and can be dangerous to the person using them if they are not properly used. For these reasons, insecticides and fungicides should only be used when other methods fail.

A combination of good cultural and management techniques and the limited use of pesticides can lead to an integrated system of control which can be far more effective and lasting than the application of chemicals only.

When it becomes necessary to use pesticides, the following guidelines should be noted:

(1) Only pesticides which are recommended for use on a particular crop and for a specific pest problem should be used. Applying the wrong pesticide many damage the crop, fail to control the pest or even leave a toxic residue on the leaves or within the edible portion of the crop.
(2) Directions on the pesticide label should be followed exactly. All amounts should be carefully measured with measuring equipment which is used only for pesticides.
(3) Where several pesticides are available for the same purpose, choose one which has no, or very low, human toxicity to minimize any possible damage to users.
(4) Persons applying pesticides should be trained in their proper application so that both the upper and lower surfaces of leaves are thoroughly sprayed.
(5) Operators should be fully briefed on safety regulations and supervisors should ensure that they are closely followed. All spraying equipment should be kept in good working order at all times.
(6) Despite worker resistance, operators using specified chemicals must wear rubber boots and overalls when applying pesticides.
(7) The safety period before harvest, at which time pesticide applications should stop, must be strictly adhered to. The safety period varies between pesticides and crops but is generally in the range of between seven and 21 days. If safety periods are not followed, toxic residues may remain in or on the crop, thus endangering the health of consumers.
(8) The person applying chemicals should always thoroughly wash after spraying and should not eat or smoke before doing so.
(9) All containers used for mixing chemicals and all equipment used in applying them must be correctly labelled and be thoroughly washed after use. Containers used for concentrated chemicals should not be reused but should be buried. All chemical containers must be kept away from children at all times and should be kept in a locked store when not in use.

Insect control

Insects can be roughly divided into two groups based on their feeding habits. Biting insects have chewing mouthparts and may entirely consume parts of the plants on which they feed, resulting in holes in leaves, fruits, and stems. Grasshoppers, beetles, and caterpillars are examples of this group. Sucking insects possess piercing mouthparts which penetrate inside the plant, allowing the insect to feed on the plant sap. Aphids, mealybugs and scale insects are examples of sucking insects. The sensitivity of insects to insecticides is partly influenced by their feeding habits so it is important to classify an insect at least into a chewing or sucking type before attempting to control it with insecticides.

Insecticides are also classified into groups which are based on the way in which they kill insects. Thus there are stomach poisons, which kill when the chemical is eaten, and contact poisons which kill when the insect is either sprayed directly or crawls on a sprayed surface. Since stomach poisons can only kill an insect if the poison is eaten they are not effective against sucking insects unless the insecticide penetrates into the sap of the plant. Since sucking insects do not feed on the surface of the leaf those stomach poisons which remain on the leaf surface are ineffective. Some stomach poisons, however, have *systemic* activity which means that they can penetrate into the plant sap and will therefore kill sucking insects. Contact insecticides will kill both sucking and chewing insects, provided that the parts of the insect body which are capable of absorbing the insecticide come into contact with the spray. Some insects such as scales and mealybugs are difficult to kill with contact insecticides since they possess an outer waxy covering which repels the insecticide. Sucking insects are therefore best controlled by systemic stomach poisons but may also be affected by contact poisons. Chewing insects are controlled effectively by both contact and stomach poisons.

As a general principle, it is more economical to apply an insecticide at the first sign of attack rather than to wait until the infestation becomes severe and significant crop damage occurs. Prompt spraying may result in a high level of control but delayed action may lead to the need for repeated applications.

Nematodes or eelworms are often grouped with insects although they are not in any way related. Nematodes can extensively damage the roots and other parts of plants and may become serious where crops are not rotated at frequent intervals. Control measures are outlined in Chapter 2. Details of insect pests of selected vegetables, including control measures, are given in Appendix Table 14.

Disease control

Diseases are classified into groups according to the pathogen which causes them. The most common diseases are caused by fungi but those due to bacteria and viruses may be serious. There is also a class of diseases called physiological diseases; these are not caused by pathogens but are due to various internal disorders of the plant which may be associated with the environment, nutrition or even irregular watering, as with 'blossom end rot' of tomato and related fruits.

Diseases caused by fungi can be prevented by the application of fungicides to plants prior to the establishment of the disease. Most fungicides, in fact, are not effective after the plants have become heavily infected. Protective sprays should therefore be applied whenever environmental conditions are favourable for the development of disease.

During rainy periods, sprays once or twice a week may be required to cover the surfaces of leaves with a protective layer of fungicide. Systemic fungicides such as benomyl, which need not be used as frequently, are now available and are becoming more widely used.

Diseases caused by bacteria are not as common in vegetable crops as are fungal diseases but they are far more difficult to control. Control is primarily through the adoption of preventive measures although sprays of copper and other materials may sometimes be effective. Some bacteria enter the plant system mainly via wounds and therefore the reduction of damage to leaves and stems by minimizing cultural operations may sometimes reduce the incidence of bacterial infection.

Viral diseases cannot be economically controlled in vegetable crops so they must be prevented by prohibiting smoking while working with plants in the Solanaceae family such as tomato and pepper, requiring workers to wash their hands before handling susceptible plants, roguing infected plants and controlling insect vectors.

Physiological diseases can be controlled by determining their causes and then correcting them. This generally involves changing watering and fertilizer practices but may also mean growing cultivars of some crops which are known to be relatively tolerant to a wide range of conditions and are therefore less susceptible to physiological disorders. Some of the more common diseases of vegetables with recommended control measures are given in Appendix Table 15.

Weed control

Weeds can seriously reduce the yields of vegetable crops by competing with the crops for nutrients, water, soil oxygen, sunlight and, in some cases, by secreting toxic substances into the soil. All weeds must therefore be controlled if vegetables are to produce optimum yields.

In tropical regions, weeds are generally grouped into those which complete their life cycle and die in one growing season, called annuals, and those which live from year to year, called perennials. Annual weeds are readily controlled by hoeing, hand weeding, and mechanical cultivation. These methods, however, are ineffective when used to control most perennial species and, in many instances, may increase the problem. Most perennial species are capable of reproducing by means of rhizomes, stolons, tubers or other vegetative plant parts which are merely spread by ordinary mechanical means of control. Many perennial weeds also produce large amounts of seed, as do annual weeds.

Perennial weeds must therefore be controlled either by digging and complete removal from the holding or killing them by chemical means. The most effective method is to uproot the weeds manually and either burn them or place them on the compost heap. The alternative is to control them by the use of certain herbicides. Annual weeds reproduce by seed rather than by vegetative means and can be most easily controlled by cultivation, mulching, and by removing all weeds before they produce seed.

Herbicides or weed killers are classified as either *pre-emergence* or *post-emergence*. Pre-emergence herbicides will kill weeds prior to or during the germination process. They generally have little effect on established weeds. Post-emergence herbicides, on the other hand, kill weeds after they have become established. In addition, herbicides are classified according to their ability to distinguish between a crop species and a weed species if applied correctly. A herbicide which can kill weeds in a crop without harming the crop is said to be *selective*. A herbicide which kills both weed and crop is *non-selective*.

Herbicides are not, however, the total answer to a weed problem, they are more of an aid to controlling weeds which are difficult to control by other means or when labour for weeding is unavailable. It is seldom possible to justify economically the exclusive use of herbicides, unless the above factors are applicable. If the decision to use a herbicide is made, the herbicide must be used strictly according to the directions given on the container.

A good example of a situation in which herbicides can be useful is when land is so infested with a perennial weed that growing a crop is not feasible. In this case, non-selective, post-emergence herbicides may be useful in killing the weed so that subsequent planting can be carried out in a relatively weed-free environment.

Another instance of the value of herbicides is when labour is unexpectedly unavailable for weeding. In this instance, a pre-emergence selective herbicide should be used shortly before planting the vegetable crop. The resulting decrease in weed germination will greatly reduce the need for labour. Although the cost of herbicides is high, the value of the vegetable crop and the scarcity of sites suitable for vegetable production in many areas may justify its use in these and similar instances.

Harvesting, handling, and storage of vegetables

Harvesting is usually carried out very early in the morning with the objective of maintaining the full turgidity of leaves and other fleshy parts of the plant. Since transpiration is normally at a minimum during the hours of darkness and early in the day, this is the optimum time at which to harvest fleshy or succulent crops which ideally will remain fresh until they are either consumed or sent to market.

All leafy crops should be cut from the root or removed from the stem with a sharp knife. Pulling and tearing may damage the remaining plant tissues which, in some cases, will produce new shoots or leaves. Legumes should be broken from the parent plant at the pedicel or remains of the flower stalk; there is often a node or joint at this point which separates easily. Root crops should be lifted carefully, without damaging the root; this operation is often best performed by inserting a spade or fork below the root and lifting gently until it is clear of the soil.

The cucurbits and the solanaceous crops are both often referred to as 'fruit-vegetables'; they are generally sensitive to handling and may bruise easily in the ripe state. When harvesting, therefore, maximum care is required to ensure that the surface of the fruit is not scratched or bruised; detachment from the plant by cutting with a sharp knife is therefore recommended for the cucurbit crops. Tomatoes and peppers can usually be twisted so that the 'joint' below the calyx will easily break. With some fruit vegetables the stage of maturity of the crop at harvest will determine its keeping qualities; immature or overripe tomatoes will rarely gain the best market price.

Handling and cleaning   Once the crop has been harvested, it may require some treatment before it can be consumed or marketed. Many root crops, for example, may have to be washed to remove the soil adhering to their roots. The tops of some vegetables may have to be removed and leafy crops may need to have outer leaves removed if they are discoloured or partially eaten by insects.

Washing vegetable crops is advisable since this will tend to keep them in a fresh, unwilted condition. Where crops are to be sent to market, they may have to be graded or sorted to remove inferior, diseased, or damaged samples and, crops such as shallots, radishes, amaranth, bologi, and other leaf crops are normally tied in bundles of a size normally accepted on the local market.

Containers of uniform size are rarely used; they are often made locally from materials readily available such as baskets woven from grass or palm leaves. Such containers are rarely ideal for transport, since they allow for crushing and bruising and the produce may overheat if it is in transit for any length of time. There is a general need, in most tropical climates, for well-designed and properly constructed containers for transporting perishable produce. Good market presentation is rarely considered of importance but fresh, well-presented, uniform produce will almost always attract the best prices; the extra effort given to good handling, care in transporting and presenting produce will often be well repaid.

Short-term storage   The main objective with perishable crops is to preserve quality by retaining them in a fresh condition. Exposure to sun, rain or wind is therefore to be avoided, and whenever possible the produce should be kept in humid conditions at as low a temperature as possible. Cool storage at temperatures in the region of 10–12°C will keep many leafy vegetable crops in good condition for several days, but this is rarely practical in most areas. However, the provision of a well-ventilated but shaded structure into which the harvested crops can be placed before they have acquired 'field heat' by exposure to the sun and regular sprinkling with water will assist in maintaining freshness and quality for some time. Crushed ice, sprinkled on the surface, is widely used in some areas, particularly for harvested leaf crops packed in crates or baskets.

For some root crops such as yams, special structures known as 'barns' have traditionally been used and underground storage pits, sometimes lined with polyethylene, can be successfully used for storing root crops for weeks or even months. Onions rapidly deteriorate if kept at moderate ambient temperature conditions, even if they have been carefully 'cured' by being exposed to the sun immediately after harvesting. However, if kept for a short period at temperatures in the range of 30–35°C at a humidity level of about 65 per cent before being stored, the storage life of many cultivars can be considerably extended. Details of optimum storage conditions for selected vegetables are given in Appendix Table 17.

Seed production and storage

As a general rule, where commercially produced vegetable seed of good quality is available, it is preferable to sow this rather than for the farmer to try to save his own seed. This is because commercial seed, produced under controlled conditions, is likely to be higher yielding and is less likely to harbour pathogens than locally produced seed. Also, by purchasing seed it is possible to buy high-yielding, disease-resistant hybrids which are impossible for the farmer to produce.

For most local vegetables, however, seed must be saved by the farmer from the previous year's crop. Seeds of some introduced vegetables can also be saved if they cannot be purchased. The best plants, which should always be selected as seed parents, should be identified early and the seeds collected as they mature; if seed is taken only from the remaining few plants at the end of the harvesting period they are likely to prove inferior and may well be affected by diseases and pests. Seeds should be gathered when fully ripe, threshed and winnowed as necessary and then dried. Seeds from moist fruits, such as the tomato, will require washing before being dried. Seeds should be stored in as cool a location as possible (7–8°C is ideal, but rarely practicable) and should be kept dry. Tins or jars with tight lids are ideal for seed storage if they are kept tightly closed. One method of keeping seeds cool is to bury the airtight container in moist, shady soil to a depth of 60–90 cm. If large amounts of seeds are to be stored, they should be dusted with an insecticide powder to control weevils or other insects which may be present; the use of a desiccator, with an appropriate moisture-absorbing chemical, is recommended for storing large quantities of seeds, particularly if they have been purchased at a high cost.

Questions for study and review

1. List the main vegetables consumed in your area and classify them into (a) indigenous and (b) imported types.
2. By referring to Table 10 in the Appendix, draw a table of the nutritional values of the vegetables you have listed in Question 1.
3. What are the major factors to be considered in planning a vegetable garden?
4. Assume that you are required to plan a vegetable garden of 0.5 hectares in area for producing crops to sell in the local market. List the crops you would select and indicate what area each crop would occupy.
5. Write brief notes on soil nematodes (eelworms) and, by reference to Table 14 in the Appendix, indicate what control measures you would use for a nematode-infested soil.
6. Write brief notes on the ways in which local growers of vegetable crops market their produce. (A brief market survey in your area will help you to collect the required information.)
7. What is the value of windbreaks in vegetable cultivation? Suggest various types of windbreaks which would be suitable for your area.
8. Write brief notes on the value of vegetable compost and draw a sketch to illustrate the method of construction of a compost heap.
9. Outline the various ways in which vegetable crops can be irrigated
10. Write brief notes on the importance of rotation in vegetable growing.
11. What do you understand by the term 'intercropping'? Suggest some intercropping combinations suitable for your area.
12. Write a brief account of vegetable seed sowing (a) in containers, (b) by direct sowing in beds.
13. What special precautions are required in transplanting vegetable seedlings?
14. What is meant by 'mulching'? What materials are available for mulching in your area?
15. Write a brief essay on chemical fertilizers and their use in vegetable growing.
16. List the safety precautions which should be taken when using chemicals for insect or disease control.
17. What are the main types of insecticide in common use?
18. Write notes on weeds in vegetable crops and methods of controlling them.
19. What are the main principles to be considered in harvesting and handling vegetables?
20. Write a brief account of the principles to be followed in seed saving and storing.

CHAPTER 8

Brassica or Cole Crops

General characteristics

These crops belong to the family Cruciferae (Brassicaceae). The flower is 'cruciform' or cross-shaped, with four often brightly coloured petals and six stamens. The fruit of the Cruciferae is pod-like, usually long and narrow but with a central cross-wall or septum. The seeds are attached to the wall and the two halves of the pod split apart when the seeds are mature. The seeds are round and yellow to black. All members of the group have taproots which are often fleshy. The lower leaves are often divided and have large terminal lobes.

Ecological factors

Most Brassicas produce large leaf areas and soils which are well supplied with nitrogen and other major and trace elements are recommended. Some members of the family are sensitive to molybdenum and boron deficiencies. Medium loam soils are considered suitable and good moisture retention is essential.

In general, Brassicas require a cool period before flowering can be initiated. For this reason seeds can only be produced at higher elevations. The heading of Brassicas such as cabbage is also stimulated by low temperatures although some cultivars will head at high temperatures. Brussels sprouts are only successful at high elevations in relatively cool conditions. Some cultivars of cauliflower have been selected for head formation in high temperatures and short daylengths. Most Brassicas are adapted to short daylengths and many appear to be daylength neutral.

Economic importance and distribution

Some forms of Brassica are widely grown in many tropical regions and are used as cooked vegetables. Cauliflower is becoming more widely grown in many areas and various forms of cabbage are now widely grown in both lowland and highland locations. Broccoli, turnip, and kohlrabi are of minor importance in most parts of Africa, mainly due to the fact that cultivars for high-temperature conditions are not generally available, but kohlrabi (knol-kohl) is widely grown in India and many other tropical and subtropical areas. The various forms of chinese cabbage are grown in many parts of South-east Asia as are the numerous forms of leaf mustard (*Brassica juncea*).

Nutritional value of brassica crops

See Table 10 in the Appendix.

Brussels sprouts (*Brassica oleracea* var. *gemmifera*)

General characteristics

The crop is biennial with a vigorous main stem which becomes woody at the base. The large leaves are produced on well-developed petioles. The edible parts are the axillary buds on the main stem and the lower sprouts normally mature first. An ideal sprout is large and firm. Flowers are rarely produced in high-temperature conditions.

Environmental factors

Altitudes of more than 1000m are required for successful production of sprouts; a daily variation in temperature is necessary, with night temperatures of 16°C or lower. Most cultivars tolerate high rainfall and heavy loams are generally suitable. Optimum pH is 6.5.

Cultural techniques

Soils with a high organic matter content are preferable; rolling or firming the soil is recommended before seeding. Seeds are sown in containers or nursery beds and seedlings are transplanted to rows 70−85cm apart with plants 60−70cm apart. Removal of the growing point of plants which have begun sprout development will stimulate the production of larger sprouts, the older lower leaves may also be removed to assist in bud development. Irrigation should be regular during dry periods.

A complete fertilizer should be applied before planting, followed by dressings of nitrogenous fertilizer at regular intervals. Excessive nitrogen may, however, cause the sprouts to become 'blown' or loosely packed.

Sprouts may be harvested 100−130 days from transplanting by pulling them from the main stem with a downward twist or by cutting with a knife. Yields of 8−12t/ha (1kg/m^2) may be obtained.

Pests and diseases

These are similar to those of cabbage and cauliflower. Descriptions of the pests and diseases of Brassicas, with recommended control measures, are given in Tables 14 and 15 in the Appendix.

Cabbage (*Brassica oleracea* var. *capitata*)

General characteristics

Cabbage is a biennial herb with a short, thickened stem surrounded by a series of overlapping expanded leaves which form a compact head. Head shape may be pointed or round and leaf colour and shape are variable. 'Savoy' cabbages have deeply wrinkled dark green leaves (Figure 8-1).

Environmental factors

Soils should be well provided with organic material and have good moisture-holding properties. A pH of approximately 6.5−6.8 is considered best.

Figure 8-1
Cabbages (*Brassica oleracea* var. *capitata*) grown on smallholding

Although many cabbage cultivars are tolerant of varying soil conditions, acid soils are not generally suitable.

Cabbages can be successfully grown at elevations of more than 800m, but cultivars which form heads at low elevations are available. Some cultivars can withstand temperatures in excess of 30°C but head formation is more likely to occur at temperatures lower than 24°C. A difference of approximately 5°C between day and night temperatures appears to be necessary for adequate head development.

Cultural techniques

Seeds are normally sown in nursery beds and seedlings are transplanted to rows 60−75cm apart with plants 45−60cm apart. The wider spacings are required for later-maturing cultivars. Approximately 300−400g/h (0.5g/10m²) of seed is required for transplanted seedlings. Direct seeding followed by thinning is not normally successful in tropical areas. Cuttings from decapitated stumps may be rooted under nursery conditions and transplanted.

Regular watering is necessary throughout the growing period to avoid any check in development.

Organic manure and NPK fertilizer should be used before transplanting and dressings of nitrogenous fertilizers applied as top dressings at intervals of 14−21 days until the heads begin to develop.

Under lowland conditions, heads may be harvested in 70−90 days; at elevations over 1000m the growth period may be 80−110 days, depending on the cultivar. Harvesting should be carried out selectively as individual heads mature. The heads should be severed just below the bottom leaves using a sharp knife.

Yields range from 12−35t/ha (1−3.5kg/m²) depending on environmental conditions and the period to maturity required by the cultivar. Cultivars are generally classed as 'early', 'medium' or 'late', based on their length of time to reach maturity.

Pests

Aphid (*Aphis* spp.), Table 14 (117).
Cabbage moth (*Crocidolomia binotalis*), Table 14 (14).
Cutworm (*Spodoptera littoralis*), Table 14 (34).
Diamond back moth (*Plutella xylostella*), Table 14 (35).

Diseases

Alternaria or leaf blight (*Alternaria brassicae*), Table 15 (17).
Black leg fungus (*Phoma lingam*), Table 15 (18).
Black rot (*Xanthomonas campestris*), Table 15 (19).
Cercospora leaf spot (*Cercospora brassicicola*), Table 15 (20).
Club root (*Plasmodiophora brassicae*), Table 15 (21).
Downy mildew (*Peronospora parasitica*), Table 15 (10).

Figure 8-2
Cauliflower (*Brassica oleracea* var. *botrytis*) with curd protected from the sun by outer leaves

Cauliflower (*Brassica oleracea* var. *botrytis*)

General characteristics

Cauliflower is a biennial, grown as an annual. The edible portion is called the head or curd and consists of a white mass of abortive flowers on enlarged branches produced at the top of a short, thick stem (Figure 8-2).

Environmental factors

In general, cauliflowers require more fertile soils than cabbages. Cultivars suitable for growing in lowland tropical areas have been selected, but cool conditions such as are found at elevations of more than 1000m are normally required for optimum growth and development. There are two main groups, classed according to curd development: those which will form curds and flower in hot weather conditions and the later-maturing types which take a long time to develop curds and also require cool conditions for curd development. Dry season cultivation with irrigation is preferable since the curds are likely to decay in wet conditions. Optimum soil pH is 6.0−7.0.

Cultural techniques

Seeds are sown in nursery beds and the seedlings are transplanted to rows 60−75cm apart with plants 45−60cm apart. Approximately 0.25g of seed is required for 10m^2.

Soils with a high organic content are preferable and an application of NPK should be made before planting. Dressings of nitrogenous fertilizer are beneficial up to the formation of the curds. A shortage of nitrogen during early growth may lead to the condition known as 'buttoning' which results in the formation of a curd with an irregular and misshapen surface. Since cauliflower has a high demand for magnesium, boron, and molybdenum, deficiencies may occur and can be corrected by the application of dolomitic limestone, borax or ammonium molybdate at the rates recommended by the local Extension Service.

Most cultivars of cauliflower will require blanching when the curd begins to develop. Blanching, which involves shading the developing curd by tying the upper leaves over the curd, protects the curd from sunburn and prevents it from turning green.

Cauliflowers can normally be harvested 80−120 days from transplanting, depending on the cultivar. The heads are severed in one piece with a sharp knife. Yields varying from 12−25 t/ha (1−2.5 kg/m^2) may be obtained if growth conditions are favourable.

Pests

Aphid (*Aphis* spp.), Table 14 (28).
Cutworm (*Agrotis* spp.), Table 14 (29).
(see also pests of cabbage, page 434.)

Diseases

Black rot (*Xanthomonas campestris*), Table 15 (19).
Damping-off (*Pythium aphanidermatum = P. ultimum*), Table 15 (1).
Stem rot (*Thanatephorus cucumeris = Rhizoctonia solani*), Table 15 (39).
Club root (*Plasmodiophora brassicae*), Table 15 (37).
(See also diseases of cabbage, page 448−449.)

Chinese cabbage (*Brassica chinensis*)

Alternative names are flowering cabbage, Pekin cabbage, pe-tsai, celery cabbage, white cabbage, spoon cabbage, and Shantung cabbage. There are two main forms: paak tsoi and pe-tsai, both of which are widely grown in South-east Asia.

General characteristics

Chinese cabbage is a biennial herb grown as an annual. Basal leaves are broad, 20−50 cm in length, with thick white petioles. Flowers are yellow and approximately 1 cm in length. Fruits are slender and 3−6 cm in length (Figure 8-3).

Figure 8-3
Chinese cabbage (*Brassica chinensis*)

Environmental factors

Chinese cabbage is tolerant of a wide range of soil conditions although excessively well-drained soils are unsuitable for this crop, which matures rapidly. The crop is normally grown at elevations up to 1500 m and withstands periods of high rainfall although it requires full exposure to the sun for optimum development. Flowering is delayed by high temperatures but low temperatures of less than 16°C promote flowering, particularly in daylengths of 13 hours or more. A day-night variation of 5−6°C appears to increase the vigour of the plant. High-yielding, firm-headed crops are produced at high elevations during cool weather; at lower elevations heading is less likely to occur.

Cultural techniques

Seeds are sown in nursery beds or containers and seedlings are transplanted when 6−8 cm high to rows 40−45 cm apart with plants 30−40 cm apart. Seeds may also be sown directly and the seedlings thinned. Approximately 1−1.5 kg seed is required per hectare (1−1.5 g/10 m²) for transplanted seedlings.

Irrigation may be required to maintain a constant rate of growth. A complete fertilizer should be applied before sowing or transplanting. Nitrogenous top dressings are frequently required to obtain full leaf development (see also Appendix Table 13).

Headed plants are harvested 50−80 days from sowing seed and may attain a height of 40–50cm. Either the complete plant or individual leaves may be harvested, preferably early in the day. Heads and leaves should be kept cool & moist to prevent wilting. Yields vary and are normally between 10–20t/ha (1–2kg/m^2) although higher yields may be obtained with good cultural methods.

Pests

Aphid (*Aphis* spp.), Table 14 (33).
Cutworm (*Prodenia litura*), Table 14 (34).
Diamond back moth (*Plutella xylostella*), Table 14 (35).
Flea beetle (*Phyllotreta* spp.), Table 14 (36).

Diseases

Black rot (*Xanthomonas campestris*), Table 15 (19).
Cercospora leaf spot (*Cercospora brassicicola*), Table 15 (20).
Downy mildew (*Peronospora parasitica*), Table 15 (10).

Kale or Collard (*Brassica oleracea* var. *acephala* or *B. oleracea* var. *viridis*)

Alternative names are borecole, green cabbage, curly kale, or Chinese kale. Collard is often regarded as a variety or form of kale. The Chinese broccoli or 'kai laan' of South-east Asia is *Brassica alboglabra*.

General characteristics

Several types of kale exist: 'Thousand Headed', 'Marrow Stem', and 'Curled' but the latter is not often used as a vegetable. The selections referred to as collards have smooth leaves in a rosette at the stem apex and can withstand higher temperatures than the other forms of kale.

Kale is a perennial grown as an annual and stems may reach up to 1m. The edible leaves are divided or oval with long petioles and the plant is non-heading. Leaf colour is blue-green, the stems are robust and the flowers large and pale yellow.

Environmental factors

Elevations over 500m are most suitable for growth although some cultivars are adapted to the climatic conditions existing at lower elevations. Crops grown during dry periods with irrigation often produce optimum yields. A high level of soil organic material is desirable.

Cultural techniques

Seeds are sown in containers or nursery beds and seedlings are transplanted to rows 60–75 cm apart with plants 60 cm apart. Propagation may also be by cuttings made from side shoots.

Regular irrigation is required since the crop has a high demand for water due to the extensive leaf area. A complete fertilizer should be applied before planting followed by top dressings of a nitrogenous fertilizer at regular intervals. (See also Table 13 in the Appendix). Harvesting of leaves and shoots may begin 60–85 days from transplanting. Lateral shoots develop when the terminal shoots have been removed. Flowering shoots should be removed to encourage leaf development. Yields are variable but may be in the region of 20 t/ha (2 kg/m^2).

Pests and diseases

These are similar to those affecting other Brassicas. Descriptions of the pests and diseases of Brassicas, with recommended control measures, are given in the Appendix, Tables 14 and 15.

Kohlrabi or Knol-kohl (*Brassica oleracea* var. *gongylodes*)

An alternative name is turnip-rooted cabbage.

General characteristics

A biennial plant grown as an annual (Figure 8-4). The edible swollen stem-base, with attached leaves, has a superficial resemblance to a root or tuber. These stems may be up to 10 cm in diameter and may be white, green or purple when fully developed. The leaves often have a waxy appearance.

Environmental factors

Kohlrabi is a fairly tolerant crop as far as soil type and rainfall pattern are concerned although good drainage is essential. Elevations of more than 800 m with some variation between day and night temperature are favourable to stem development.

Cultural techniques

Seeds are sown in containers or nursery beds and seedlings transplanted when about 8–10 cm high to rows 40 cm apart, with plants 24–30 cm apart in the row. Regular irrigation is required and a complete fertilizer should be applied before planting, followed by regular dressings of a nitrogenous fertilizer. Boron, molybdenum and nitrogen deficiencies may cause physiological disorders.

Plants normally mature within 10 weeks from transplanting and should be harvested before the swollen stems become fibrous and woody. Yields may be up to 20 t/ha (2 kg/m^2).

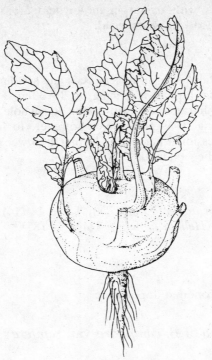

Figure 8-4
Kohlrabi (*Brassica oleracea* var. *gongylodes*)

Pests and diseases

These are similar to those of other Brassicas. Descriptions, with recommended control measures, are given in the Appendix, Tables 14 and 15.

Mustard or Leaf mustard (*Brassica juncea*)

Alternative names are Chinese mustard, Indian mustard, mustard greens, and white mustard. Kai choy, widely grown in South-east Asia, is *Brassica juncea* var. *rugosa*.

General characteristics

Mustard is an erect annual herb growing to 160cm in height. The stems are glabrous and the leaves are thin with irregular margins. Basal leaves are often divided but the upper leaves are only slightly lobed. The flowers are borne in racemes.

Environmental factors

Mustard is tolerant of moderate amounts of rainfall but is mainly suited to dry season cultivation with irrigation in lowland areas. A well-drained, fertile soil is required for optimum yield.

Cultural techniques

Seeds are sown in containers or in nursery beds and seedlings are transplanted to beds when 10−12cm high at a spacing of 20−30cm × 20−30cm.

The plants may be harvested by uprooting the whole plant at about 50 days from transplanting. It is preferable, however, to harvest only single mature leaves in successional harvesting lasting 4−6 weeks. Yields of up to 50t/ha ($5kg/m^2$) may be obtained.

Pests

Cabbage moth (*Crocidolomia binotalis*), Table 14 (14).
Cabbage budworm (*Hellula undalis*), Table 14 (85).

Diseases

These are rarely of economic importance.

Turnip (*Brassica rapa* or *B. campestris* var. *rapifera*)

General characteristics

The turnip is a biennial herb with a swollen taproot which is normally white. The leaves are light green with large terminal lobes. The flowers are yellow. Seeds are round, black or red-brown, 1.5−2mm in diameter. The crop is grown both for its edible leaves and its roots (Figure 8-5).

Environmental factors

Few cultivars are suitable for growing at low altitudes but satisfactory yields are obtained from crops at elevations of 1000m or more. Light shade is sometimes beneficial for plants grown for leaf consumption. Most cultivars are sensitive to excessive rainfall and poor soil-drainage conditions. Cultivation is most successful during dry periods. Soils with a pH in the region of 6.0−6.8 appear to be most suitable.

Cultural techniques

Seeds are sown directly at a depth of 1−2cm in rows 25−30cm apart and thinned to 10−15cm apart in the row. Phosphorus and nitrogen should be applied to the soil before sowing. Seed requirement is approximately 2.5kg/ha ($2.5g/10m^2$).

Irrigation should be regular, since low water availability may lead to the development of small roots. The rooting zone of turnips is shallow and

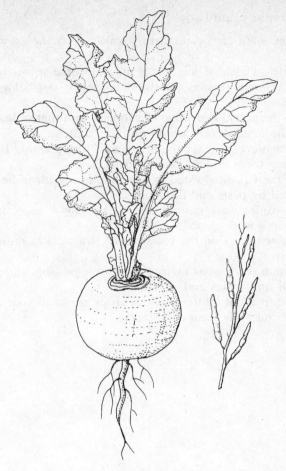

Figure 8-5
Turnip (*Brassica rapa*) with inset seedhead

irrigation may therefore have to be applied frequently in dry weather to keep this zone moist.

Roots of early maturing cultivars may be harvested about 60−70 days from sowing; the leaves may be harvested as they become mature. Yields in the range of 20−25 t/ha (2−2.5 kg/m^2) may be obtained under good cultural conditions.

Pests and diseases

These are similar to those of other Brassicas. Descriptions, with recommended control measures, are given in the Appendix, Tables 14 and 15.

Questions for study and review

1. Discuss the main ecological factors which affect the growth of Brassica crops.
2. What particular mineral deficiencies may affect the growth of cauliflowers?
3. Outline the main reasons for Brussels sprouts not being grown at low elevations.
4. Compare the nutritional value of cabbage with that of leaf mustard and indicate the main differences in nutrient content.
5. Give the main reasons for the practice of transplanting Brassicas rather than sowing them directly in seed beds.
6. Which of the Brassicas listed in this chapter appear to be most likely to be attacked by pests and diseases?
7. Why are cauliflowers more likely to be grown successfully in a dry season than a wet season?
8. Comment generally on the reactions of Brassicas to either low or high temperatures.
9. Try to obtain seeds of as many Brassicas as possible and write notes on their main similarities and differences.
10. List the cultivars of all the Brassica crops grown in your area and write brief notes on their main characteristics.

CHAPTER 9

Leaf Crops

Leaf crops are used frequently in the tropics as boiled vegetables. They are often added to soups and stews and many have high nutritional value. Some leaf crops grow rapidly and may be harvested within a few weeks of sowing. Others mature more slowly but provide leaves over a long period. In addition to cultivated vegetables, an appreciable percentage of tropical leaf vegetables are still gathered as wild or naturalized plants. In some areas, however, this valuable natural nutritional resource is neglected as cash crops become more popular.

Economic importance and distribution

The leaf crops belong to many different families and have few similarities. Their ecological requirements also differ. Since most leaf crops wilt and deteriorate rapidly within a few hours of harvesting, they are mainly grown for domestic consumption although some are commonly found in markets.

Their importance varies in relation to the food habits and preferences of the population and local climate. The leaf vegetables consumed in lowland humid areas are generally different from those eaten at higher altitudes. However, there are some crops common to both regions such as the Indian or Ceylon spinach and some of the amaranths.

The leaves are not the only food produced by many crops. The leaves of sweet potato, for example, are an important and nutritious vegetable but the tubers are an important source of carbohydrate. Other examples of the dual value of some crops as leafy vegetables and also as fruit or tuber-producing crops are pumpkin, cassava, cocoyam, and several leguminous crops.

Nutritional values

There is an increasing awareness of the value of leafy vegetables in contributing to a balanced diet, particularly in areas where animal protein is deficient. In addition to their iron content, leafy vegetables contribute significant amounts of beta-carotene and ascorbic acid (Vitamin C), protein, minerals (particularly calcium) and carbohydrate. Nutritional values are given in the Appendix in Table 10.

Uses

The *low elevation leaf crops* in this group are used as cooked vegetables and are commonly added to soups and stews. Crops such as *Amaranthus* spp. *Celosia* and *Corchorus*, or long-fruited jute, have relatively small leaves and whole shoots are harvested. Exceptions are crops such as cassava with large leaves. These are often harvested singly at intervals since successional harvesting does not significantly reduce the growth of the plants and the alternative food supply in the root tubers. In addition to the use of the leaves as a cooked vegetable, the seeds of some cucurbit crops are pounded and added to soups.

The young shoots of roselle are eaten either raw or cooked; the swollen calyces of the flowers are also cooked and may be used in the preparation of beverages. The leaves and also the flowers of waterleaf (*Talinum*) are both cooked. In many parts of Africa the leaves of vegetables such as cassava are dried, powdered and stored for later use.

The *medium to high elevation crops* are also used as cooked vegetables, exceptions being celery, endive, and lettuce which are usually eaten raw in salads although they are sometimes cooked and added to soups and relishes.

LOW ELEVATION LEAF CROPS

Amaranth (*Amaranthus* spp.)

Alternative names are African spinach, Chinese spinach, Indian spinach, bush greens, green leaf and spinach greens.

There are many species of amaranth in cultivation but *A. caudatus*, *A. dubius*, *A. cruentus* (= *A. hybridus*), *A. lividus*, and *A. tricolor* (= *A. oleraceus*, *A. gangeticus*) are the most widely grown species (Figures 9-1 and 9-2.) Numerous hybrids between species and varieties exist, some of which have been designated as species or subspecies.

General characteristics

All the above species are short-lived annuals growing to 1 m in height. The stems are erect, often thick and fleshy and are sometimes grooved. Leaves are variable in shape, green or purple and normally alternate. The inflorescence is an axillary or terminal spike. The insignificant flowers are numerous, regular and unisexual and the seeds are small, shiny, and black or brown.

Environmental factors

Soils with a high organic content and with adequate nutrient reserves are

Figure 9-1
Amaranth (*Amaranthus caudatus*)

Figure 9-2
Amaranth (*Amaranthus tricolor*)

required for optimum yields, although some species are tolerant of a wide range of soil conditions. Optimum pH range is 5.5−7.5 but some cultivars will tolerate more alkaline conditions.

Surface dressings of nitrogenous fertilizers are normally required during active growth. In some areas, additional applications of potassium may also be required.

Most species are tolerant of high temperatures and thrive within a temperature range of 22−30°C. Areas with elevations below 800 m are most suitable for cultivation, but the crop can be grown at higher altitudes.

Cultural techniques

Seeds are small and so are often mixed with dry sand to ensure uniform distribution. They are normally directly sown broadcast on prepared beds at a rate of 1.5−2 kg/ha (2 g/m^2). However, they may also be sown on nursery beds and the seedlings transplanted to rows 20−30 cm part with 10−15 cm between plants. Very vigorous species or cultivars may be transplanted to 30−40 × 30−40 cm. Broadcast-sown seedlings are thinned to 15−22 cm apart or more liberal spacings if they have a very vigorous habit. A grass mulch is sometimes used for covering seeds to protect them from heavy rain. This mulch should be removed after germination.

Amaranthus is grown during both wet and dry seasons and irrigation is normally required for dry season crops. Frequency of irrigation will be determined by the stage of growth of the crop and the moisture-retaining capacity of the soil.

A complete fertilizer and organic material should be applied to the beds before sowing directly or transplanting (see also Table 13 in the Appendix).

Most species and cultivars of *Amaranthus* grow rapidly and may be harvested from 30−50 days from sowing when they are 15−20 cm tall. Either the whole plant may be uprooted, or plants may be cut back to 15 cm to encourage lateral growths for successive harvests. If the entire plant is harvested, yields may be 20−25 kg/10 m^2. If the lateral shoots are harvested, 30−60 kg/10 m^2 is an average yield.

Pests

Stink bug (*Aspavia armigera*), Table 14 (1)
Cletus capensis, Table 14 (2)
Leaf caterpillar (*Hymenia recurvalis*), Table 14 (9)
Psara bipunctalis, Table 14 (4)
Stem borer (*Lixus trunculatus*), Table 14 (5)

Diseases

Leaf spot (*Cercospora beticola*), Table 15 (2).
Leaf and stem wet-rot (*Choanephora cucurbitarum*), Table 15 (3).
Leaf spot (*Cladosporium variabile*), Table 15 (4).
Damping-off (*Pythium aphanidermatum*), Table 15 (1).

Cassava (*Manihot esculenta* or *M. utilissima*)

Alternative names are manioc, manihot, taro, tapioca.

Figure 9-3
Cassava (*Manihot esculenta*).
(Photograph by Rex Parry)

General characteristics

Cassava is a short-lived shrubby perennial up to 5 m in height (Figure 9-3). Leaves vary in size and shape but are often palmate with from five to seven lobes. The flowers are small with yellow sepals and no petals. Male and female flowers are on the same plant. The fruit is a capsule with three seeds.

The stems and leaves contain latex and the tubers a toxic glucoside which is destroyed by boiling. The glucoside content divides cassavas into two main groups, 'sweet' and 'bitter'.

Environmental factors

Cassava tolerates a wide range of environmental variation. Tuber size is often related to a combination of climate and soil factors. Sandy or sandy loam soils with a pH of 5.5−8 are preferable but most soils, even those of low fertility, are suitable provided they are not waterlogged. Growth is severely restricted at temperatures below 16°C. Many cultivars are drought-resistant but high humidity will encourage growth even when soil moisture is low. Elevations of up to 1500 m are acceptable.

Cultural techniques

The sweet forms are used for leaf production since the glucoside content is minimal. Cuttings are selected from mature portions of the stem of virus-free parent plants. They are normally about 25 cm in length and are inserted to about half that length in prepared beds or ridges 75—90cm apart with 30—40cm between cuttings. They are usually inserted at an angle in double rows 20—30cm apart when planted on beds.

Although mature plants are often drought-resistant, cuttings should be watered during dry periods until they are established. Fertilizers are rarely required, except for potassium and nitrogen on some soils deficient in these elements, but organic material in the soil encourages early leaf production (see also Table 13 in the Appendix).

Mature leaves are formed 50—70 days from cutting insertion and are detached with a knife when they are fully developed. Yields of 20t/ha (2 kg/m^2) per year have been recorded.

Pests

Red spider mite (*Tetranychus cinnabarinus*), Table 14 (25).
Tobacco white fly (*Bemisia tabaci*), Table 14 (26).
Variegated grasshopper (*Zonocerus variegatus*), Table 14 (27).

Diseases

Anthracnose (*Glomerella manihotis*), Table 15 (30).
Bacterial blight (*Xanthomonas manihotis*), Table 15 (31).
Brown streak virus, Table 15 (32).
Cercospora leaf spot (*Cercospora* spp.), Table 15 (33).
Mosaic virus, Table 15 (34).
Root rot (*Sclerotium rolfsii*), Table 15 (98).

Cock's comb (*Celosia argentea*)

General characteristics

Cock's comb is an erect short-lived annual herb up to 150cm in height (Figure 9-4). The leaves are alternate, light green, 2 × 6cm long; the leaves on flowering shoots are slightly longer. The flowering spikes are pink, becoming white when the seeds reach maturity. The seeds are 1mm in diameter and are normally black.

Both green and red forms are widely grown; the red form is similar to the green form but is taller and more spreading. Leaves of the red form have a distinct purple marking and the leaves on flowering shoots may be very large.

Environmental factors

This plant is tolerant of a wide range of soil conditions although high levels of organic material are required for the production of good yields, particularly of the green form. Low elevations with a temperature range of 25—30°C are

Figure 9-4
Cock's comb (*Celosia argentea*)

suitable for both forms and stable high temperatures are desirable. Heavy rainfall will not limit growth but the red form is sensitive to drought.

Cultural techniques

Green form: for leaf production, seeds are sown broadcast on seedbeds which may be temporarily mulched with dried grass. This should be removed after germination. Organic material and complete fertilizer should be added to the soil before sowing or transplanting. Dressings of a nitrogenous fertilizer at intervals will stimulate rapid leaf development (see also Table 13 in the Appendix). Irrigation is required during dry periods.

The seedlings are harvested 30–40 days after sowing by thinning out the tallest plants after they reach a height of 15 cm. Harvesting continues until the plants are about 30 cm in height. For seed production, broadcast-sown seedlings are transplanted to rows about 70 cm apart with plants 40–45 cm apart or on a square planting 60×60 cm. Seeds are produced 80–90 days from sowing.

Red form: this form requires wider spacing. Seedlings for leaf production are transplanted to 15–30 cm apart. For seed production, a spacing of 90–100 cm each way is required.

Leaf yields from the green form may be 10–15 kg/10 m^2. The red form yields about 25 kg/10 m^2.

Pests and diseases

These are similar to those described for amaranth (see also Tables 14 and 15 in the Appendix).

Indian or Ceylon spinach (*Basella alba*, *B. rubra*, or *B. rubra* var. *alba*)

Alternative names are vine spinach, Gambian spinach, Malabar nightshade, and Malabar spinach.

General characteristics

Ceylon spinach is a short-lived perennial with a twining habit. The stems are fleshy, growing to 6 m in length, and the leaves are alternate, fleshy and almost triangular in shape (Figure 9-5). Flowers are small, pink or white and arise from the axils of the leaves.

Mature fruits are round, succulent, dark purple and measure about 8 mm in diameter. The form referred to as *B. rubra* has purple stems, purple leaves and pale pink flowers.

Environmental factors

Most cultivars are tolerant of a range of soil conditions but grow best in

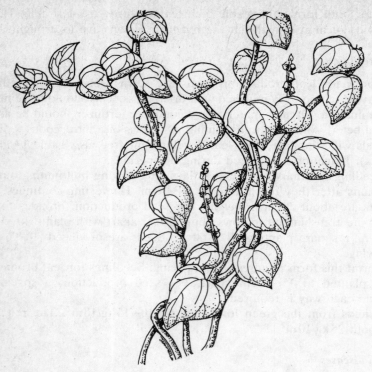

Figure 9-5
Indian or Ceylon spinach (*Basella alba*)

moist, fertile soils which are well supplied with organic material and essential elements. *Basella* is adapted to high-temperature conditions; low temperatures may reduce the growth rate. Plants are tolerant of high rainfall and some cultivars have drought resistance. Growth is likely to be limited at altitudes greater than 500 m.

Cultural requirements

Propagation is by seed or, more frequently, by cuttings up to 25 cm in length. With seed propagation, the whole fruit is often sown in nursery beds and seedlings are transplanted when 10–12 cm high. Seedlings or rooted cuttings are transplanted to beds on a square planting of 40–50 × 40–50 cm or, alternatively, to rows 60–70 cm apart with 25–30 cm between plants. Support is required and pruning side shoots prolongs vegetative growth. Irrigation is required during dry periods to continue leaf development.

Pre-transplant or sowing application of a complete fertilizer is recommended with subsequent dressings of a nitrogenous compound (see also Table 13 in the Appendix).

Harvesting of the young shoots normally begins 55–70 days from transplanting and continues over several months. Yields are variable. Depending on cultivar and frequency of harvest a single plant may yield up to 1.5 kg of fresh leaves and young shoots over 180 days.

Pests

These are rarely serious but infection by root-knot nematode (*Meloidogyne* spp.) may occur, Table 14 (74).

Diseases

A species of *Cercospora* can cause leaf spots.

Jute mallow or Long-fruited jute (*Corchorus olitorius*)

Alternative names are bush okra, Jew's mallow, tossa jute, and West African sorrel.

General characteristics

Jute is a woody, short-lived perennial growing to 1.5 m in height (Figure 9-6). The leaves are ovate. Flowers occur in groups of two or three and are small and yellow. The fruits are long, rounded, and up to 8 cm in length. Seeds are small and dark green.

Environmental factors

Most selections are tolerant of a wide range of soil and climatic variation, and plants grow well in both dry and wet seasons. Plants grown at altitudes over 700 m are liable to produce low yields. Alluvial soils with a good moisture retaining capacity are generally suitable. Regions with moderately high rainfall and a temperature range of 22–35°C provide ideal growing conditions.

Cultural techniques

Seeds are usually broadcast on beds or ridges and thinned to 15–22 cm each way. Alternatively, they are grown in the nursery and transplanted to rows 40–50 cm apart with 30–40 cm between plants. Irrigation may be required in dry periods particularly in the early stages of growth.

A complete fertilizer should be applied to the beds before sowing or transplanting and later nitrogenous dressings given at regular intervals (see also Table 13 in the Appendix).

Pests

Beet army worm (*Spodoptera exigua*), Table 14 (76).

Figure 9-6
Jute mallow or long-fruited jute (*Corchorus olitorius*)

Leaf beetle (*Podagrica sjostedti, P. uniforma*), Table 14 (77).
Root-knot nematode (*Meloidogyne* spp.), Table 14 (78).
Sweet potato butterfly (*Acraea terpsichore*), Table 14 (79).

Diseases

Wilt (*Sclerotium rolfsii*), Table 15 (98).

Kang kong or Water spinach (*Ipomoea aquatica*)

Alternative names are swamp cabbage, water convolvulus.

General characteristics

There are two main forms of this vegetable, which is widely grown in South-east Asia; one is a semi-aquatic perennial which grows in moist places and

Figure 9-7
Kang kong — semi-aquatic variety (*Ipomoea aquatica*)

Figure 9-8
Kang kong — dryland variety (*Ipomoea aquatica*)

often floats in water (Figure 9-7), and the other is the dryland form which is generally an annual (Figure 9-8). Both have white, pink or red flowers, and the aquatic type has entire leaves similar to those of sweet potato. The leaves of the dryland form are narrow and oval with pointed tips.

Environmental factors

Both forms grow well in moist soils with high levels of fertility; clay soils are generally preferable. The high temperatures and short daylengths typical of the lowland tropics are required for optimum growth.

Cultural techniques

Semi-aquatic form   Cuttings with between six and eight nodes taken from sections of stem will root readily when inserted in moist soil at a spacing of 4−8cm between plants and 16cm between rows. Moist soil or moving water is essential for the satisfactory growth of these plants.

Dryland form   May be grown from either seeds or cuttings. Seeds are sown on raised beds and thinned to 15−20cm apart or transplanted to rows 4−5cm apart with 15cm between plants in the row. Cuttings should be 15−25cm in length and inserted to half their length. Nitrogenous fertilizer will stimulate the growth of both the dryland and the semi-aquatic types of kang kong.

   Both the semi-aquatic and dryland plant forms can be harvested from six to eight weeks after planting; three or four successional harvests can be obtained from one planting − about 10cm of main stem should remain after each harvest for the production of axillary shoots. Yields from the dryland form may be in the region of 20−30t/ha (2−3kg/m^2).

Pests and diseases

Similar to those listed for sweet potato (Tables 14 (134−137) and 15 (128−131)).

Katuk or Chekor manis (*Sauropus androgynus*)

General characteristics

A perennial shrub, growing to 3m high, with ovate, pinnate leaves up to 7cm long, slightly drooping (Figure 9-9). The leaves are exceptionally high in vitamin A content; the vitamin C and protein contents are also well above the average for most green-leaved vegetables.

Environmental factors

Katuk is well adapted to growing in low elevation areas with moderate to high levels of rainfall and high temperatures. Medium loams or clay soils are suitable if adequately drained.

Figure 9-9
Katuk (*Sauropus androgynus*)

Cultural techniques

Propagation is mainly by cuttings. Seeds may be sown in containers and transplanted when about 10 cm high. Cuttings up to 20 cm long root very easily and are often inserted around the edges of raised beds to prevent erosion. They may also be planted in the centre of a bed to provide light shade. Little cultivation is required and the plants are pruned regularly to maintain lateral branching. The young tips and individual leaves can be harvested at regular intervals from about 10 weeks after the insertion of the cuttings; they are used mainly as a cooked vegetable.

Pests and diseases

These are rarely serious although root-knot nematode (*Meloidogyne* spp.) may occur in some soils, Table 14 (119).

Roselle (*Hibiscus sabdariffa* var. *sabdariffa*)

Alternative names are Indian sorrel, Jamaican sorrel, red sorrel, and sour-sour.

General characteristics

Roselle is an herbaceous, upright plant growing to 2 m. The habit is variable and the leaves also vary in shape and size. The flowers are yellow, sometimes with dark red centres. The edible calyx swells and becomes fleshy. The fruits are up to 2.5 cm in length and the seeds contain a high percentage of oil.

Environmental factors

The crop is adapted to a wide range of soil conditions. While it is often grown on infertile soils, economic yields are only obtained on soils supplied with organic material and essential nutrients. Most selections are tolerant of high temperature.

Some cultivars give satisfactory yields in wet humid areas although most are partially drought-resistant.

Cultural techniques

For the production of the calyces, seeds are sown broadcast and thinned to $60-80 \times 60-80$ cm apart or sown on seed beds and transplanted to rows $80-100$ cm apart with $60-80$ cm between plants. Alternatively, plants may be propagated from cuttings. The seed requirement is $5.5-7.5$ kg/ha $(5-7 \text{g}/10 \text{m}^2)$.

Irrigation is required during dry periods and fertilizers should be applied where the organic content of the soil is low (see also Table 13 in the Appendix).

The inflated and ripened calyces should be ready for harvesting $80-110$ days from transplanting. Leaves and shoots may be removed when fully developed. Yields of approximately 1.5 kg of calyces per plant per annum are obtained.

Pests

Cotton stainer (*Dysdercus superstitiosus*), Table 14 (120).
Flea beetle (*Podagrica* spp.), Table 14 (121).
Root-knot nematode (*Meloidogyne* spp.), Table 14 (119).
Spiny bollworm (*Earias biplaga*, *E. insulana*), Table 14 (123).

Diseases

Anthracnose (*Colletotrichum hibisci*), Table 15 (117)
Fruit rot (*Phytophthora parasitica*), Table 15 (118)
Leaf blight (*Phyllosticta hibisci*), Table 15 (119)

Sunset hibiscus or Aibika (*Hibiscus manihot*)

General characteristics

Approximately eight different forms of this variable branching shrub have been described, the differences among them being mainly due to the

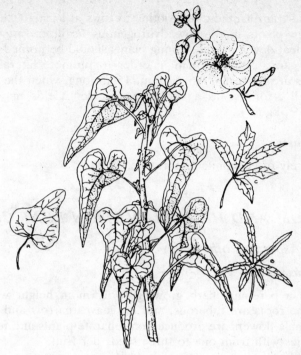

Figure 9-10
Sunset hibiscus (*Hibiscus manihot*)

variation in leaf shape, size and colour. Leaf form may be entire or deeply divided; leaf size varies from 10—60 cm in length and width. The leaf colour may be light or dark green; some leaves have a red tinge and both smooth and hairy leaf types occur. Plants grow to about 5 m high and produce large yellow terminal flowers, often with a purple spot at the base of the petals (Figure 9-10). The fruit is a capsule and the seeds are round or kidney-shaped, brown or black, with a diameter or 3—4 mm.

Environmental factors

Fertile soils with a good reserve of organic material are required for satisfactory growth, and optimum yields are obtained from plants grown in areas with stable high temperatures with a well distributed rainfall above 1000 mm per annum. Irrigation is necessary during dry periods. Full exposure to sun is recommended for full leaf development to occur.

Cultural techniques

All forms of sunset hibiscus are normally propagated from cuttings up to 20 cm in length; these are inserted in well prepared cutting beds or in containers and transplanted after rooting to rows at least 100 cm apart with 25 cm between plants in the row. Nitrogenous fertilizers are essential to promote full leaf development. Young plants should be pruned by reducing the main axis to a height of 1 m in order to promote lateral branching. Young shoots are harvested when about 15 cm long, when the leaves have developed fully, for use as a cooked vegetable. Yields may vary from 40−60 t/ha (4−6 kg/m^2).

Pests and diseases

These are rarely of economic importance.

MEDIUM AND HIGH ELEVATION LEAF CROPS

Asparagus (*Asparagus officinalis*)

General characteristics

Asparagus is a perennial herb growing to 1.5 m in height with a fleshy rootstock. The roots are tuberous, and the leaves narrow and needle-like. Male and female flowers are produced on separate plants and female plants produce berries with from one to three seeds per fruit.

Environmental factors

Sandy, silt loam or alluvial soils are preferred but a high level of organic material in the soil is essential. A pH of 6−6.7 is suitable.

A temperature range of 16−24°C is appropriate for most of the growing season but lower temperatures are necessary for 60−90 days of the year when the plants are dormant. Elevations of over 1000 m are generally required for economic yields.

Cultural requirements

Seeds are often soaked in water for 24 hours before being sown 3−4 cm deep in nursery beds in rows 30 cm apart with 4−6 cm between seeds. Germination is best at temperatures of 25−30°C. The seedlings are transplanted to ridges after a period of between seven and 12 months, preferably at the beginning of a wet season. The ridges should be 75−90 cm apart with 50−60 cm between plants in the row.

Alternatively, crowns of established plants may be divided to provide planting material when they are about two years old. These are planted at the same spacing as seedlings but at a depth of 10−15 cm. Crowns should be mulched and given an annual surface dressing of fertile topsoil for blanching of the young shoots or 'spears'. This covering of the young shoots excludes light and therefore they become white or cream since no chlorophyll develops.

Consumer preference is for blanched spears rather than green ones which are unblanched, however there is a growing preference in European markets for green spears which are unblanched. Irrigation is necessary to maintain a high soil moisture content and top dressings of nitrogenous fertilizer are beneficial (see also Table 13 in the Appendix).

The spears are harvested when they reach 10–15cm. Harvesting may begin 13–20 months from establishment of the seedlings. The shoots are removed by inserting a long-bladed knife close to the base of the shoot and cutting it and breaking it at the base.

Yields of 2.5–4.0t/ha (2.5–4kg/10m^2) may be obtained from established beds. These should be replanted from every three to five years, since the crowns will become unproductive in a relatively short time where the dormant season is inadequate. Male plants give the highest total yields, although female crowns produce larger individual spears.

Pests and diseases

These are rarely serious.

Endive (*Cichorium endivia*)

An alternative name is escarole.

General characteristics

Endive is an annual or biennial herb with leaves in a dense rosette. The young leaves are added to salads; the mature leaves are sometimes cooked. Leaf form is variable, being divided, curled, or broad. Leaves of most forms have a slightly bitter taste. The flowers are blue but are rarely formed in tropical areas.

Environmental factors

Plants respond to high levels of organic material in the soil and pH should ideally be in the range 5.8–6.5.

Temperature is an important factor influencing growth. Leaves become fibrous in high temperatures although endive is, in general, more tolerant to high temperatures than lettuce. Altitude also has an important effect on growth. More satisfactory growth is obtained from plants grown at elevations greater than 400m than from those in lowland conditions.

Cultural techniques

Seeds are sown in seed beds or containers and seedlings transplanted when 5–7cm in height to rows 30–36cm apart with 20–25cm between plants. Approximately 0.5kg of seed is required to plant one hectare (0.5g/10m^2). Blanching, which reduces the bitter flavour of the leaves, is induced by tying the leaves together for a period of 10−14 days when the plants are approaching maturity. However, during wet weather, internal leaves are liable to rot and so blanching is only attempted during relatively dry periods.

The water requirement is moderately high. Plants grown during the dry season require irrigation. In addition to a pre-planting addition of a complete fertilizer to the beds, additional dressings of a nitrogenous fertilizer should be made during early growth.

Plants normally mature within 70−85 days from transplanting, depending on factors such as soil fertility and the application of irrigation during dry periods. Yields of up to 5 t/ha (0.5 kg/m^2) may be obtained under average cultural conditions.

Pests and diseases

These are rarely serious.

Fennel (*Foeniculum vulgare*)

General characteristics

A very erect perennial plant growing up to 2m in height; the stems are ribbed and the leaves finely divided into hair-like segments with wide leaf bases. The small yellow flowers are produced in compound terminal umbels.

Environmental factors

Fertile soils, with adequate mineral reserves, are required for optimum growth although plants are tolerant of a wide range of climatic conditions. Dry season cultivation at altitudes of more than 500 m is generally successful.

Cultural techniques

Seeds are sown in containers or in seedbeds and are transplanted to well prepared beds when about 5−8 cm in height. The spacing is 45−60 × 45−60 cm each way, or in rows 60−70 cm apart in which 35−45 cm is allowed between plants in the row. Regular irrigation and occasional dressings of potash and nitrogen are recommended. The young leaves are normally harvested between four and six weeks from transplanting; the swollen leaf bases are sometimes blanched and used in salads. The swollen stem bases may be harvested 12−16 weeks from planting for use in salads or as a cooked vegetable. The seeds are used for flavouring soups and sauces.

Pests and diseases

These are rarely of economic importance.

Lettuce (*Lactuca sativa*)

General characteristics

Three main varieties of the European type of lettuce commonly grown are:
  (a) Cabbage or head lettuce (var. *capitata*). The compact rosette of leaves forms a solid head. The leaves are broad and the midrib branches into small veins (Figure 9-11).
  (b) Cos or romaine lettuce (var. *longifolia*). The upright leaves are oblong and coarse in texture; the midrib is prominent. Most of the leaves are self-folding, forming loose heads (Figure 9-12).
  (c) Leaf lettuce (var. *crispa*). The leaves are in a loose rosette similar to cabbage lettuce but the plant is non-heading. Some cultivars have curled or coloured leaves.

Indian lettuce (*Lactuca indica*) is popular in some South-east Asian countries. It is an erect perennial, up to 130 cm high, with yellow flowers. The stems contain latex. It is raised from seed or root cuttings and plants are established at a 30 × 30 cm space interval. The leaves are harvested individually.

Environmental factors

Most forms and cultivars are tolerant of a wide range of climatic and soil conditions, although a pH of 6.0−6.8 is preferable. Cultivars adapted to tropical environments have been selected. Many tolerate high day temperatures of up to 30°C and short days although cooler temperatures of 15−20°C are preferred. High temperatures often result in stunted growth and bitter leaves. At low-elevation high-temperature areas some cultivars do not form solid heads; they tend to flower early without forming the normal number of leaves. Elevations above 1000 m provide optimum growth conditions.

Cultural requirements

Propagation is by seed which germinates four or five days after sowing. Seeds require a period of dry storage before sowing and some cultivars which have been stored at high temperatures have a requirement for exposure to light before dormancy can be broken. Optimum germination temperature is 25°C. Above this temperature germination percentage falls rapidly. Viability of seeds is reduced under moist conditions and high temperatures.

Seeds may be sown in drills 25−30 cm apart and the seedlings thinned initially to 8−10 cm apart and later to 20−25 cm apart. Lettuce is, however, more frequently sown in containers or a nursery bed and transplanted when between four and six weeks old. Irrigation is required at frequent intervals, particularly until the seedlings are established. Dry soil conditions induce premature flowering.

A good response is obtained from a high soil organic matter content and top dressings of nitrogenous fertilizers. Phosphate fertilizers encourage the

Figure 9-11
Lettuce, cabbage type (*Lactuca sativa* var. *capitata*)

Figure 9-12
Lettuce, cos type (*Lactuca sativa* var. *longifolia*)

production of firm heads (see also Table 13 in the Appendix).

Most heading cultivars mature within 60–85 days from transplanting but the leaf types may be harvested 35–50 days from planting. Harvesting during the early part of the day is preferable, particularly in hot weather. Yields of approximately 10–15 t/ha (1–1.5 kg/m^2) may be obtained.

Pests

Aphid (*Aphis* spp.), Table 14 (45).
Root-knot nematode (*Meloidogyne* spp.), Table 14 (119).
White fly (*Bemisia tabaci*), Table 14 (26).

Diseases

Bottom rot (*Thanatephorus cucumeris* or *Rhizoctonia solani*), Table 15 (89).
Cercospora leaf spot (*Cercospora lactucae*), Table 15 (90).
Downy mildew (*Bremia lactucae*), Table 15 (91).
Lettuce mosaic virus, Table 15 (92).
Root rot (*Sclerotium rolfsii*), Table 15 (98).

Mint (*Mentha* spp.)

Spearmint, peppermint and round-leaved mint are the names of the different species which are widely grown in warm climates. *Mentha javanica* is probably the most commonly grown species.

General characteristics

Widely used for flavouring, this perennial herb grows up to 60 cm high and has underground stolons or modified stems which readily form roots. The leaves are generally oval, sometimes hairy and often with serrated margins. They have numerous oil glands which give the flavour typical of the various species when crushed.

Environmental factors

Mint is very tolerant of a wide range of climatic and soil conditions but will grow best in fertile, well-drained soils with high moisture-retaining capacities. Shading the plants may be an advantage in positions exposed to direct sunlight and high temperatures.

Cultural techniques

The various species of mint are readily propagated by division of the main rootstock, the rooted underground stems being planted at a spacing of 15–20 cm × 15–20 cm. Irrigation is required during dry periods. Shoots 4–6 cm in length can normally be harvested from six to eight weeks from planting; they should be severed with a sharp knife, leaving 10−12cm of the main stem for the production of lateral shoots.

Pests

These are rarely serious.

Diseases

Mint rust (*Puccinia menthae*) Table 15 (100).

New Zealand spinach (*Tetragonia tetragonioides* or *T. expansa*).

General characteristics

This crop is a vigorous, spreading annual growing 30−40cm in height but with semi-prostrate stems. The leaves are dark green, triangular, succulent and approximately 12cm long and 5cm across (Figure 9-13). The flowers are small, green-yellow and borne in the leaf axils. The fruits are hard, triangular, 8−10cm in length and contain several seeds.

Environmental factors

Moderately fertile soils are suitable for this crop and soils with a fairly high organic matter content and an adequate reserve of essential minerals will produce satisfactory yields. Well-drained, sandy soils are preferable. Most cultivars tolerate short periods of drought and many also withstand heavy rainfall without reduction in yield.

The crop is adapted to high temperatures up to 35°C due to the succulent leaves which do not transpire readily. Altitudes of over 1000m provide conditions for satisfactory growth and development. At low elevations the crop is much less vigorous and is grown with some difficulty.

Cultural techniques

New Zealand spinach is propagated by seeds. These are normally soaked in water for 24 hours before sowing directly in rows 65−75cm apart with 30−40cm between plants. Germination is sometimes slow and erratic, so two or three seeds are often sown per planting station or hill and later thinned to one. Alternatively, seeds may be sown in containers and the seedlings transplanted when 8−10cm high. Approximately 15kg/ha (15g/10m^2) of seeds are required for direct sowing.

Irrigation is generally necessary during prolonged dry periods and nitrogenous fertilizer, applied at intervals of 20−30 days, may be necessary to stimulate leaf production (see also Table 13 in the Appendix).

Terminal shoots are harvested when they are about 8−20cm in length and harvesting may continue for several months. The removal of terminal shoots

Figure 9-13
New Zealand spinach (*Tetragonia tetragonioides*)

stimulates the development of lateral branches. Yields are approximately $2.5\,kg/m^2$ per annum.

Pests and diseases

These are rarely serious.

Parsley (*Petroselinum crispum*)

General characteristics

Two main forms are commonly grown: the tall, upright, flat-leaved Hamburg or Italian; and the shorter, serrated-leaf form known as 'moss-curled'. Parsley is a biennial or short-lived perennial herb. The taller forms grow to 30—60 cm high, while the moss-curled type rarely exceeds 20 cm in height. The roots are fleshy and aromatic.

Environmental factors

Well-drained soils with good reserves of essential minerals are required for optimum growth; acidic, poorly drained soils are generally unsuitable. Elevations of 500 m and above are preferable, although plants will grow adequately at lower elevations, particularly the more upright forms. Light shading may be required in areas exposed to direct sunlight.

Cultural techniques

Seeds are sown in containers or on nursery beds and may take several weeks to germinate. The seedlings are transplanted to nursery beds or plastic containers when about 6 cm high; container growing is often more successful at low elevations. Seedlings grown in beds should be planted in rows spaced at 30—38 cm, with 15 cm between plants. Regular irrigation in dry periods and top dressings of a nitrogenous fertilizer are required for optimum growth. The leaves may be harvested about 10—14 weeks from transplanting and successional harvesting continued over a period of between six and 12 months. The leaves should be removed by cutting the base of the petiole with a sharp knife. Fresh leaves can be kept in good condition at low temperatures for a considerable period; dried crushed leaves can also be stored in an airtight container.

Pests and diseases

These are rarely serious.

Swiss chard (*Beta vulgaris* var. *cicla*)

Alternative names are seakale beet, leaf beet, silver beet and spinach beet.

Figure 9-14
Swiss chard (*Beta vulgaris* var. *cicla*)

General characteristics

These are similar to beetroot. Swiss chard is a form of beetroot which has been selected for the production of large, fleshy leaf stalks and broad leaf blades with pronounced white or red midribs (Figure 9-14).

Cultural techniques

The cultural requirements of Swiss chard are similar to those of the garden beet except that a wider spacing of 60–75 cm between rows and 20–30 cm between plants is used. Swiss chard is tolerant of high temperatures and slightly saline soils. Fertilizer requirements, particularly for nitrogen, are higher than those of beetroot. The outer leaves are harvested singly as they mature, starting after 50–60 days from sowing. Yields of up to 10 t/ha (1 kg/m^2) are obtained.

Pests and diseases

These are similar to those described for beetroot.

Questions for study and review

1. Examine the nutritional contents (Table 10 in the Appendix) of the leaf crops described in this chapter and select the three which have the highest overall nutritional value.
2. From all the leaf crops described in this chapter how many are grown in your area? Try to discover reasons for the remainder not being grown.
3. How many crops in this chapter provide not only leaves but also other food materials? What do you consider to be the most important dual-purpose vegetable grown in your area?
4. Compare the yields, in kg/m^2, of all the crops listed in this chapter. Which one would you grow to give the highest total yield?
5. How many leaf crops are likely to be attacked by leaf spots? How would you control these diseases?
6. What are the main advantages in growing cassava, compared with other leaf crops?
7. If *Amaranthus* or *Celosia* are grown in your area, list and describe the main cultivars or forms grown and write notes on the local methods of cultivation which are used.
8. List those leaf crops which are sown directly in beds and those which are normally transplanted. Attempt to find reasons for the different methods used.
9. Do leaf crops in general appear to require more of one type of fertilizer than any other? If so, what is this fertilizer and how often would you apply it to the most important leaf crops grown in your area?
10. Write an account of the methods used in lettuce cultivation in your area, including all the operations from seed sowing to harvesting.

CHAPTER 10

Cucurbit Crops

General characteristics

Most of the crops of the family Cucurbitaceae are annuals or are grown as annuals. A few are short-lived perennials. The habit of all species grown as food crops is trailing and many must be supported to prevent the fruits touching the soil, and to make the crop more convenient to harvest.

The root system is generally fibrous and shallow, and most species are sensitive to excessive soil water. Stems are generally soft, ridged, hairy or covered with stiff prickles and some become woody at the base. Many species have tendrils.

The leaves are variable in shape and size. The leaf edges are toothed, the leaf blade is thin, simple or lobed, often hairy and may grow to a very large size in some cultivated forms.

The flowers are borne in the axils of leaves and are unisexual, with male and female flowers either on the same plant (monoecious) or on separate plants (dioecious). In some instances, fertilization of the female flowers is undesirable since this imparts a bitter taste to the fruit; this may be prevented by removal of the male flowers.

The fruit is a berry which normally has a firm outer layer or rind. The seeds vary in number and are normally flattened. Their colour varies from white to black .

Ecological factors

Fertile soils are generally considered essential for the production of high cucurbit yields since they grow rapidly and require readily available sources of nutrients. It is customary to excavate trenches or pits which are then filled with well decomposed organic material before transplanting seedlings or sowing the seed of some cucurbits. Good drainage is, however, essential since excessive soil water encourages the development of harmful soil organisms.

Most cucurbits grow in high rainfall areas but growth is often more vigorous in dry periods because high humidity encourages leaf diseases, particularly mildew. The water requirement of most species is high due to the rapid growth rate and the large leaf surface. Watering should therefore be regular during dry periods.

Many cucurbits thrive only at low elevations and are therefore widely grown in the lowland humid areas. All are sensitive to cold and may be killed or injured by low temperatures. There are, however, cultivars or forms which have been selected for their adaptation to high elevations.

Cucurbits, in general, are adapted to short daylengths, and very few species are sensitive to variation in length of day.

Economic importance and distribution

The wide cultivation of the cucurbits indicates their importance, particularly in South-east Asian and African countries. In many areas, the leaves are consumed in addition to the fruits. The rapid rate of growth of most forms therefore provides a readily available source of food. In some areas, the fruits of various species are valuable as containers for liquids and for food storage.

Production of the cucurbits is labour-intensive. The soil has to be well prepared and many species require stakes or trellises. Weed control and spraying to prevent diseases and pests involve a high level of labour input.

In general, cucurbits are grown on a domestic scale and are used as intercrops, but some commercial production has developed particularly in countries where the small marrow or courgette has become popular as an export crop.

Nutritional value

In general, the fruits of cucurbits do not have a high nutritional value although the seeds of some melons have a relatively high nutrient content. Nutritional values of some cucurbits are listed in the Appendix, Table 10.

Uses

The leaves of many cucurbits are edible and are used as a cooked vegetable; examples are bottle gourd, cho-cho, pumpkin, and watermelon. The main source of food, however, is usually the fruit which may be boiled or roasted in either the immature or mature state. In some species, the main source of nutrition is in the seeds which are either roasted or used in soups; watermelon seeds are also made into flour.

LOW ELEVATION CUCURBITS

Bitter gourd or Balsam pear (*Momordica charantia*)

Alternative names are bitter cucumber, bitter melon.

General characteristics

Bitter gourd or cucumber is a climbing annual plant, the vines growing to 4m in length (Figure 10-1). The stems are furrowed, have tendrils, and the leaves are lobed, varying in diameter from 5–15cm. They have a foetid odour when crushed. The flowers are yellow, up to 3cm in diameter, and the oblong fruits are pendulous with numerous wart-like protuberances on

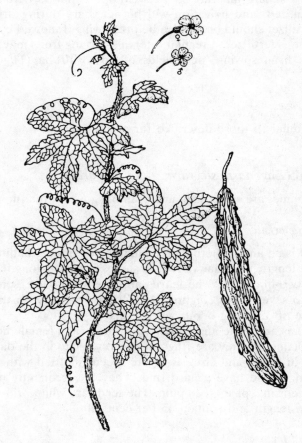

Figure 10-1
Bitter gourd (*Momordica charantia*)

the surface. They are up to 25 cm long and 8 cm in diameter when mature and orange-yellow in colour. The seeds are brown, flattened and oval, containing 32 per cent of oil.

Environmental factors

This crop is widely grown in lowland humid areas and is well adapted to high temperatures and high rainfall conditions at elevations up to 500 m. Fertile soils with a high level of organic content are required for the production of optimum yields.

Cultural techniques

Seeds are sown on well prepared beds or ridges with 60–75 cm between rows or ridges and 30–40 cm between plants in the row. Alternatively, they may be sown on a square planting 50 × 50 cm apart. Support from a trellis or stakes is required, and irrigation will be necessary during dry periods. A complete fertilizer should be applied before sowing, followed by applications of a nitrogenous fertilizer at regular intervals. Young fruits may be harvested 8–10 weeks from sowing, and yields of up to 10 t/ha (1 kg/m^2) may be obtained.

Pests and diseases

These are similar to those described for cucumber.

Bottle gourd (*Lagenaria siceraria* or *L. vulgaris*)

Alternative names are calabash gourd, trumpet gourd, white-flowered gourd.

General characteristics

Bottle gourd is an annual herb with a climbing or trailing habit. Vines grow to 3–4 m in length with hairy, robust stems furrowed longitudinally. The tendrils are two-branched. The leaves are 10–40 cm wide, sometimes with from three to seven lobes (Figure 10-2). They are hairy on the lower surface and give off a strong odour when bruised.

Flower petals are white and woolly. Both male and female flowers are up to 10 cm in diameter; they are fragrant and open late in the day. The fruits are variable in shape and size. They are green, mottled with white, up to 100 cm in length and have a hard rind. They are frequently bottle-shaped with many seeds in a pale green pulp. The seeds are white or brown, ridged, up to 2 cm in length and contain 45 per cent oil.

Environmental factors

Economic yields are obtained at elevations up to 500 m. Most cultivars are

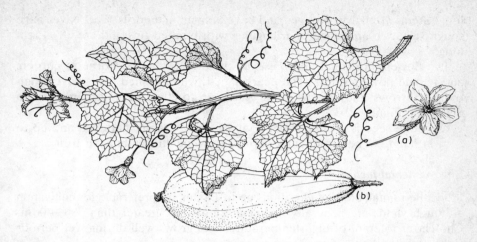

Figure 10-2
Bottle gourd *Lagenaria siceraria* or *L. vulgaris* (a) female flower, (b) fruit

tolerant of a wide range of rainfall but normally produce higher yields under moderate rainfall. Bottle gourds may be cultivated in either a wet or dry season. For optimum growth, soils should be amended with organic material.

Cultural techniques

Seeds are sown in mounds, ridges, or prepared holes 75−90cm apart and are later thinned to single seedlings. Plants should be supported by stakes and terminal shoots removed to encourage branching. Seed requirement is approximately 0.5−2.0kg/ha (0.5−2g/10m^2).

Irrigation is required at regular intervals and a complete fertilizer should be applied to the soil before planting. Liquid manure or nitrogenous fertilizer may also be applied at intervals until the onset of flowering.

Immature fruits may be harvested 70−90 days from sowing; mature fruits may be left until they reach a length of 30−45cm. Yield is approximately 20t/ha (2kg/m^2).

Pests and diseases

These are similar to those of cucumber.

Cho-cho (*Sechium edule*)

Alternative names are chayote, christophine, choko, chow-chow and sou-sou.

General characteristics

This cucurbit is a vigorous, climbing perennial herb with vines up to 12m long. Stems are hairy, grooved and have branching tendrils. The leaves have rough surfaces, are lobed or triangular with pointed tips and are 10–15 cm long.

The flowers are small and unisexual with male flowers borne in clusters and female flowers singly. The pear-shaped, white-fleshed fruits have longitudinal furrows; they are white or green, up to 20 cm in length and 8 cm in diameter (Figure 10-3).

Only one seed is produced per fruit. Seeds are large, flattened and up to 5 cm in length. They frequently germinate while still within the fruit.

Environmental factors

Moderate rainfall areas at elevations up to 1500 m are preferable for cultivation although yields are generally higher at elevations greater than 300–400 m. Cho-cho is tolerant of high temperatures and grows well during wet periods since moist soil conditions favour growth.

Soils should be supplied with organic material and should have a good moisture-retaining capacity. Slightly acid soils with a pH of 5.5–6.5 are preferred.

Cultural techniques

The most usual method of propagation is by planting a mature fruit containing the germinating seed. The fruits are planted at a depth of two-thirds of their

Figure 10-3
Cho-cho (*Sechium edule*)

length in prepared planting holes with the wide end pointing downward. Rows should be 100–120cm apart with plants 60–75cm apart in the row. Propagation by stem cuttings 15–20cm in length is sometimes used. Cuttings are inserted under shade and kept moist. Plants require support by poles or a trellis. Irrigation should be frequent to keep the soil moisture content at a high level.

A complete fertilizer should be applied before planting and this may be followed by dressings of nitrogenous fertilizer or liquid manure at intervals until fruit formation (see also Table 13 in the Appendix).

Fruits may be harvested 100–120 days from planting and are produced over a long period. Individual fruits may weigh up to 0.5kg each.

Pests and diseases

These are rarely of economic importance.

Cucumber (*Cucumis sativus*)

General characteristics

The cucumber is a climbing or trailing annual herb with an extensive, shallow root system. Stems are four-angled and hairy with unbranched tendrils. The leaves are triangular, oval, 7–20cm in length and have sharply pointed tips.

The flowers are yellow, 3–4cm in diameter with male flowers predominating. The female flowers are usually single and are borne on short, thick pedicels. The pendulous fruits are variable in shape and size. The flesh is pale green and many white seeds are produced by each fruit.

Environmental factors

Cucumbers are adapted to warm climates but will grow at lower temperatures than melons; the optimum range is 20–25°C. Water requirement is high, but high humidity encourages the development of leaf diseases and may affect flower production. Soils should be well drained with a high organic content. Sandy loams with a pH of 5.5–6.7 are generally suitable for early-maturing crops. Elevations of up to 1200m are suitable for most cultivars.

A high light intensity increases the number of male flowers produced; lower light levels result in more female flowers being formed. Shading is sometimes beneficial and can also protect crops from damage by high winds.

Cultural techniques

Seeds are sown directly in ridges, mounds, or prepared planting holes 60–70cm apart. Alternatively, plants may be raised in containers and transplanted when 8–12cm high. Lateral shoots may be pruned after one fruit has formed to restrict excessive leaf and flower production. Support should be provided for some trailing cultivars.

Irrigation is required at frequent intervals and a high level of soil moisture should be maintained, A complete fertilizer should be applied before sowing or transplanting, followed by applications of liquid manure every 14−21 days until the fruits form (see also Table 13 in the Appendix).

Fruits may be harvested 40−60 days from sowing or planting, when 15−20cm in length. Yields of approximately 5t/ha (0.5kg/m^2) may be obtained.

Pests

Aphid (*Aphis* spp.), Table 14 (117).
Epilachna beetle (*Epilachna* spp.), Table 14 (143).
Melon fruitfly (*Dacus cucurbitae*), Table 14 (51).
Root-knot nematode (*Meloidogyne* spp.), Table 14 (52).

Diseases

Cucumber mosaic virus, Table 15 (59).
Downy mildew (*Pseudoperonospora cubensis*), Table 15 (112).
Fusarium wilt (*Fusarium oxysporum*), Table 15 (61).
Powdery mildew (*Erysiphe cichoracearum*), Table 15 (62).

Pumpkin or Squash gourd (*Cucurbita maxima*)

Alternative names are Chinese pumpkin and crookneck squash.

General characteristics

Pumpkin is an annual vine growing to 3m in length. The stems are slightly hairy and have branched tendrils. The dark green leaves are rarely lobed; they sometimes have white markings and are 15−30cm in diameter (Figure 10-4). The flowers are unisexual with yellow-orange female flowers 15cm in diameter. Male flowers are smaller.

The fruits are large and vary in shape. They may be round or oblong, are often covered with small raised spots and weigh up to 5kg. The rind is sometimes brightly coloured and the flesh is yellow. The seeds are white or brown, flattened, and contain 35−40 per cent oil and 30 per cent protein.

Environmental factors

Most cultivars require a high temperature above 25°C during the growing period and fairly low humidity. Some cultivars are tolerant of cool conditions. Dry periods favour rapid growth and excessive humidity may be harmful due to the encouragement of leaf diseases.

Figure 10-4
Pumpkin (*Cucurbita maxima*)
(Photograph by Alan Thomas)

Pumpkins tolerate a range of soil conditions and some cultivars are tolerant of slightly acid soils. A high organic matter content is preferable and altitudes of up to 2000 m are suitable for most cultivars.

Cultural techniques

Seeds can be sown directly 2 cm deep in prepared holes or beds. Bush types may be sown in rows 20 cm apart with 60−100 cm between plants; trailing types should be sown 1−3 m apart each way. Seeds may also be sown in containers and transplanted when 8−10 cm in height. Seed requirement is about 3−4 kg/ha (3 g/10 m^2).

Irrigation is required at regular intervals, particularly during dry periods. A complete fertilizer should be applied before sowing or transplanting, followed by applications of a nitrogenous fertilizer or liquid manure at intervals of 10−14 days until the fruits develop (see also Table 13 in the Appendix).

Fruits may be harvested from 80−140 days from sowing or planting. The optimum harvesting stage is before the seeds ripen and when the skin of the fruit begins to harden. Yields of between three and six fruits/plant may be expected.

Pests

Aphid (*Aphis* spp.), Table 14 (117).

Cucurbit leaf beetle (*Aulacophora* spp.), Table 14 (113).
Leaf beetle (*Copa* spp.), Table 14 (126).
Leaf-eating beetle (*Lagria villosa*), Table 14 (115).
Lesser melon fly (*Dacus ciliatus*), Table 14 (116).

Diseases

Downy mildew (*Pseudoperonospora* spp.), Table 15 (112).
Leaf and stem wet-rot (*Choanephora cucurbitarum*) Table 15 (3).
Leaf blight, brown spot (*Alternaria cucumerina*), Table 15 (114).

Pumpkin or Winter squash (*Cucurbita moschata*)

This species is closely related to *C. maxima* and these two forms of pumpkin are sometimes difficult to distinguish.

General characteristics

The crop is an annual vine growing to 3 m and generally lacking hairs on leaves and stems which mainly distinguishes it from *C. maxima*. The stems are moderately hard with tendrils which are three- or four-branched. The leaves are palmately divided, up to 20 × 30 cm with rounded lobes and sometimes with white markings.

The flowers are unisexual; male flowers are borne on long stems and the

Figure 10-5
Winter squash (*Cucurbita moschata*)

female flowers on shorter stems. The large fruits are variable in shape and size and are borne on stiff, angular, grooved stalks which are expanded at the junction with the fruit (Figure 10-5). The flesh is yellow to orange. The white or brown seeds have a broad rim and contain 40–50 per cent of oil and 30 per cent of protein.

Environmental factors

C. moschata is well adapted to the environment of the lowland tropics. Soils with high organic content are essential for optimum yields.

Cultural techniques

These are as described for *C. maxima* except that seeds are sown directly at a spacing 2 × 3 m or seedlings may be raised in containers and transplanted.

Pests and diseases

These are similar to those affecting *C. maxima*.

Ridged gourd or Angled loofah (*Luffa acutangula*)

Alternative names are silk gourd, angled gourd, vegetable gourd and Chinese okra.

General characteristics

This is a vigorous, climbing, annual herb, the vines growing to a length of several metres (Figure 10-6). The stems are furrowed, with tendrils, and the leaves are mainly oval with slight lobes and pointed tips. They have a rough texture and emit a strong odour when rubbed. The separate male and female flowers are yellow. The male flowers are formed in clusters and the female flowers are single; they are larger than the male flowers and may be up to 8 cm in diameter. The fruits are club-shaped with 10 clearly defined ridges, and fully developed fruits may be up to 60 cm long and 12 cm in diameter. The numerous seeds are black, flattened, and have a rough surface; they contain an edible oil.

Environmental factors

This species grows well in lowland areas with relatively high temperatures and regular rainfall, although excessive rain may reduce flowering and fruiting. Soils with a high organic content and a pH value of 6.5–7.5 are preferable, although satisfactory crops may be grown on sandy loams if well supplied with essential nutrients.

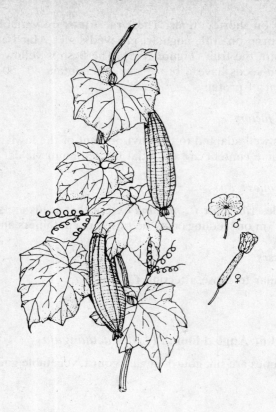

Figure 10-6
Ridged gourd (*Luffa acutangula*)

Cultural techniques

Seeds may be sown in containers and transplanted or sown direct in well prepared soil on mounds or ridges. These should be 75−100 cm apart and the plants established at 45 cm apart in the row. Alternatively, a square planting of 60−90 × 60−90 cm may be used. Plants are normally supported on a trellis up to 3 m high. Irrigation is required during dry periods and particularly before flowering occurs. Liquid manure or a nitrogenous fertilizer should be applied at regular intervals up to the flowering period. Fruits may be harvested 6−10 weeks from planting and 15−20 fruits may be produced per plant.

Pests and diseases

These are rarely of economic importance.

Figure 10-7
Snake gourd (*Trichosanthes cucumerina* or *T. anguina*)

Snake gourd (*Trichosanthes cucumerina* or *T. anguina*)

An alternative name is club gourd.

General characteristics

Snake gourd is a climbing annual herb with leaves which have a strong smell when damaged. The leaves are hairy, normally with five to seven lobes, 10–25 cm in length and 15 cm in diameter.

The unisexual flowers are axillary, white, and up to 5 cm in diameter. Male flowers are borne on long racemes but female flowers are borne singly. The flowers have frilled edges, and the inner surfaces of the petals and the ovary are hairy.

The waxy fruits are cylindrical, slender and 40–120 cm in length (Figure 10-7). The rind is orange when ripe. The brown seeds have a rough surface and are 1–1.5 cm long.

Environmental factors

This crop is adapted to low elevations in high rainfall areas. It does not withstand drought and requires a plentiful reserve of moisture in the soil.

Cultural techniques

Seeds are sown in planting holes or ridges 100−150 cm apart with 60−75 cm between plants. Mature plants require support from poles or a trellis. Weights can be attached to the tips of the fruits to prevent them from curving.

Young fruits may be harvested 50−70 days from sowing when 30−60 cm in length. Picking may be continued for 30−60 days. Single fruits may weigh 1 kg; one plant is likely to produce from six to 10 fruits per year.

Pests

Cucurbit leaf beetle (*Aulacophora* spp.),Table 14 (113).
Leaf beetle (*Copa* spp.), Table 14 (126).
Leaf-eating beetle (*Lagria villosa*), Table 14 (115).

Diseases

Storage or black rot (*Botryodiplodia theobromae*), Table 15 (121).
Other diseases of snake gourd are similar to those described for cucumber.

Sponge gourd or Smooth loofah (*Luffa cylindrica*)

Alernative names are dish-cloth gourd, loofah, rag gourd, vegetable sponge.

General characteristics

This gourd is a vigorous, annual, climbing herb with a five-ridged stem and long tendrils up to 10 cm long (Figure 10-8). The leaves have a rough surface, are mainly oval with from five to seven lobes, and may be up to 25 cm in length and width. The male and female flowers are yellow and are produced together in the same axil. The single female flowers may be up to 10 cm in diameter; the smaller male flowers are in clusters. The fruits have 10 slight furrows on the surface and may be 30−60 cm long and 8−12 cm in diameter when fully developed. The seeds are black, flat and smooth and contain 46 per cent oil and 40 per cent protein.

Environmental factors and cultural techniques

These are essentially the same as those outlined for the ridged gourd, although the lateral shoots of young plants are often pruned to encourage early flowering. Hand pollination of the female flowers early in the morning

Figure 10-8
Sponge gourd (*Luffa cylindrica*)

is also recommended to stimulate fertilization and fruit production. Mature fruits are produced 14−16 weeks from planting and immature fruits, eaten either raw or as a cooked vegetable, may be harvested 10−12 weeks from planting when about 15−20 cm long. Yields of 30−40 t/ha (20−25 fruits per plant) may be obtained under optimum conditions.

Pests

Fruitfly (*Dacus* spp.), Table 14 (128)

Diseases

Powdery mildew (*Erysiphe cichoracearum*), Table 15 (62)
Downy mildew (*Pseudoperonospora cubensis*), Table 15 (112)

Watermelon (*Citrullus lanatus, C. vulgaris, Colocynthia citrullus*)

General characteristics

The watermelon is an annual herb with slender, hairy stems growing to 5 m. The stems are angular and grooved, with two- or three-branched tendrils. The large leaves, 5–20 × 2–12 cm, have three or four pairs of lobes which are subdivided and toothed. Petioles are 1–10 cm in length. The unisexual flowers are axillary with more male than female flowers normally being formed. The flowers are 2–3 cm in diameter with from three to five stamens and a hairy ovary.

The globular fruits are up to 60 cm or more in length (Figure 10-9). The rind is smooth, green or cream and may be striped or mottled. The flesh is

Figure 10-9
Water melon (*Citrullus lanatus*)
(USDA Photo)

red, green, yellow, or white and is usually sweet. The numerous flattened seeds may be black, brown, red, or yellow and have a protein content of 25−32 per cent. The oil content is also high.

Environmental factors

Growth and fruit production in most areas are generally satisfactory during dry periods. Excessive rainfall and humidity reduce productivity by affecting flower production and encouraging leaf diseases. Soils should be well drained, rich in organic content and with high moisture-retaining capacity. Well-drained sandy loam soils are ideal and pH values as low as 5.0 can be tolerated by some cultivars. Elevations up to 1000m provide a suitable climate and stable day-night temperatures promote rapid growth.

Cultural techniques

Seeds are sown 2−4cm deep in trenches, on mounds or in prepared planting holes at a spacing of 1.2−2.0m each way. Seedlings are later thinned to one per station. Alternatively, seedlings may be raised in containers and transplanted when 10−14cm high. The seed required per hectare is approximately 2.5−3kg (3g/10m^2).

Irrigation should be regular throughout the growing period. A dressing of NPK fertilizer should be applied to the soil before sowing or transplanting and followed by nitrogenous fertilizer applied at regular intervals up to flowering.

Developing fruits can be protected from soil insects by being raised on pads of grass or straw. First fruits may be harvested about 80 days from sowing with harvesting continuing for a further 40−50 days. Maturity is indicated by the withering of the tendril nearest the fruit, and the weight of the fruit. Yields of 10−20t/ha (1−2kg/m^2) may be obtained depending on the cultivar.

Pests and diseases

These are mainly as described for cucumber.

Wax or Ash gourd (*Benincasa hispida*)

Alternative names are white gourd, hairy wax gourd, tallow gourd, Chinese watermelon.

General characteristics

The wax or ash gourd is a vigorous, climbing, annual herb, with ribbed stems which may grow to several metres in length (Figure 10-10). The leaves are hairy, are dark green with long petioles, and may measure 10−25cm in length and 10−20cm in width; they are lobed and give off a foetid odour when bruised. Both male and female flowers are yellow and approximately

Figure 10-10
Wax gourd (*Benincasa hispida*)

10cm in diameter. The fruits are either rounded or oblong and may be 25−30cm long and 15−20cm in diameter; they are hairy in the young state but become covered with a waxy, grey deposit when mature. The seeds are flat with a narrow base and may be yellow, white or pale brown.

Environmental factors

This crop is well adapted to the relatively high temperatures of the lowland areas, but excessive rainfall may reduce yield and areas of moderate rainfall are therefore preferable. However, the water requirement is fairly high in the early stages of development due to a rapid rate of growth. Although the crop will grow satisfactorily in a wide range of soils, a good organic content is necessary for optimum development; soil pH should be in the range 5.5−6.5.

Cultural techniques

Seeds are sown in planting holes which have been partially filled with decayed manure or compost mixed with topsoil; the holes may be in rows 90–120 cm apart with 60–80 cm between plants in the row. Regular irrigation is required, particularly in the early stages of growth, and a regular dressing of nitrogenous fertilizer at an interval of 10–14 days is generally beneficial. Plants may be supported on a trellis or with stakes 2–3 m long, although they are allowed to trail on the ground in some areas. Only one fruit should be allowed to develop on each lateral shoot.

Hand-pollination may be required to obtain a satisfactory level of fertilization and fruits may be harvested 12–20 weeks from the date of sowing. Individual fruits may weigh 14–20 kg and yields of up to 20 t/ha (2 kg/m^2) may be obtained.

Pests

Aphid (*Aphis* spp.), Table 14 (117)
Fruitfly (*Dacus* spp), Table 14 (116)
Root-knot nematode (*Meloidogyne* spp.), Table 14 (119)

Diseases

These are rarely of economic importance.

MEDIUM AND HIGH ELEVATION CUCURBITS

Melon (*Cucumis melo*)

Alternative names are sweet melon, cantaloupe, muskmelon, honeydew melon.

General characteristics

Melon is a hairy annual herb with a trailing habit and ribbed stems covered with fine hairs. The alternate leaves have between three and seven lobes and are 8–15 cm in diameter. Leaf margins are dentate, leaf under-surfaces hairy, and the vine produces unbranched tendrils.

The male flowers are borne in clusters but female flowers are borne singly; some flowers are hermaphrodite. Flower colour is yellow, the corolla is five-partite, and the petals are round and 2 cm in length.

The fruits vary in size and shape. They may be globular or oblong, smooth or furrowed, pale to deep yellow, yellow-brown or green (Figure 10-11). The flesh is yellow, pink, or green and the fruits have many seeds.

The white-to-buff seeds are flat, smooth, and 5–15 mm in length. The edible kernel contains approximately 46 per cent oil and 36 per cent protein.

Figure 10-11
Melon (*Cucumis melo*)

Environmental factors

Melons are very tolerant of a wide range of environmental conditions and are grown in both temperate and tropical areas. Optimum growth is obtained in dry periods with high temperatures with little diurnal variation, but in many cultivars fruit quality, particularly the sugar content, is increased with lower night temperatures. Full exposure to sun is recommended and growth is best in moderate humidity; humid conditions encourage leaf diseases, reduce flower production and adversely affect fruit quality. Soils should be well supplied with organic material and have a high moisture-retaining capacity. Acid soils are unsuitable and a pH value of 6.0−6.7 is preferred.

Cultural techniques

Seeds may be sown 1−2 cm deep in prepared holes or on mounds or ridges in rows 1.3−2 cm with 60−90 cm between plants. Between one and three seeds are normally sown together and thinned to one. Seedlings may also be raised in containers and transplanted when 10−15 cm tall. When plants have become established, earthing-up may be necessary to protect the roots from excessively high temperatures. The seed required per hectare is approximately 1.5−2.5 kg ($1.5−2.5\,g/10\,m^2$) for transplanted seedlings; direct sowing requires 2−4 kg per hectare ($2−4\,g/10\,m^2$).

Irrigation should be frequent since plants have a high demand for water until the fruits have almost reached maturity. A complete fertilizer should be applied before planting, followed by regular applications of liquid manure during active growth.

Fruits may be ready for harvesting in about 80−120 days from sowing or planting. Yield is about 8−12 t/ha ($10\,kg/m^2$).

Pests and diseases

These are similar to those listed under cucumber and pumpkin.

Vegetable marrow (*Cucurbita pepo*)

Three varieties are recognized by some authorities: field pumpkins (var. *pepo*); vegetable marrows or courgettes (var. *medullosa*); bush and summer squash (var. *melopepo*). Alternative names are marrow, summer squash, zucchini.

General characteristics

Vegetable marrows are annual herbs which are bushy or, more commonly, trailing. The stems have rough hairs and are often five-angled with tendrils on the trailing forms. The leaves are deeply lobed with bristly hairs and sometimes with white markings.

The flowers are unisexual with yellow or orange petals. The elongated fruits are variable in size; no cork is present at the point of fruit attachment. The flesh is white or yellow and coarse-textured. The smooth seeds are white or brown.

Environmental factors

The vegetable marrow is a dry season crop which responds adversely to high temperatures and excessive humidity. Elevations exceeding 500 m are more suitable to growth than lowland areas. Some forms are tolerant of relatively low temperatures. Well-drained soils with a high content of organic material give optimum growth, but crops will also grow adequately on moderately fertile soils. A pH range of 5.5−6.8 is suitable.

Cultural techniques

Seeds of the bush types are sown directly on ridges or mounds 75−90 cm apart each way. For the trailing types, mounds or prepared holes 2.0−2.5 m apart each way should be used. The plants are thinned to two per station when established. Alternatively, seeds may be sown in containers and seedlings transplanted when 10−15 cm high. The seed required per hectare is 3−4 kg (3−4 g/10 m^2).

Irrigation is required at regular intervals during dry weather to promote rapid growth. A complete fertilizer should be applied before sowing and followed by applications of liquid manure or dressings of a nitrogenous fertilizer at intervals of 10−14 days (see also Table 13 in the Appendix).

Immature fruits may be harvested 40−70 days from sowing or planting. Mature fruits may take 80−100 days to ripen. Yields of 6−8 t/ha (0.6−0.8 kg/m^2) may be obtained for immature fruits (courgettes) and up to 20 t/ha (2 kg/m^2) for mature fruits.

Pests and diseases

These are similar to those given for cucumber and pumpkin.

Questions for study and review

 1. By reference to Table 10 in the Appendix, identify the three most nutritious cucurbit crops from those listed.
 2. What main structural characteristics do most of the cucurbit crops have in common?
 3. How does high rainfall affect the growth of cucurbits? What can be be done to prevent this?
 4. How does daylength affect the growth of cucurbit crops?
 5. Identify three cucurbit crops which produce not only fruits but also other food materials.
 6. Collect as many seeds as possible of different cucurbit crops grown in your area and draw them in detail; include descriptions of size, shape, and colour.
 7. List the cucurbit crops which are used both as immature and mature fruits.
 8. Name the cucurbit plant which has a single large central seed which is surrounded by a fleshy covering.
 9. What are the two main types of mildew which affect cucurbit crops grown in both dry and wet seasons?
10. Which of the cucurbit crops grow best when they are supported on stakes or a trellis?
11. What do you consider to be the most widespread pest of cucurbit crops?
12. Which is the largest of the cucurbit fruits described in this chapter?

CHAPTER 11

Solanaceous Crops

General characteristics

The vegetables included in the family Solanaceae are mainly herbs. Although some, such as the egg-plant and the hot peppers, are woody and may be regarded as short-lived perennials, most are grown as annuals.

The roots of most species are fibrous, relatively shallow and prone to invasion by nematodes. The leaves vary but are generally alternate and oval. The flower structure is regular, the petals form a corolla with five lobes, although there are examples of six or more lobes being present as in tomato. The calyx has five sepals and there are normally five stamens as well. The superior ovary develops into a berry, often with two carpels. Most solanaceous fruits produce many seeds which are often covered with fine hairs.

Ecological factors

Most crops in this family grow well in moderately fertile soil, but the addition of organic material and fertilizers promotes optimum growth. Slightly acid soil conditions are suitable for most Solanaceae and good drainage is essential. Few vegetables in this group are tolerant of drought.

Heavy rainfall and the accompanying high humidity will promote damage by pests and diseases, and the cultivated forms of this family are particularly sensitive to leaf diseases. Nematodes and bacterial wilt become serious pathogens in soils where rotations have not been followed.

Temperature has a significant effect on most solanaceous vegetables. Some thrive where there is a day-night (diurnal) variation, for example the tomato. Very high temperatures can induce pollen sterility in some crops but high night temperatures are necessary for fertilization in others.

Although well adapted to humid lowland areas, many crops in this family give good yields at elevations over 400m. These areas are most likely to have diurnal temperature variation and lower humidity than lowland areas.

Most members of the Solanaceae are daylength neutral. There are exceptions, however. The Irish potato forms tubers best when daylengths shorten to about 12 hours.

Economic importance and distribution

Many of the vegetables listed have fruits which are either consumed fresh or are cooked. Some also have edible leaves but are included in this section because the fruit produced is generally the most important product. Although many of these crops are produced on a limited scale in warm climates and are frequently grown as intercrops, some details are given for their commercial cultivation because some, such as garden eggs, tomatoes and peppers are grown for export. Since many of them require regular spraying, staking and successional harvesting, they are classified as labour-intensive crops.

Nutritional values

These are given in the Appendix, Table 10.

Uses

All of the crops listed in this chapter produce berries of varying size, shape, and colour. These are usually added to soups and stews but may also be fried or roasted. Some fruits, such as those of tomato and peppers, are mainly used for flavouring; hot peppers are commonly used in sauces.

Fruits which are eaten raw include tomato and sweet pepper. Potato tubers are boiled or roasted; they may also be dehydrated and made into flour; the berries produced by flowering cultivars of potato are poisonous.

LOW AND MEDIUM ELEVATION SOLANACEOUS CROPS

Capsicum or Sweet pepper (*Capsicum annuum, C. annuum* var. *grossum, C. annum* var. *acuminatum*)

Alternative names are Cayenne pepper, bell pepper, cherry pepper, chili, chilli pepper, green pepper, paprika, pimento and red pepper.

General characteristics

Capsicum is a short-lived herb normally grown as an annual. The plant is sometimes woody at the base and may grow to 1.5 m tall. Leaves vary in size and are mainly oval with pointed tips. The flowers are borne singly and arise terminally but appear to be axillary. The calyx is ribbed and usually enlarges to enclose the base of the fruit. The fruit is a berry which is variable in size, shape, colour and pungency (Figure 11-1).

Figure 11-1
Capsicum (*Capsicum annuum*)
(Photograph by Alan Thomas)

Environmental factors

Fertile loam soils with a high level of organic material and adequate reserves of essential elements are preferable. Good drainage is important since water-logging is likely to cause leaf drop. Plants are tolerant of slightly acid soils but a pH of 5.5−6.6 is preferable. Most cultivars are adapted to growing at altitudes up to 2000 m.

Sweet peppers are adapted to high temperatures, but excessively hot, dry weather may produce infertile pollen and thereby reduce fruit set. Temperatures above 32°C, with a fairly low relative humidity, may also cause excessive transpiration, resulting in the dropping of buds, flowers and fruits. A drop in temperature to below 16°C at flowering may result in poor fruit set or seedless fruits.

A rainfall level of 600−750 mm is adequate, but excessive rainfall affects flowering and fruit set and may encourage fruit rot. Mulching is an advantage in both dry and wet seasons.

Cultural techniques

Seeds may be sown in containers or nursery beds, and seedlings transplanted when 8−10 cm high to beds in rows 60−75 cm apart with 35−45 cm between plants. Square planting of 50−60 × 50−60 cm each way is equally acceptable, and plants are sometimes established in double rows 1−1.5 m apart with 30 cm between the double rows and 25−30 cm between plants in the row.

In addition to a pre-planting application of a complete fertilizer, dressings of nitrogen or potash are generally required during growth (see also Table 13 in the Appendix).

Fruits may be harvested 60–80 days from transplanting and picking can continue for 60 or more days. Yields of up to 20t/ha ($2\,kg/m^2$) may be obtained with good culture and irrigation during dry periods.

Pests

Aphid (*Aphis* spp.), Table 14 (45).
Fruitfly (*Ceratitis capitata*), Table 14 (21).
Root-knot nematode (*Meloidogyne* spp.), Table 14 (119).
Thrips (*Scirtothrips dorsalis*),Table 14 (23).

Diseases

Anthracnose (*Colletotrichum nigrum*), Table 15 (23).
Bacterial wilt (*Pseudomonas solanacearum*), Table 15 (136).
Leaf spot, fruit rot (*Colletotrichum capsici*), Table 15 (25).
Powdery mildew (*Leveillula taurica*), Table 15 (26).
Virus diseases, Table 15 (145–147).

Egg-plant (*Solanum melongena*)

Alternative names are garden egg, aubergine, brinjal, melongene.

General characteristics

Egg-plant is a perennial woody herb growing to 1.5m tall. The leaves are alternate, oval, and hairy. The corolla has five or six segments which are normally purple in colour and approximately 1cm in length. The fruits may be 15cm long and vary in size and shape but are often oval; they may be yellow, purple, white or black and contain many small brown seeds. (Figure 11-2).

Environmental factors

Well-drained sandy soils with good moisture-retaining properties are suitable. The root system is sensitive to excess water and deep cultivation prior to planting is advisable. A pH from 5.5–6.8 is satisfactory.

Diurnal variation in temperature is not essential and may limit development. The most satisfactory environmental conditions are normally found in lowland coastal areas with stable, high temperatures varying from 25–32°C. High soil temperatures are often injurious to the root system and cultural treatments such as mulching can be applied to reduce variation in soil temperature.

The egg-plant is adapted to both wet and dry season cultivation but excessive rainfall will check vegetative growth and flower formation. Elevations of up to 800m are suitable.

Figure 11-2
Egg-plant or garden egg

Cultural techniques

Seeds can be soaked for 24 hours to hasten germination and are sown in
containers or nursery beds. Seedlings are transplanted when 8−10 cm in
height to ridges or raised beds. For bed planting, rows are 75−90 cm apart
with plants 60−75 cm apart in the row. In square planting, a spacing of
50−60 cm each way is suitable. For ridge planting there should be 75−90 cm
between ridges and 50−60 cm between plants. The terminal growing point

may be removed when plants are established to encourage branching. Tall-growing cultivars will require support.

Fruits may normally be harvested 80–120 days from transplanting, depending on the cultivar and rate of growth. Most cultivars continue fruit production into a second year. Yields vary with cultivar; from eight to 14 fruits per plant may be expected on an average, with fruit size varying from 0.25–0.4 kg per fruit.

Pests

Bollworm, corn earworm, gram caterpillar (*Heliothis armigera*), Table 14 (140).
Cutworm (*Spodoptera littoralis*), Table 14 (34).
Epilachna beetle (*Epilachna hirta*), Table 14 (143).
Red spider mite (*Tetranychus* spp.), Table 14 (46).
Root-knot nematode (*Meloidogyne* spp.), Table 14 (119).

Diseases

Bacterial wilt (*Pseudomonas solanacearum*), Table 15 (136).
Fruit rot (*Phytophthora parasitica*), Table 15 (68).
Fusarium wilt (*Fusarium oxysporum*), Table 15 (75).
Phomopsis rot (*Phomopsis vexans=Diaporthe vexans*), Table 15 (70).

Hot pepper (*Capsicum frutescens*)

Alternative names are bird chilli, bird's-eye pepper, goat pepper, red pepper, spur pepper, tabasco pepper.

General characteristics

A shrubby perennial similar in structure to *C. annuum*. The main difference is the production of two or more fruits instead of a single fruit which is typical of the sweet pepper. These fruits are small and narrow, up to 1 cm in length and 7 mm in diameter. They are red or yellow when ripe and are extremely pungent.

Environmental factors

Hot peppers are similar to *C. annuum* in their reaction to environment except that they are generally more tolerant of extremes of temperature and rainfall than sweet peppers.

Cultural techniques

These are also practically identical to those used with *C. annuum*. However, since *C. frutescens* is a short-lived perennial rather than an annual and will tolerate slightly less fertile soil conditions, it requires more generous spacing and will have greater demands for fertilizer and irrigation than the annual pepper.

The first fruits are produced 80−100 days from transplanting. Yields of between 1000−1250 kg/ha (120 g/m^2) of fresh pods (350−500 kg/ha or 40−50 g/m^2 of dried fruits) may be obtained.

Pests and diseases

These are similar to those listed for *C. annuum*.

Tomato (*Lycopersicon lycopersicum* or *Lycopersicum esculentum*)

The cherry tomato (var. *cerasiforme*) is now widely grown in some areas; it is less susceptible to some diseases which affect tomato.

General characteristics

The tomato is an annual herb up to 2 m tall (Figures 11-3 and 11-4). The stems are hairy with a strong odour. The terminal bud often becomes an inflorescence and growth is continued by an axillary bud. The leaves are spirally arranged, up to 30 cm long and 10−15 cm across. The leaf blade is lobed and divided.

The flowers, which are up to 2 cm in diameter, are borne in inflorescences of between four and 12 flowers. The calyx is short and remains green when the fruit ripens. The six petals are yellow and up to 1 cm in length.

The fruit is a fleshy berry, ripening to become red or yellow and may be smooth or with longitudinal furrows. The hairy seeds are oval and light brown.

Environmental factors

Well-drained, fertile soils with a good moisture-retaining capacity and a relatively high level of organic material are best, although many cultivars tolerate a range of soil conditions. Slightly acid soils with a pH of 5−6.5 are suitable. Low soil temperatures retard the growth of seedlings and absorption of minerals.

High air temperatures above 27°C can cause pollen sterility and high night temperatures adversely affect flower initiation. Night temperatures of about 18−20°C are considered ideal for most cultivars. A diurnal variation of at least 5−6°C is considered necessary for optimum growth and development. High temperatures, combined with low relative humidity, can seriously affect fruit setting. Both high and low temperatures can affect fruit quality, particularly the colour of the fruits.

Excessive rainfall can harm a tomato crop, particularly if it is unstaked,

Figure 11-3
Staked (indeterminate) tomato plants

Figure 11-4
Determinate form of tomato

due to the spread of leaf diseases in humid conditions. Uneven levels of water application, combined with a lack of calcium or potassium in the soil water, may lead to a physiological disorder of the fruit known as blossom-end rot. Erratic irrigation may also produce cracking and splitting of the fruit skin. Fruits rarely ripen fully during wet periods and production is generally higher during the dry season with irrigation.

Elevations of up to 2000 m are suitable for tomato cultivation and yields are generally higher at elevations over 500 m. Many cultivars, however, are well adapted to cultivation at low elevations, although yields are generally low mainly due to a lack of diurnal temperature variation and to high humidity which encourages leaf diseases.

Cultural techniques

Seeds are sown in containers or a nursery bed, preferably in sterilized soil, and transplanted when 8−10 cm height to a prepared bed or ridges. Shelter may be required to protect the seedlings from wind, excess sun, and heavy rainfall. Plants may be established in single rows 50−60 cm apart with 30−38 cm between plants or in double rows 45−60 cm apart with 45 cm between plants in the row and 75−90 cm between the double rows. For ridge cultivation, ridges should be 75−90 cm apart with 45−60 cm between plants. In areas where bacterial wilt and root-knot nematode are present, direct sowing of tomatoes in prepared beds or planting holes has been recommended as a means of reducing pests and diseases by eliminating infection from nursery beds. In some areas, yields from direct-sown crops may be significantly higher than those obtained from transplanting. The seed requirement is approximately 0.5 kg seed/ha (0.5 g/10 m^2) for transplanted seedlings.

Liquid fertilizer can be applied at planting and repeated after about 40 days. Soils low in potassium should be dressed with potash at intervals of about 21 days until flowering begins. Excessive nitrogen may, however, promote leaf growth instead of flower initiation and nitrogen should therefore be applied sparingly after the second flower cluster has set. Trace element deficiencies may occur in some soils and micronutrients may have to be added if deficiencies are severe. Mulching is advantageous in both dry and wet seasons (see also Table 13 in the Appendix).

The removal of axillary shoots is usual if plants are staked. This increases quality and early yield but results in a reduction of total yield compared with unstaked and unpruned plants. Staked and pruned plants are also more liable to sun scald on the fruits.

Fruits can normally be harvested from indeterminate (tall-growing) cultivars 70−100 days from transplanting. Determinate (dwarf) cultivars begin fruiting after about 60 days. During wet periods green fruits are often harvested and ripened by being kept in warm, dry conditions for several weeks. A yield of 10−15 t/ha (1−1.5 kg/m^2) is normal. This can be exceeded with proper management and favourable environmental conditions.

Pests

Bollworm, corn earworm, gram caterpillar (*Heliothis* spp.), Table 14 (140).
Cotton leaf roller (*Sylepta derogata*), Table 14 (141).
Cutworm, click beetle (*Spodoptera littoralis*), Table 14 (34).
Epilachna beetle (*Epilachna* spp.), Table 14 (143).
Leaf hopper (*Empoasca* spp.), Table 14 (144).
Mite (*Hemitarsonemus latus*), Table 14 (145).
Red spider mite (*Tetranychus* spp.), Table 14 (46).
Root-knot nematode (*Meloidogyne* spp.), Table 14 (119).

Diseases

Bacterial wilt (*Pseudomonas solanacearum*), Table 15 (136).
Collar rot (*Sclerotium rolfsii*), Table 15 (137).
Early blight (*Alternaria solani*), Table 15 (139).
Grey leaf spot (*Stemphylium solani*), Table 15 (140).
Late blight (*Phytophthora infestans*), Table 15 (141).
Leaf mould (*Cladosporium fulvum*), Table 15 (142).
Septoria leaf spot (*Septoria lycopersici*), Table 15 (143).
Target leaf spot (*Corynespora cassicola*), Table 15 (144).
Tomato wilt (*Fusarium oxysporum*), Table 15 (148).

Virus diseases

Cucumber mosaic virus, Table 15 (59).
Tomato bunchy top virus, Table 15 (145)
Tomato double streak virus, Table 15 (146).
Tomato (and tobacco) mosaic virus, Table 15 (147).

MEDIUM & HIGH ELEVATION SOLANACEOUS CROPS

Irish potato (*Solanum tuberosum*)

General characteristics

The potato is an annual growing to 1 m high and producing edible underground tubers when mature. The roots are fibrous and the tubers arise separately from the main root system. The stems are angular, branched, and bear compound, alternate leaves up to 30 cm long (Figure 11-5). The flowers, which are produced in clusters or cymes, are yellow, white or purple but are rarely produced under conditions in which daylengths are short and temperatures high. The fruits are globular berries and contain poisonous alkaloids.

Figure 11-5
Potatoes, with flowers forming

Environmental factors

A well-drained, fertile soil with a pH of 5.0−6.0 is necessary. Deep cultivation and a high reserve of organic material and major elements are also essential. Although excess nitrogen encourages leaf and stem growth instead of tuber formation, low levels of nitrogen sometimes encourage tuber formation at high temperatures.

High soil temperatures may restrict tuber initiation and high air temperatures may limit tuber development in some cultivars. The optimum soil temperature for tuber formation in most cultivars is 16−24°C, but cultivars adapted to tropical areas may withstand higher temperatures.

Excessive rainfall spreads leaf diseases and therefore dry season cultivation is generally practised in high rainfall areas. Irrigation during dry weather can have a beneficial effect on tuber formation by preventing excessive soil temperatures.

An altitude of over 1000m is required for successful growth and the production of economic yields. Most cultivars respond to relatively short day lengths by forming tubers. Long days prolong the vegetative growth phase at the expense of tuber formation.

Cultural techniques

Soil cultivation to a depth of 25cm is advisable, with organic material and fertilizers being applied to the soil before planting. Tubers should be certified to be free from diseases such as virus and are either planted whole or cut into 'setts', each sett having one or more dormant buds or 'eyes'. Tubers may be placed in boxes, with the end bearing most buds facing upwards, for two weeks before planting so that young shoots can develop. They should be sprayed with water daily during this period.

The sprouted tubers are planted 10−15cm deep in shallow trenches spaced 60−75cm apart with 25cm between plants. They are then lightly covered with soil and irrigated if necessary. Filling in of the trenches is desirable when the stems are 10−15cm in height to ensure that the tubers develop in loose soil and to prevent exposure to light which turns the tubers green and makes them unfit for human consumption. Nitrogen and potash may be applied as surface dressings every 20−30 days until the tubers form.

The propagation of potatoes from seed is now being practised in some tropical countries; this technique is particularly effective in reducing the incidence of virus diseases.

Potatoes may be harvested 80−120 days from planting. The tubers should be lifted carefully using a garden fork, if available, to avoid damage. Tubers which are not required for immediate use may be left in the soil where they will remain in a good condition for some time. They may be stored in well-ventilated rooms or barns for several months, where they should be placed either on layers of dry sand or on slatted shelves. Pit or clamp storage is also possible; these are trenches dug up to 75cm deep and lined with dried grass and leaves. The tubers are placed in the clamp and covered with leaves and a top layer of soil. Ventilation holes should be made to prevent condensation; short tubes of bamboo cane are sometimes used. Yields of 10−15t/ha $(1-1.5 kg/m^2)$ may be obtained under favourable conditions.

Pests

In addition to the pests listed as attacking tomato, the following may also affect the potato:

Aphid (*Aphis* spp.), Table 14 (117).

Cutworm (*Agrotis ipsilon*), Table 14 (29)

Diseases

Most pathogens attacking tomato affect the potato.

Questions for study and review

1. What major effects may high or low temperatures have on solanaceous crops such as tomatoes and peppers?
2. Most solanaceous crops are sown in containers or seedbeds and are transplanted. Suggest reasons for this practice.
3. Why is a mulch frequently used with crops such as egg-plant or peppers?
4. What do you consider to be the most serious disease of solanaceous crops grown in the humid tropics?
5. List the crops in this family which benefit from support by stakes or a trellis.
6. How would you attempt to prevent virus infection of solanaceous crops, particularly tomatoes?
7. What is the main cause of physiological disease known as blossom-end rot? How can it be prevented?
8. What is meant by 'determinate' and 'indeterminate' tomato cultivars?
9. List three of the most important insect pests of solanaceous crops and state how you would prevent or control them.
10. What are the best climatic conditions for potato cultivation? Write brief notes on the preparation of potato tubers for planting.

CHAPTER 12

Root and Tuber Crops

General characteristics

The crops in this section are adapted to varying elevations and belong to different botanical families. Although most are herbaceous and are reproduced from parts of tubers, others are temperate zone biennials, grown from seed as annuals. Their main feature in common is that the edible part is produced underground.

Low elevation crops:
    Chinese yam (*Dioscorea esculenta*) — Dioscoreaceae.
    Cocoyam (*Colocasia esculenta*) — Araceae
    Sweet potato (*Ipomoea batatas*) — Convolvulaceae
    Tannia (*Xanthosoma sagittifolium*) — Araceae
    Yam bean (*Pachyrrhizus erosus*) — Leguminosae
Medium and high elevation crops:
    Beetroot (*Beta vulgaris*) — Chenopodiaceae
    Carrot (*Daucus carota*) — Umbelliferae
    Radish (*Raphanus sativus*) — Cruciferae

Ecological factors

Due to their diversity, few of these root and tuber crops have ecological requirements in common. However, like most root crops, they generally require deep, well-drained soils which are well prepared before planting so that the underground portion can develop without restriction. Many of the lowland root and tuber crops are grown on ridges due to high rainfall conditions. Higher altitude crops tend to be grown on raised beds. Many intercropping patterns exist, but only details of row-crop and ridge planting are given here.

    High-temperature tolerance is a feature of many root and tuber crops although high soil temperatures are generally damaging. Many crops in this group tolerate high rainfall, particularly the lowland crops. Irrigation in dry periods is essential for the medium and high elevation crops to maintain a regular growth rate.

Nutrient requirements vary with each species but a high level of organic material in the soil is necessary for the successful production of most crops. Nitrogen and potassium are most in demand and additional dressings in the early stages of growth are generally recommended (Table 13 in the Appendix).

Economic importance and distribution

Most important crops in this section are grown in the lowland humid areas and provide an essential part of the carbohydrate content of the average diet. Most lowland area crops are grown for domestic consumption or sale in local markets. The possibilities for export are limited by their short post-harvest life, although, in some warm climate countries improved post-harvest techniques are increasing the export potential of some crops.

The medium and high elevation crops, on the other hand, have a longer storage life and can be transported for considerable distances if adequately packed in suitable containers. They may also be stored if kept at a low temperature and high humidity. Post-harvest diseases and pests can be serious on both lowland and higher elevation crops, and storage hygiene is important if stored tubers or roots are to remain in good condition.

Nutritional values

The nutritional values of the major crops in this group are listed in the Appendix, Table 10.

Uses

The tubers and corms of the crops included in this chapter are used as boiled vegetables; many of them (such as cocoyam and tannia corms) are also roasted, baked, or fried. The young shoots and leaves of cocoyam, tannia, and sweet potato are frequently used as boiled vegetables.

The roots of beetroot and carrot are also cooked although young carrot roots may also be eaten raw in salads, as are radish roots. Beetroot, carrot, and radish leaves are all edible and are used as cooked vegetables in some areas.

LOW ELEVATION ROOT AND TUBER CROPS

Chinese yam or Lesser yam (*Dioscorea esculenta*)

Alternative names are Asiatic yam, lesser Asiatic yam.

General characteristics

The stems of Chinese yam may reach a length of 3m. They twine and are spiny, often with a purple colouration at the base (Figure 12-1). The leaves are smooth, up to 12cm long and 15cm wide with pointed tips; the bases of the petioles are often spiny. Flowers are rarely formed. The tubers are small compared with those of most other species of yam. They are oval, up to 20cm long 6–8cm in diameter, and each plant produces 5–20 tubers. The flesh is yellow or white and the average tuber weight varies from 0.25 to 1kg.

Environmental factors

Chinese yams do not have the same requirement for water as do other yams, such as the water yam or white yam. Well-drained soils with a high organic content are preferable. Optimum growth is obtained from plants grown in

Figure 12-1
Chinese yam (*Dioscorea esculenta*)

hot climates with an evenly distributed rainfall of 850—1000 mm per annum. Elevations of more than 500 m are considered unsuitable since yields are reduced at higher altitudes. The tubers normally have a short storage life due to a limited dormancy period.

Cultural techniques

The normal method of propagation is by planting tubers or small sections of tubers with dormant buds. These are inserted in prepared mounds or ridges with the buds pointing upwards; the ridges are 80—100 cm apart, and the tubers are planted at a depth of 4—8 cm spaced 45—55 cm apart. Propagation from stem cuttings is also possible. The plants are supported on stout stakes 3—4 m long. Since they are normally planted at the beginning of the wet season, irrigation is rarely necessary.

The nutrient requirements of Chinese yam are normally met by an application of NPK fertilizer to the planting holes and one or two subsequent top dressings of potash and nitrogen.

Tubers may be harvested approximately 28—40 weeks from planting, depending on the climate and soil conditions; they are sensitive to rough handling. Yields are in the region of 8—20 t/ha (0.8—2 kg/m^2); higher yields may be obtained under ideal growing conditions.

Pests

Yam shoot beetle (*Lilioceris livida*), Table 14 (37)
Root-knot nematode (*Meloidogyne* spp.), Table 14 (119)
Yam beetle (*Heteroligus meles*), Table 14 (39)
Yam nematode (*Scutellonema bradys*), Table 14 (40)

Diseases

Anthracnose (*Glomerella cingulata*), Table 15 (46)
Leaf mosaic Table 15 (47)
Tuber rot (*Botryodiplodia theobromae*), Table 15 (48)
Tuber rot (*Fusarium oxysporum*), Table 15 (49)

Cocoyam (*Colocasia esculenta, C. antiquorum, C. antiquorum* var. *esculenta*)

Alternative names are Chinese eddoe, dasheen, eddo.

General characteristics

Cocoyam is an erect herbaceous perennial. The West Indian or Trinidadian dasheen (*Colocasia esculenta* var. *esculenta*) normally has a large main corm with a few side cormels. The eddo (*C. esculenta* var. *antiquorum*) has a relatively small main corm but produces many small oval cormels at the base

Figure 12-2
Cocoyam (*Colocasia*)

of the stem. Cormels of the eddo mature earlier than those of the dasheen.

The leaves are pale green and oval, with long petioles. At least two forms exist, one with green and the other with purple petioles and leaves. Cocoyams rarely flower (Figure 12-2).

Environmental factors

The cocoyam produces optimum yields when planted in fertile soils with a high water-retaining capacity. It is traditionally planted along streams or rivers but some forms tolerate upland conditions.

Cocoyam is adapted to high-temperature areas with high humidity and most forms respond well to stable temperatures. Crops are normally grown in lowland areas at elevations up to 1000 m.

In most growing areas, crops receive sufficient moisture from rainfall but some forms of eddoes, which are more adapted to drier soil conditions, require irrigation during dry periods.

Cultural techniques

Propagation is by corms or sections of corms, preferably planted during the wet season after they have begun sprouting. In some areas corms or cormels are preferred, but tops and young suckers are also used as planting material.

For bed planting, rows should be 90–100 cm apart with 30–45 cm between plants in the row. For ridge planting, the ridges should be 1 m apart with 45–60 cm between plants. A complete fertilizer should be applied in addition to manure or compost before planting. Dressings of nitrogen and potash during the growing period are also beneficial (see Table 13 in the Appendix). Plants are often earthed up after they have become established to encourage corm development.

Most forms of cocoyam mature in 240–300 days from planting but the eddo matures in 180–210 days. The corms are lifted by hand and the main tuber is often harvested with the smaller corms left to develop later.

Yields are variable but may range from 4–6 t/ha (500 g/m^2). Yields of 15 t/ha have been recorded.

Pests

Aphid (*Aphis* spp.), Table 14 (45).
Red spider mite (*Tetranychus* spp.), Table 14 (46).
Root-knot nematode (*Meloidogyne* spp.), Table 14 (119).

Diseases

Leaf spot (*Phyllosticta* spp.), Table 15 (57)
Tuber rot (*Sclerotium rolfsii*), Table 15 (58).

Sweet potato (*Ipomoea batatas*)

General characteristics

Sweet potato is a perennial herb cultivated as an annual with trailing or twining stems 1–5 m in length. The stems are mainly prostrate, sometimes twining and light green to purple. The leaves are spirally arranged and either simple or deeply lobed. They are up to 15 cm long with pointed tips and may be green to purple.

The root system is extensive and roots grow from the stem nodes where stems contact the soil. Tuber structure is mainly globular and smooth or ridged. The tuber surface may be white, yellow, orange, purple or brown and the flesh is white, yellow, orange, red or purple (Figure 12-3).

The flowers may be single or in clusters (cymes); the calyx is five-lobed, the corolla is funnel-shaped or tubular and the petals are purple with pale margins.

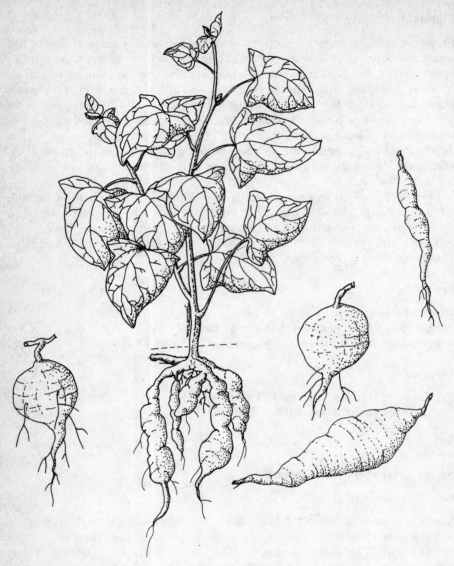

Figure 12-3
Sweet potato (*Ipomoea batatas*)

Environmental factors

Sweet potatoes are grown between 40°N and 32°S latitude at elevations up to 2500 m. The optimum average temperature is 24–25°C with a well-distributed rainfall in the range 75–125 cm per year. Full exposure to sun is essential. The sweet potato is a short-day plant and a photoperiod of less than 11 hours induces flowering.

The plants tolerate a range of soil conditions, but a well-drained sandy loam with a pH of about 6.0 is preferable. The crop is sensitive to water-logging.

Cultural techniques

Seedlings vary in performance and propagation from tubers and tuber shoots is rare in tropical countries. Stem cuttings are therefore most frequently used for propagation. These are from 25−45 cm in length and are obtained from terminal shoots since plants from terminal cuttings produce higher yields than those propagated from basal or mid-stem cuttings. The lower leaves are removed and cuttings are inserted to half their length on ridges or mounds. Ridges are constructed about 45 cm high and 75 cm apart, and the cuttings are inserted 26−30 cm apart on the ridge.

Irrigation is rarely necessary since most cuttings are inserted during the wet season. If soil moisture content is not adequate, furrow irrigation may be used.

Sweet potatoes respond well to manure although too much nitrogen may promote excess stem and leaf growth instead of tuberous root formation. Potash is sometimes required for the formation of large well-shaped tubers (see also Table 13 in the Appendix).

Most cultivars mature in 80−160 days from planting. At maturity, the leaves yellow and fall; tubers become firm and do not discolour when cut. Yields of 15−20 t/ha $(1.5−2.0 \text{ kg/m}^2)$ may be obtained but yields average 8−12 t/ha $(0.8−1.2 \text{ kg/m}^2)$.

Pests

Root-knot nematode (*Meloidogyne* spp.), Table 14 (119)
Stem borer (*Megastes grandalis*), Table 14 (135)
Sweet potato hawk moth (*Agrius convolvuli=Herse convolvuli*), Table 14 (136).
Sweet potato weevil (*Cylas formicarius* and *C. brunneus*), Table 14 (137).

Diseases

Black rot (*Ceratocystis fimbriata*), Table 15 (128).
Sclerotial wilt (*Sclerotium rolfsii*), Table 15 (129)
Soft rot (*Rhizopus nigricans* and *R. stolonifer*), Table 15 (130).
Storage or black rot (*Botryodiplodia theobromae*), Table 15 (131).

Tannia (*Xanthosoma sagittifolium*)

Alternative names are yautia, elephant's ear, tannier. *X. brasiliense* is known as belembe or calalou; it produces small corms and is grown mainly for the young leaves and shoots.

General characteristics

Tannia is an herbaceous perennial 1.3–2.5 m in height, closely related to the cocoyam. The main underground stems are corms and may be white, pink, or yellow; smaller offshoots, referred to as cormels, are produced by the main corm.

The leaves are large with a marginal vein and a long petiole joined at the margin of the lamina (Figure 12-4). This is unlike the cocoyam where the petiole joins near the centre of the leaf blade. Cultivars differ in leaf size, colour, venation and the length of their petioles; also in the size, shape and the flesh colour of the corms. Flowers are rare in cultivated forms.

Environmental factors

Heavy soils are unsuitable for this crop which is often grown where the soil moisture content is too low for *Colocasia*. A high level of soil organic material is essential for early maturity of the corms and cormels.

Plants are not sensitive to high temperatures and will tolerate shade, although growth is optimum in full sunlight. Tannia is normally grown as a rain-fed crop in fairly moist climates where rainfall is regular and evenly distributed. Altitudes up to 800 m are considered suitable. The crop is usually restricted to lowland areas where climatic conditions favour rapid growth.

Figure 12-4
Tannia (*Xanthosoma sagittifolium*)

Cultural techniques

Propagation is normally by means of small corms, although the tops of young plants are also often removed and inserted as cuttings. Rooted cuttings or corms are planted on low ridges 75—90 cm apart with 90 cm between plants. More vigorous forms require wider spacing. Earthing up during the early part of the growing period is advantageous, followed by mulching during hot weather.

In addition to pre-planting applications of organic material and complete fertilizer, surface dressings of nitrogen may be required at intervals during active growth (see also Table 13 in the Appendix).

After a period of 240—420 days large corms, to which smaller cormels 15—22 cm in length are attached, are produced from the original seed corms or cuttings. These should be harvested before they produce new shoots. Successional harvesting of these mature cormels may be continued for 500 days or more, leaving the main corm undisturbed. Yields of 6—12 t/ha (1 kg/m^2) of corms are often obtained.

Pests

These are rarely of economic importance.

Diseases

Root rot (*Thanetophorus cucumeris*) Table 15 (134)
Tuber rot (*Sclerotium rolfsii*) Table 15 (135)

Yam bean or Potato bean (*Pachyrrhizus erosus*)

General characteristics

This climbing legume is unusual in that it produces both edible tubers and pods, although the mature pods, leaves and seeds are poisonous due to the development of a glucoside (Figure 12-5). The stems may grow to 6 m in length and are covered with brown hairs. The large leaves are trifoliate, the leaflets oval and sometimes lobed. The flowers, which are mauve or white, are borne in clusters of between one and five florets. The pods, up to 14 cm long and 2 cm in diameter, contain from six to 12 flat, almost square, yellow seeds; the roots are large and tuberous, light brown in colour and up to 20 cm in diameter.

Environmental factors

Sandy, well-drained soils are suitable since the yam bean is sensitive to wet soil conditions. Moderate rainfall and elevations up to 1000 m are generally regarded as being favourable for growth. Plants exposed to daylengths of

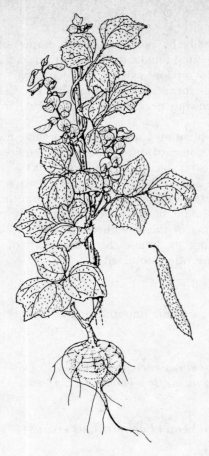

Figure 12-5
Yam bean (*Pachyrrhizus erosus*)

14–15 hours do not produce tubers although vegetative growth is normal under these conditions.

Cultural techniques

Seeds and tubers may be used in propagation, although tuber propagation is relatively rare and is only used to maintain trueness to type of particular cultivars. For pod production, seeds are sown on beds or ridges 60–75 cm apart, with 30–40 cm between plants in the row. If tuber production is required, the wider spacings should be used. Plants for pod production should be supported on a trellis or by staking to a height of 2 m; plants for tuber production require 1 m stakes. Removal of the growing point and flowering shoots when plants are 90 cm high is likely to increase tuber size and quality. Irrigation is required during dry periods, and top dressings of fertilizers containing potassium (and possibly nitrogen) should be used at intervals of 14−21 days. Immature pods may be harvested 12−14 weeks from the date of sowing, and tubers will become fibrous unless they are harvested 20−24 weeks from sowing. Tuber yields of 2 kg/plant may be obtained.

Pests and diseases

These are rarely of economic importance although Mosaic virus may be prevalent in some areas.

MEDIUM AND HIGH ELEVATION ROOT AND TUBER CROPS

Beetroot (*Beta vulgaris*)

Alternative names are garden beet, beet, red beet.

General characteristics

Beet is a biennial herb grown as an annual. The enlarged roots are usually red and rounded or tapering. The leaves are in a rosette and may be dark red or green (Figure 12-6). The flowers are borne in groups of three or four and are formed in the second year if environmental conditions are favourable.

Environmental factors

Beetroot is rarely grown successfully at low elevations with high temperatures. Elevations over 600 m are normally required for proper growth and root development. Moist soil with adequate drainage is necessary and sandy loam soils are best. The crop is tolerant of alkaline soils but sensitive to acid conditions; a pH of 6−6.8 is considered suitable. A high content of soil organic material is also desirable.

Cultural techniques

Propagation is by direct seeding in drills 25−30 cm apart and a complete fertilizer should be applied before sowing. The seedlings are later thinned to 10 cm apart.

Irrigation is required at regular intervals to maintain rapid growth and give tender roots. Applications of nitrogen fertilizer at intervals are also recommended. Boron deficiency can occur in soils high in calcium, causing the formation of black spots on the roots.

Plants normally mature within three months from sowing, developing roots 5−8 cm in diameter. These are lifted by hand since the outer surface of

Figure 12-6
Beetroot (*Beta vulgaris*)
(Photograph by Mary Waltham)

the root is easily damaged. Yields of 15−18t/ha (1.5−1.8kg/m^2) may be obtained with irrigation.

Pests

Flea beetle (*Phyllotreta* spp.), Table 14 (118)
Leaf caterpillar, beet web-worm (*Hymenia recurvalis*), Table 14 (9)

Diseases

Bacterial soft rot (*Erwina carotovora*), Table 15 (28)
Downy mildew (*Peronospora parasitica*), Table 15 (10)
Powdery mildew (*Albugo candida*), Table 15 (11)

Carrot (*Daucus carota*)

General characteristics

Carrot is an erect biennial grown as an annual, with a swollen orange taproot which varies in shape and size. The stem is condensed, forming a rosette plant and the leaves are divided (Figure 12-7). The inflorescence is a compound umbel. The flowers are white and small, and the seeds are oval.

Figure 12-7
Carrot (*Daucus carota*)
(Photograph by Alan Thomas)

Environmental factors

Well-drained soils with reserves of organic material and essential minerals are suitable; poorly-drained stony soils encourage malformed roots of inferior quality. Optimum soil pH is 5.8−6.5.

The crop is fairly sensitive to temperature; the optimum air temperature is 16 -24°C. High soil temperatures can result in the production of short fibrous roots and may inhibit germination. Many cultivars tolerate a range of rainfall but excessive rainfall may adversely affect root colour.

Altitudes of more than 500m are necessary for economic yields. Some cultivars will grow in lowland areas but rarely give high yields.

Cultural techniques

The soil should be cultivated to a depth of 20−25cm and seeds sown 1−2cm deep in rows 30−40cm apart. Mulching the seedbed is recommended to encourage good germination. The seedlings are thinned to 5cm apart when 5—8cm high. Later thinning should leave the seedlings 8—10cm apart. Seedlings may be earthed up when the roots begin to swell to protect them from excess heat. In very hot weather, light shading may be required, but additional fertilizer is rarely necessary.

Roots may be harvested 70—85 days from sowing, depending on the size required and the cultivar. Yields of 8.5—12t/ha (approximately 1 kg/m^2) may be obtained.

Pests

Root-knot nematode (*Meloidogyne* spp.), Table 14 (119)

Diseases

Bacterial soft rot (*Erwinia carotovora*), Table 15 (28)
Leaf spot, leaf blight (*Alternaria dauci*), Table 15 (29)

Jerusalem artichoke (*Helianthus tuberosus*)

An alternative name is sunchoke.

General characteristics

A perennial herb up to 3m in height with fleshy roots. The leaves are opposite or spirally arranged and hairy; the stems are also hairy and often grooved. The flowers are in large yellow heads of disc and ray florets. The tubers are formed from stolons and are irregular (Figure 12-8). Their skin may be red, purple, white or brown.

Environmental Factors

The crop tolerates a range of soil conditions but well-drained soils with a high organic content and mineral reserves are most suitable. Temperatures above 27°C may reduce yield. The water requirements of the crop are not high and rainfall of more than 150cm per year may reduce growth and rot the tubers. Tuber yield is reduced at low elevations and an altitude of more than 500m is generally required for satisfactory yield.

Cultural Requirements

Propagation is by tubers and the size of the tuber often relates directly to the yield. Tubers are planted 2.5—4cm deep, usually on ridges or raised beds in rows 45—75cm apart with 30—45cm between plants. Supplementary applications of nitrogen and potash may be required on poor soils (see also Table 13 in the Appendix).

Tubers normally mature 80—120 days from planting. Removal of the flower heads in the bud stage will encourage tuber development. The stems

Figure 12-8
Jerusalem artichoke

are normally cut and removed before the tubers are lifted. Yields of 12−25 t/ha (1.2−2.5 kg/m^2) are often achieved.

Pests and Diseases

These are rarely of economic importance.

Radish (*Raphanus sativus* var. *hortensis*)

The Chinese or Japanese radish is *R. sativus* var. *longipinnatus* and is widely grown in South-east Asia. The roots are large, up to 2 kg in weight, and normally white. Cultural requirements are similar to those of var. *hortensis*.

General characteristics

Radish is an annual herb with a tuberous red or white root. Root shape may be round, cylindrical, or tapering and the internal tissue white or red. The leaves are divided into several minor lobes and are up to 25cm in length. The flowers are white or pink and small. The fruit is a pod-like capsule (siliqua) 3–7cm in length containing from six to 12 seeds.

Environmental factors

Radish is tolerant of varying climatic conditions. Growth is little affected by temperature and daylength variation, although cool conditions promote optimum growth. White-fleshed cultivars flower under short daylengths at low elevations but red-fleshed cultivars normally require long days or elevations of 500m or more for flower production. White-fleshed radishes are more pungent than red-fleshed.

Light, well-drained soils favour early root production but adequate organic material is also necessary for good yields. Optimum pH is 6.0–7.0. Crops grown on infertile soils or at high temperatures may be very pungent.

Cultural techniques

Seeds are sown directly into prepared beds in drills 20–30cm apart. Seedlings are later thinned to 2–4cm apart. For small areas, seeds are often broadcast and lightly raked into the soil. Intercropping radish with salad crops is popular in many areas and light shading may improve the root quality during hot, dry weather.

A complete fertilizer should be applied to the soil before sowing and surface dressings of nitrogen at intervals, until the first roots mature, are often beneficial (see also Table 13 in the Appendix).

Roots are ready for harvesting 30–50 days from sowing depending on the cultivar and the climate. The leaves are often used as a vegetable and are harvested when mature. Yields or 7–10t/ha ($0.7-1 kg/m^2$) are often obtained.

Pests

Aphid (*Aphis gossypii*), Table 14 (117)
Flea beetle (*Phyllotreta* spp.), Table 14 (118)
Root-knot nematode (*Meloidogyne* spp.), Table 14 (119)

Diseases

Downy mildew (*Peronospora parasitica*), Table 15 (10)
Leaf spot (*Cercospora brassicicola*), Table 15 (116)

days favour bulb development although cultivars which form bulbs in short days are available.

Economic importance and distribution

Although not of high nutritional value, the onion and related crops are extremely popular throughout the tropics as a flavouring. They are largely consumed locally but export potential is developing in several regions. This is because, if properly dried and packed, many forms of onion, shallot, and garlic can be transported for considerable distances without deteriorating. If suitable temperatures can be maintained, storage for several months is also possible.

Dehydration techniques are now being developed commercially in some areas, particularly for onions, and cultivars which have the characteristics required for successful dehydration have been developed.

Nutritional values

These are listed in the Appendix, Table 10.

Uses

In addition to bulb formation, as in onions and shallots, some alliums have leaves which are used as a flavouring; examples are shallots and chives. Garlic bulbs or cloves are also used as a flavouring and condiment. The blanched lower portions of alliums which do not form bulbs, such as leeks, are used as a cooked vegetable; the upper portions of leek leaves are rarely eaten.

Onion and shallot bulbs may be boiled or fried but may also be eaten raw in salads. They are also sometimes preserved in the form of pickles.

In parts of South-east Asia, the flowering shoots of some alliums are used as a cooked vegetable, e.g. Chinese chives (*Allium tuberosum*).

Bunching onion (*Allium fistulosum*)

Alternative names are: Japanese bunching onion, green bunching onion, Chinese small onion.

General characteristics

There are two main forms of this type of onion, tall- and short-leaved types (Figure 13-1). A bulb is rarely formed, although the basal portion of the plant is slightly swollen. The leaves are hollow, similar to those of the onion (*Allium cepa*) and may grow from 12–30 cm in length. Lateral buds at the base of the plant expand to form several offshoots. The yellow flowers are borne in an umbel.

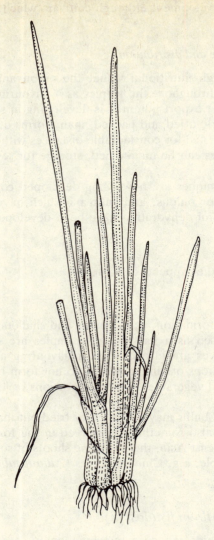

Figure 13-1
Bunching onion (*Allium fistulosum*)

Environmental factors

Well-drained medium-loam soils, with a high level of organic material, are generally suitable; most forms are well adapted to high rainfall conditions. Temperatures in excess of 25°C may reduce yields, and elevations over 1000 m promote vigorous growth resulting in higher yields than are obtained in lowland areas.

Cultural techniques

This type of onion may be propagated by either seeds or by division of the basal shoots. Seeds are sown either in containers or in seedbeds and seedlings may be transplanted when 12–15cm high to well prepared beds. Planting distances of 24–30cm between rows with plants 12–20cm apart in the row are generally suitable. Alternatively, the lateral buds produced at the base of the plant may be removed and transplanted when large enough at the spacing recommended for seedlings. A dressing of NPK fertilizer before sowing or planting, followed by top dressings of potash and nitrogen at intervals of 14–21 days, is normally adequate. The lower parts of the plants may be blanched by covering them with soil to a height of about 15cm.

Harvesting may begin 8–12 weeks from planting and can be extended over a considerable period by detaching the outer leafy shoots, leaving the main cluster of the parent plant intact. Alternatively, the whole plant may be lifted and divided into the individual offshoots. An average yield of 20t/ha ($2 kg/m^2$) may be obtained.

Pests

Cotton leafworm (*Spodoptera littoralis*), Table 14 (13).
Beet armyworm (*Spodoptera exigua*), Table 14 (76).
Onion thrip (*Thrips tabaci*), Table 14 (107).

Diseases

These are similar to those listed for onion.

Chives (*Allium schoenoprasum*)

General characteristics

Chives are a perennial herb growing to 30cm tall with many shoots arising from the base of the parent plant. The leaves are hollow and up to 25cm long. Bulbs do not develop but are present as a swelling at the base of the plant. The pink flowers are borne in an umbel.

Environmental factors

Plants tolerate a wide range of soils but fertile loams are best. Sandy soils are also satisfactory if supplied with organic material and nutrients.

Cultural techniques

Propagation is by division of existing clumps or by seeds sown in containers or on a nursery bed. Seedlings or divided plants are transplanted to prepared beds when 12–15cm high to a square planting 15–20 × 15–20cm. Liquid manure is beneficial during early growth.

Shoots can usually be harvested beginning 70–100 days from planting. The outer shoots should be removed at the base of the plant, leaving the clump intact. Clumps should be replanted every two to three years.

Pests and diseases

The main pest affecting chives is *Thrips tabaci* (onion thrips). Other pests and diseases are as listed under onion.

Chives, Chinese (*Allium tuberosum*)

An alternative name is Chinese leeks.

General characteristics

This type of chive does not form bulbs or offshoots but has a rhizomatous habit and develops in closely grouped clumps (Figure 13-2). Plants can grow up to 40 cm in height; the leaves are long, up to 5 mm wide and have a flattened upper surface. The flowers are white with a green or red stripe and are produced in umbels on solid flowering stalks up to 45 cm long.

Figure 13-2
Chinese chives (Allium tuberosum)

Environmental factors

These are generally similar to those described for chives.

Cultural techniques

Plants may be propagated from seed, sown in rows 30 cm apart, and thinned to 4—6 cm apart in the row. The more usual practice is to divide the clumps, replanting the offshoots or bulbils in rows 30 cm apart, 8—10 cm apart in the row. The leaves are harvested when mature, and the flowers when they reach the bud stage. The lower parts of the leaves may be blanched by covering them with soil.

Pests and diseases

These are similar to those listed for bunching onion.

Garlic (*Allium sativum*)

General characteristics

Garlic is an erect biennial up to 60 cm tall. The bulbs have a flattened conical stem from which arise several cloves or individual sections consisting of thickened storage leaves and a growing point. The leaves are flat, solid, and 2.5 cm wide; the flowers are pink.

Environmental factors

Fertile well-drained sandy or silt loams are preferred with organic material added to improve water-retaining capacity. High temperatures are required for optimum bulb development but cooler conditions in the early stages favour vegetative growth. Excessive humidity and rainfall are detrimental to vegetative growth and bulb formation. Garlic is normally grown in low rainfall areas with irrigation during the early vegetative stage. Elevations from 500—2000 m provide suitable growing conditions.

Long days favour bulb development. In areas with a seasonal variation in length of day, it is preferable to plant during short photoperiods to encourage maximum vegetative growth before bulbing begins.

Cultural techniques

The cloves are separated and planted on firmed, prepared beds, preferably towards the end of the wet season. The cloves are almost covered with fine soil after planting. Rows should be 30—40 cm apart with 10—15 cm between plants in the row. Approximately 700—1000 kg of cloves are required to plant one hectare (1 kg/m^2). Surface dressings of nitrogen are often beneficial during the growing period (see also Table 13 in the Appendix).

The bulbs should be mature in 90–120 days from planting when the leaves turn brown and dry. Leaves for flavouring can be cut before they mature. Yields are in the range of 5–10t/ha (0.5–1kg/m^2).

Pests and diseases

These are similar to those of onion.

Leek (*Allium ampeloprasum* var. *porrum, A. porrum*)

General characteristics

Leek is a biennial with no bulb formation. The leaves are flat and grow up to 100cm long (Figure 13-3).

Environmental factors

Leeks have a high water requirement. Deeply cultivated soils which are capable of moisture retention, but are also well drained, are necessary for optimum growth. Soils should be rich in organic material and well supplied with the major elements. Cool conditions favour optimum growth, but large plants may be obtained at high temperatures if they are well watered. Growth will continue as long as adequate soil moisture is available. Elevations of more than 500–700m are preferable but adequate yields may be obtained at sea level from local or specially selected cultivars.

Cultural techniques

Seeds are sown at almost any time of the year in containers or nursery beds. The seedlings are transplanted when 15–20cm high to beds or trenches 30–40cm wide. Trenches should be dug 30–35cm deep and loosely refilled to half the depth with topsoil. Seedlings are inserted into holes 7–10cm deep in two rows 20–22cm apart with 20–22cm between plants in the row. If beds rather than trenches are used, the seedlings are planted in rows 30–36cm apart with 15–20cm between plants in the row. They will require earthing up to ensure blanching of the lower portion of the leaf bases. Since consumer demand is generally for leeks with the lower half of the plant white due to blanching, trench planting is preferred, particularly for small-scale cultivation.

Irrigation is required during dry weather and supplementary nitrogen and potash may be required (see also Table 13 in the Appendix).

Mature plants may be harvested 120–150 days from transplanting, by which time they will have grown to 25–30cm in length and have a lower stem diameter of 3–4cm. Yields of 15–20t/ha (1.5–2.0kg/m^2) may be obtained.

Figure 13-3
Leek (*Allium ampeloprasum* var. *porrum*, *A. porrum*)
(Photograph by Alan Thomas)

Pests and diseases

These are similar to those listed for onion.

Onion (*Allium cepa*)

Alternative names are common onion or bulb onion for bulb-forming types; spring onion or green onion for onions which are harvested in the early stages of growth.

General characteristics

Onion is a biennial herb normally grown as an annual. The bulb is formed from the bases of thickened food-storage leaves (Figure 13-4). The hollow foliage leaves are produced from a flattened, basal, conical stem. Groups of 5–10 flowers are produced in umbels and the whole inflorescence may grow 100 cm tall.

Figure 13-4
Onion, bulb-forming type (*Allium cepa*)

Environmental factors

A high level of organic material is required for optimum growth and development; alluvial or sandy loams are suitable. Reserves of the major elements, particularly phosphate and potash, should be available throughout the growing period but excess nitrogen may delay bulb formation. A pH of 6−6.8 is ideal. Bulb-forming onions normally only bulb at high temperatures, although it is preferable that vegetative growth is completed during cool weather.

Seedlings are fairly tolerant of high rainfall and adequate soil moisture is required throughout the growing period but particularly at bulb formation. A long, dry period is required for bulb ripening after the leaves have withered. The onion is considered a long-day plant, generally requiring 14−16 hours for bulb formation. However, cultivars have been selected which form bulbs in short days (12−13 hours) in tropical areas.

Cultural techniques

Bulb onion seeds are normally sown in nursery beds or containers and seedlings transplanted to prepared beds when 8−12 cm high. They may be planted in rows 30−38 cm apart with 15−20 cm between plants or on a square 15−22 × 15−22 cm apart. Alternatively, seeds may be sown in drills 30 cm apart and thinned to 10−15 cm between plants. Small bulbs from the previous season can also be replanted as setts, in a similar manner to shallots.

Spring onions are grown from seeds sown in rows 15−20cm apart and the seedlings thinned to 5−8cm apart. The quantity of seed required per hectare for bulb or spring onions by direct sowing is about 5−8kg (5−8g/10m²), for transplanted seedlings 3−5kg (3−5g/10m²) is required.

Supplemental dressings of nitrogen and potash may be required to stimulate growth (see also Table 13 in the Appendix).

Bulbs mature 100−140 days from sowing, depending on the cultivar and the weather. They should be allowed to cure in the field with bulbs protected from the sun for several days after lifting. Spring onions are ready for harvesting 35−45 days from sowing. Bulb yields of up to 5t/ha (0.5kg/m²) may be obtained. Seeds are rarely produced from bulbs subjected to high temperatures and short days.

Pests

Beet armyworm (*Spodoptera exigua*), Table 14 (76).
Onion fly (*Hylemya antigua*), Table 14 (105).
Stem and bulb nematode (*Ditylenchus dipsaci*), Table 14 (106).
Thrips (*Thrips tabaci*), Table 14 (107).

Diseases

Downy mildew (*Pseudoperonospora destructor*), Table 15 (109).
Onion smudge, anthracnose (*Colletotrichum circinans*), Table 15 (110).
Smut, black soft rot (*Urocystis cepulae*), Table 15 (111).

Shallot (*Allium cepa.* var. *aggregatum, A. ascalonicum*)

General characteristics

Shallot is an annual with several bulbs arising from a single parent bulb. The bulbs, sometimes referred to as 'cloves', vary in shape, size, and colour and consist of a series of overlapping food-storage leaves arising from a condensed conical stem. The foliage leaves are up to 40cm in length and are slightly flattened on the upper surface (Figure 13-5). The flower stalks are up to 25cm tall and the black seeds are produced within a capsule.

Environmental factors

Shallots are tolerant of many soils. Loose, sandy soils with high organic content are preferable, although silt-clay loams are often used. The bulbs tolerate high temperatures up to 30°C and high temperatures encourage bulb development in some cultivars. Bulbs are not formed at temperatures lower than 20°C. Flowering is reduced in high-temperature conditions. Yields may be reduced during heavy rainfall, partly due to disease. A dry period is required for curing the bulbs.

Figure 13-5
Shallot (*Allium cepa* var. *aggregatum*)

Altitudes from sea level to 2500 m are suitable. Although the shallot requires long days for maximum bulb development, most tropical cultivars will form bulbs of an adequate size in short daylengths.

Cultural techniques

Propagation is mainly by planting mature bulbs. The bulbs are planted on prepared beds with a well-firmed surface, 15 × 22 cm each way or in rows 25−30 cm apart with 12−15 cm between bulbs in the row. Bulb size is related to planting density and smaller bulbs are formed at closer spacings. However, size can be increased by applications of nitrogen and potash during the growing period (see also Table 13 in the Appendix). Irrigation may be necessary during very dry periods.

Bulbs can normally be harvested 60−100 days from planting or when the leaves have become yellow. Young bulbs with green leaves can be harvested earlier for use in salads.

Yields of 9−10 t/ha (1 kg/m^2) of bulbs may be obtained. Mature bulbs which are required for propagation should be thoroughly dried and stored for at least 40 days in well-ventilated conditions before being replanted.

Pests and diseases

These are similar to those listed under onion.

Questions for study and review

1. Outline the main ecological requirements of the following crops: onions, garlic, leeks, and shallots, emphasizing any similarities or differences in requirements.
2. Compare the nutritional values of all the crops listed in this chapter using data obtained from Table 10 in the Appendix.
3. Write a brief account of the various methods of propagation used with the crops included in this chapter.
4. Compare the yields of the crops included in the family Alliaceae. Attempt to estimate the yields of any of these crops grown in your area and compare them with the data given in this chapter.
5. Write brief notes on the importance of onions and shallots in your locality and indicate any ways in which you consider local growing techniques could be improved.

CHAPTER 14

Legumes

General characteristics

The family Leguminosae is now separated into the three subfamilies; Caesalpinioideae, Mimosoideae, and Papilionoideae. Of these, the Papilionoideae is the subfamily which includes many of the annual food crops of the tropics.

The flower of the Papilionoideae has five often brightly coloured petals. The lower two are often fused to form the keel, the two side petals form the wings and the remaining large petal forms the standard. There are also five sepals, which are often fused. The 10 stamens are frequently joined at the base. The ovary is the typical pod which gives the family its name.

The root sytem of the Leguminosae has distinctive nodules along the major roots as a result of the symbiotic habitation of the root system by *Rhizobium* bacteria. Products of the bacteria, particularly nitrogen compounds, are used by the plant for producing amino acids and proteins. The requirement of the plant for nitrogenous fertilizer is therefore reduced.

Ecological factors

Legumes tolerate a range of soils and lack of nitrogen in some soils is compensated for by the *Rhizobium* bacteria. Where legumes have not previously been grown, inoculation of the seed or soil with the correct *Rhizobium* strain may be necessary.

Drainage is important for most legumes and a satisfactory level of organic material is required for optimum yields. Many tropical legumes are adapted to high temperatures but diurnal variation is beneficial at medium to high elevations.

Moisture is important in growing most legumes since a soil water deficit at flowering time will often cause the flowers to drop before fertilization. However, heavy rainfall at flowering will also frequently prevent successful fertilization.

Many tropical leguminous vegetables are either daylength neutral or are

310

adapted to a short length of day. Few are affected by minor fluctuations in daylength.

Economic importance and distribution

The tropical legumes included in the following selection are grown mainly for their immature pods and seeds and for the dried seeds, but some leaves are also used as vegetables. The high protein content in legume seeds makes them a valuable food, and both the fresh pods and dried seeds are in much demand for domestic consumption and as export crops. The commercial production of beans is now established and is partially mechanized in some areas.

Nutritional values

These are listed in the Appendix, Table 10.

Uses

The young pods of most of the legumes described in this chapter are harvested for use as a cooked vegetable. The dried mature seeds are also used in soups and stews although some seeds, such as those of the dolichos bean, jack bean and sword bean, contain alkaloids which are only made non-toxic after prolonged boiling.

The yam bean and the winged bean have, in addition to edible leaves and pods, tubers which can be used as a cooked vegetable. The leaves of many legumes, particularly the dolichos bean, Lima bean, winged bean, French bean, and garden pea have edible leaves which are sometimes boiled or added to soups. In some areas, the leaves are dried and stored.

LOW ELEVATION LEGUMES

Asparagus bean or Yardlong bean (*Vigna unguiculata* subsp. *sesquipedalis*)

Alternative names are vegetable cowpea, long bean, snake bean.

General characteristics

Both dwarf and climbing forms of this annual bean exist; the climbing forms have stems which twine and may grow up to 4 m long (Figure 14-1). The trifoliate leaves may be up to 12 cm long and sometimes have purple markings.

The yellow or violet-coloured flowers are produced in groups of between three and six and are up to 2.5 cm long. The pods are very variable in length, from 30–60 cm, and may reach up to 1.25 cm in diameter; they may be white, green or red-purple and are generally succulent with 10–30 seeds

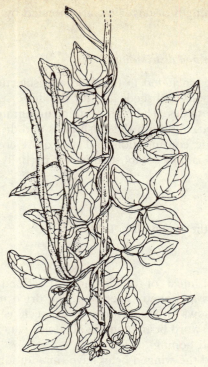

Figure 14-1
Asparagus bean (*Vigna unguiculata* subsp. *sesquipedalis*)

per pod. The seeds are up to 12 mm long and are brown or red with a white hilum.

Environmental factors

The yardlong bean is tolerant of a wide range of soil conditions; even slightly acid peat soils with a pH of 5.0−6.0 can produce adequate crops. Alkaline soils may reduce nodulation and produce chlorosis of the leaves. Most cultivars are adapted to temperatures up to 30°C. Heavy rainfall may reduce flowering, but annual precipitation in the range 1500−2000 mm is generally suitable. Most cultivars will thrive at altitudes up to 1000 m, although this is essentially a crop well adapted for the lowland humid tropics and many of the cultivars grown are daylength neutral.

Cultural techniques

Seeds are sown 1−3 cm deep in well prepared beds or ridges. The climbing cultivars are sown in rows 75−100 cm apart with plants 30−45 cm apart in the rows. They require support on a trellis or poles up to 3 m long. The dwarf or bush cultivars are sown in rows 45—60cm apart with 30—40cm between plants in the row. An NPK fertilizer should be applied as a pre-planting treatment, followed by surface dressings of nitrogen and potash.

Green pods may be harvested from early-maturing cultivars 8—10 weeks from sowing; longer duration cultivars may require 14—16 weeks to produce pods. Yields are generally variable, but 2—6t/ha (0.2—0.6kg/m^2) of fresh pods may be produced, depending on climate and soil conditions.

Pests

Spotted pod borer (*Maruca testulalis*), Table 14 (90)
Root-knot nematode (*Meloidogyne* spp.), Table 14 (119)

Diseases

Leaf spot (*Cercospora* spp.), Table 15 (33)
Anthracnose (*Colletotrichum lindemuthianum*), Table 15 (133)
Bean yellow mosaic virus Table 15 (72)
Common bean mosaic virus Table 15 (73)

Chick pea (*Cicer arietinum*)

Alternative names are Bengal gram, gram, Indian gram. At least four different races of chick pea exist, their differences being mainly based on the size and colour of the seed.

General characteristics

This annual plant grows to 50cm high when mature and both stems and leaves are covered with hairs (Figure 14—2). Plants have well developed roots which are normally covered with nodules due to the activity of the *Rhizobium* nitrogen-fixing bacterium. The leaves are pinnately divided and are up to 5cm long, each leaflet being 1—2cm in length. The fruiting pods are oblong, 2—3cm x 1—2cm with from one to two seeds per pod; these seeds are up to 10mm in diameter and may be white, yellow, red, brown or almost black.

Environmental factors

Medium to heavy loam soils, with a pH of 6—9, are generally suitable; acid soils may encourage *Fusarium* wilt. A temperature range of 18—26°C is considered suitable, preferably with a day-night variation of about 8°C. Most forms are drought resistant and are not well adapted to high rainfall areas since pod set is reduced in high humidity conditions. Semi-arid areas with an annual rainfall of 650—1000mm are usually satisfactory, and optimum seed setting is favoured in areas with a relative humidity of less than 40 per cent. Elevations of up to 1200m are suitable for the cultivation of most forms of chickpea.

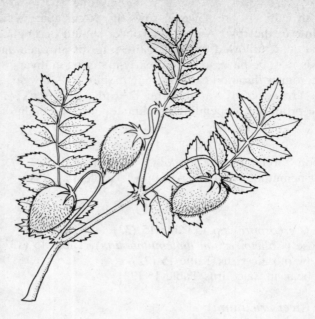

Figure 14-2
Chick pea (*Cicer arietinum*)

Cultural techniques

Seeds are sown either broadcast or in rows 25–30cm apart. A finely prepared seedbed may depress germination and a roughly prepared seedbed is preferable. Inoculation of the soil or seeds with a *Rhizobium* culture may be necessary in areas where the crop has not previously been grown. Irrigation is required in dry periods, and pinching out the terminal shoots will encourage lateral branching. Fertilizers, particularly phosphate, may be required in some areas; nitrogen application is rarely beneficial. Yields of dried seeds range from $0.5-2\,\text{kg/m}^2$; the fresh pods, used as a cooked vegetable, can be harvested 12–20 weeks from sowing.

Pests

Cutworm (*Agrotis* spp.), Table 14 (29)
Gram caterpillar (*Heliothis armigera*), Table 14 (140)
Root-knot nematode (*Meloidogyne* spp.), Table 14 (119)

Diseases

Fusarium root rot (*Fusarium oxysporum*), Table 15 (61)
Root rot (*Thanatephorus cucumeris*), Table 15 (41)

Cluster bean or Guar (*Cyamopsis tetragonoloba*)

General characteristics

The cluster bean is a bushy, upright annual which may grow to a height of 3 m. The main root is vigorous, and the lateral roots have numerous nodules. The stiff branches have white hairs and the trifoliate leaflets, which are up to 10 cm long, have long petioles. The white and pink flowers are produced in clusters of between six and 30. The pods are straight, hairy, pale green, and up to 10 cm in length; they contain from six to 12 oval, white, grey or black seeds. The young pods may contain hydrocyanic acid but this is normally absent from the mature pods.

Environmental factors

Cluster bean is generally regarded as a drought-resistant plant and grows well in alluvial or sandy-loam soils with a pH of 7.5−8.0; some forms will tolerate relatively poor soils and alkaline conditions. High temperature conditions favour growth and soil temperatures in the range 25−30°C are generally suitable for optimum growth and development. Due to its drought-tolerant properties, this crop will grow in relatively low rainfall areas of 500−750 mm per annum; below this level irrigation is required. A high level of rainfall and relative humidity at flowering time can seriously affect fertilization and pod development. Elevations of up to 1000 m are generally suitable, and most improved cultivars are daylength neutral.

Cultural techniques

Seeds are sown 2−3 cm deep in rows 40−60 cm apart with plants 20−30 cm apart in the row; alternatively a rectangular planting of 30 cm x 40 cm each way may be used. Inoculation with the required strain of *Rhizobium* bacterium may be necessary where this crop has not previously been grown. A pre-planting application of phosphate may be required, but applications of nitrogenous fertilizers are not required. Young pods may be harvested from eight to 10 weeks from sowing but should be thoroughly cooked before eating to destroy the glucoside. Yields of 6−8 t/ha (0.6−0.8 kg/m^2) of young pods may be produced.

Pests

Aphid (*Aphis* spp), Table 14 (45)

Diseases

Anthracnose (*Colletotrichum* sp.), Table 15 (133)
Powdery mildew (*Erysiphe polygoni*), Table 15 (78)
Powdery mildew (*Leveillula* sp.), Table 15 (26)
Root rot (*Thanatephorus cucumeris*), Table 15 (39)

Dolichos or Hyacinth bean (*Lablab niger*, *Dolichos lablab*)

Alternative names are bonavist bean, lubia bean.

General characteristics

The plant is a vigorous annual or perennial growing 1.5−6m tall. The leaves are alternate and trifoliate. Leaflets are oval, up to 15cm long, slightly hairy and with pointed tips. The flowers are white or purple and borne on upright racemes. The pods are generally oblong, curved and flat and up to 15cm long with wavy margins (Figure 14−3). Seed number per pod is from three to six, and seeds vary in colour from white, red, brown, or black to speckled.

Environmental factors

Dolichos beans are generally grown during the dry season since the plant is drought-resistant. Rainfall of 70−90cm is adequate but the crop may also be grown in higher rainfall areas.

Soils with average fertility can be used if well drained, but soils with high organic content will promote more vigorous growth. A pH of 5.5−6.0 is considered suitable.

Most cultivars are adapted to high temperatures and cool weather may adversely affect fertilization. Elevations of up to 2000m are suitable for the production of economic yields. The plant is sensitive to daylength and both short- and long-day cultivars exist.

Cultural techniques

Seeds are sown at the end of the wet season in prepared holes or in rows 60−75cm apart with 30−45cm between plants. Plants will require support on a trellis or stakes. Irrigation is rarely required if the plants become established during the latter part of the wet season.

A complete fertilizer should be applied before sowing; nitrogen and potassium may be applied as surface dressings to promote rapid growth (see also Table 13 in the Appendix).

Mature pods may be harvested 80−120 days from sowing, before the seeds are fully formed, for use as a cooked vegetable. Yields of 2200kg/ha $(0.2\,\mathrm{kg/m^2})$ may be obtained.

Pests

Aphid (*Aphis* spp.), Table 14 (117).
Bollworm, gram caterpillar (*Heliothis armigera*), Table 14 (140).
Root-knot nematode (*Meloidogyne* spp.), Table 14 (119).
Spotted pod borer (*Maruca testulalis*), Table 14 (90).

Figure 14-3
Dolichos bean (*Lablab niger*, *Dolichos lablab*)

Diseases

Anthracnose (*Colletotrichum lindemuthianum*), Table 15 (133).
Common blight (*Xanthomonas phaseoli*), Table 15 (64).
Halo blight (*Pseudomonas phaseolicola*), Table 15 (74).
Powdery mildew (*Leveillula tarsica*), Table 15 (26).

Jack or Horse bean (*Canavalia ensiformis*)

General characteristics

Jack bean is an erect annual herb growing 1–2 m high. The leaves are hairy
and trifoliate, smaller than those of sword bean. The flowers, in groups of
from three to five on a curved raceme, are 2–2.5 cm long and the petals are
pink or mauve. The cream-coloured pods are up to 30 cm long and 2.5 cm

wide and contain 10−20 seeds which are white, approximately 2 x 1.3cm and flat with a grey hilum.

Environmental factors

Jack bean is deep-rooted and moderately drought-resistant. Most soils can give adequate yields but a high organic content will stimulate production. Moderately high temperatures are required and elevations of up to 1500m are suitable for most cultivars.

Cultural techniques, pests and diseases

These are similar to those outlined for sword bean. Young pods may be harvested 120−200 days from sowing.

Lima bean (*Phaseolus lunatus*)

Alternative names are butter bean, Madagascar bean, and Sieva bean.

General characteristics

Lima beans are twining, perennial herbs growing to 4m tall. There are also annual and bush types which grow to 90cm. The leaves are trifoliate, and flowers are borne on a raceme up to 15cm long with from two to four

Figure 14-4
Lima bean (*Phaseolus lunatus*)

flowers at each node. Individual flowers are white, pale green, or mauve. The pods are oblong, curved, up to 12 cm long, and contain from two to four seeds (Figure. 14.4).

Both the climbing and dwarf types have large and small-seeded cultivars. The small-seeded cultivars are referred to as Sieva beans.

Environmental factors

Soils should be well drained with a pH of 6.0—6.5. Sandy loams are considered most suitable. Some cultivars are sensitive to excessive salts in the soil. Some climbing cultivars are more drought-resistant than the bush types but both are tolerant to moderately heavy rainfall. A dry period is required for the seeds to mature.

Seeds do not germinate below 16°C and temperatures above 27°C may adversely affect pollination. Most cultivars are adapted to elevations up to 1500 m and many will produce economic yields at sea level. Some small-seeded cultivars are adapted to short days; others are daylength neutral.

Cultural techniques

Seeds are sown 1—3 cm deep on beds or ridges during the latter part of the wet season. Bush types should be sown in rows 60—90 cm apart with 20—30 cm between plants. Climbing types are spaced at 75—90 cm between rows with 36—45 cm between plants; the small-seeded cultivars may be planted more closely than the large-seeded types. Climbing cultivars require support from stakes or a trellis 2—3 m high.

Irrigation is only required during exceptionally dry periods. A complete fertilizer is usually applied before sowing, followed by surface dressings of fertilizers containing potash and phosphate, if required, at intervals until flowering begins (see also Table 13 in the Appendix).

Pods of early-maturing cultivars may be harvested about 80—110 days from sowing; large-seeded cultivars take from 180—210 days to reach maturity but may be harvested over a period of several months. Yields of up to 1500 kg/ha (1.5 kg/m^2) of dried beans may be obtained.

Pests

Bean pod borer, spotted borer (*Maruca testulalis*), Table 14 (90).
Root-knot nematode (*Meloidogyne* spp.), Table 14 (119).
Stem-boring beetle (*Sagra adonis*), Table 14 (139).
Variegated grasshopper (*Zonocerus variegatus*), Table 14 (93).

Diseases

Common blight (*Xanthomonas phaseoli*), Table 15 (64).
Downy mildew (*Phytophthora phaseoli*), Table 15 (95).

Halo blight (*Pseudomonas phaseolicola*), Table 15 (74).
Root rot (*Thanatephorus cucumeris*), Table 15 (97).

Mung bean (*Vigna radiata*)

Alternative names are green gram, golden gram, Jerusalem pea.

General characteristics

Two main groups of cultivars, based on seed colour, have been identified, i.e. *golden gram* and *green gram*. Mung bean is an annual plant growing to 90 cm in height; the habit may be either upright or spreading (Figure 14-5). The root system is vigorous and the stems branched and hairy. The trifoliate leaves are also hairy and may grow to 10 cm in length. The yellow or yellow-green flowers are formed in terminal clusters, and the long, narrow pods are 6–10 cm long and 4–6 mm wide; they are hairy and contain from eight to 15 seeds which are small and rounded, and green, yellow or black in colour.

Environmental factors

This plant is fairly drought-resistant and will thrive in relatively dry areas with an annual rainfall of more than 650 mm per annum. Excessive rainfall and humidity may reduce yields. A well-drained soil with a good organic

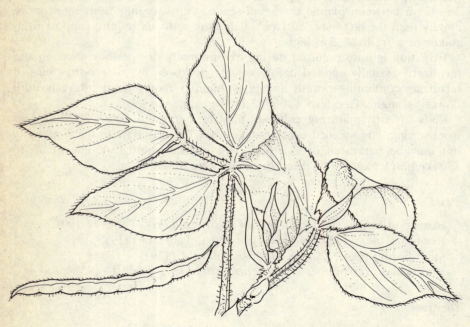

Figure 14-5
Mung bean (*Vigna radiata*)

content is preferable; some cultivars will tolerate slightly alkaline or saline soil conditions. Plants are tolerant of temperatures as high as 36°C, and altitudes of up to 2000 m are generally suitable for this crop. Both short- and long-day cultivars exist, also some which are daylength neutral.

Cultural techniques

Seeds may be sown on beds or on the flat, in rows 40–50 cm apart with 20–30 cm between plants in the row. They may also be sown at a square spacing of 25–35 cm each way. Inoculation with *Rhizobium* may be required for soils which have not previously been used for this crop. Irrigation is only required under very dry conditions but particularly before flowering.

Pre-planting fertilizer applications of both phosphorus and nitrogen are recommended in most areas. The immature pods may be harvested from eight to 10 weeks after sowing, although some late-maturing cultivars may require up to 16 weeks before harvesting. Yields of between 5–10 t/ha $(0.5–1 kg/m^2)$ of fresh pods may be obtained and up to 1 t/ha $(0.1 kg/m^2)$ of dried seeds.

Pests

Pod borer (*Heliothis* spp.), Table 14 (95).
Root-knot nematode (*Meloidogyne* spp.), Table 14 (119).

Diseases

Anthracnose (*Colletotrichum lindemuthianum*), Table 15 (133).
Collar rot (*sclerotium rolfsii*), Table 15 (137).
Halo blight (*Pseudomonas phaseolicola*), Table 15 (74).
Leaf spot (*Cercospora* spp.), Table 15 (33).
Powdery mildew (*Erysiphe polygoni*), Table 15 (78).
Root rot (*Fusarium oxysporum*). Table 15 (75).

Pigeon pea (*Cajanus cajan, C. indicus*)

Alternative names are Angola pea, Congo bean or red gram. There are two main botanical varieties: var. *flavus*, of annual habit, early maturing and of moderate height; and var. *bicolor*, which is a tall-growing, late-maturing perennial.

General characteristics

Pigeon pea is a woody, short-lived perennial shrub, growing up to 4 m in height. It has a vigorous tap root and the spirally arranged leaves are trifoliate; the leaflets are oval and hairy. The flowers are yellow, red or purple and are borne in terminal or axillary inflorescences. The flattened pods are from 6–15 cm long and contain from two to eight seeds; they have pointed tips and are normally hairy, green or dark purple when ripe. The seeds may be up to 8 mm in diameter with colour varying from white or grey to red or brown.

Environmental factors

Well-drained soils with a low level of fertility can be used, but extremely acid soils are unsuitable; a pH of 5.5—6.5 is generally required for satisfactory growth. Most forms are drought-resistant but many cultivars are adapted to growing in the wet, humid tropics. Altitudes up to 1500 m are generally suitable for growth. The optimum temperature range is from 20—30°C, but under good cultural conditions temperatures up to 35°C may be tolerated. Many tall-growing cultivars are daylength sensitive and will not flower in day-lengths longer than 12 hours.

Cultural techniques

Seeds are sown 2—3 cm deep on the flat or on ridges, in rows 90—120 cm apart with 30—120 cm between plants or 120 x 120 cm each way depending on the vigour of the cultivar. Two seeds per hill are normally sown, seedlings being later thinned to one. A dose of NPK fertilizer is sometimes applied before sowing and phosphate may increase yields on some soils. For short-duration cultivars, pods may form 100—120 days from sowing and harvesting may take place 140 days from sowing; long-duration cultivars may require 250—360 days to harvest. Yields may range from 1250—5000 kg/ha (1.2 — 5 kg/m^2) of green pods.

Pests

Aphid (*Aphis* spp.), Table 14 (117).
Gram caterpillar (*Haliothis armigera*), Table 14 (140).
Spotted pod borer (*Maruca testulalis*), Table 14 (90).
Root-knot nematode (*Meloidogyne* spp.), Table 14 (119).

Diseases

Anthracnose (*Colletotrichum lindemuthianum*), Table 15 (133).
Root rot (*Fusarium oxysporum*), Table 15 (75).
Mosaic virus Table 15 (73).

Sword bean (*Canavalia gladiata*)

General characteristics

Sword bean is a woody perennial climber with trifoliate leaves. The leaflets are up to 18 cm long and 14 cm wide with pointed tips. The flowers are 3.5 cm long and may be white or pink in groups of between three and five on

Figure 14-6
Sword bean (*Canavalia gladiata*)

a long raceme. The fruits are curved, ridged pods, up to 25 cm long and 5 cm across, containing from eight to 16 seeds (Figure 14-6). The seeds are red, 2.5−3.5 cm long, with a hilum extending almost the length of the seed.

Environmental factors

Sword bean is deep-rooted and drought-resistant. Most forms are not demanding in soil type or fertility but soils with a high organic content are preferable. The sword bean tolerates rainfall from 60−200 cm per annum and elevations up to 1000 m are favourable for most cultivars.

Cultural techniques

Seeds are sown 2−3 cm deep on the flat or on ridges 75−90 cm apart with 45−60 cm between plants. Supporting poles up to 3 m in length are required.

Irrigation is rarely necessary. A complete fertilizer should be applied before planting and occasional dressings of phosphate and potash may be beneficial (see also Table 13 in the Appendix).

The immature pods may be harvested 90−150 days from sowing when approximately 12−15 cm long. Yields of mature seeds vary from 700−1500 kg/ha ($0.7−1.5$ kg/m^2).

Pests

Stem-boring beetle (*Sagra* spp.), Table 14 (139).

Diseases

Anthracnose, root rot (*Colletotrichum lindemuthianum*), Table 15 (133).

Winged bean (*Psophocarpus tetragonolobus*)

Alternative names are Goa bean, asparagus pea, four-angled bean.

General characteristics

Winged bean is a twining, perennial herb normally grown as an annual. The ridged stems grow to 2−3 m in length and bear trifoliate leaves. Leaflets are mainly oval with pointed tips and are up to 15 cm long. The flowers are in racemes with up to 10 flowers which have white or light blue petals and are up to 4 cm in diameter. The pods are square in section, up to 30 cm long with 4 longitudinal uneven ridges (Figure 14-7). The seeds are almost round, up to 1 cm in length and white, yellow, brown, or black. The plant also produces thick, tuberous roots which develop from main lateral roots.

Environmental factors

The winged bean is adapted to humid climates although a dry period favours

Figure 14-7
Winged bean (*Psophocarpus tetragonolobus*)

fertilization and maturation of the pods. Medium loam soils are ideal although growth is satisfactory on a wide range of soils. Good drainage is essential. Elevations of up to 1000m are suitable, and a daylength of 12 hours or less is required for flower initiation, development, and pod set.

Cultural techniques

Seeds are sown on prepared beds or ridges 90−100cm apart with 45−60cm between plants. A spacing of 1−1.5 x 1−1.5m can be used between plants for long-duration crops grown for pod production. Plants require support from stakes or a trellis to a height of 2.5m.

Irrigation may be required and a complete fertilizer should be applied before sowing (see also Table 13 in the Appendix).

Immature pods may be harvested when 15−20cm in length and 2−2.5cm wide. The first pods are produced about 80 days from sowing and harvesting may continue for several months. Tubers mature from 180−240 days from sowing. If allowed to remain in the soil they will develop new shoots the following season. Yields of green pods may reach 3 t/ha (0.3kg/m^2). Seed production is in the region of 1500kg/ha (0.15kg/m^2) and tuber yields vary from 2.5−6.0t/ha (0.25−0.6kg/m^2).

Pests and diseases

These are similar to those of lima bean, but orange gall or false rust (*Synchytrium psophocarpi*) is sometimes serious, Table 15 (153).

HIGH ELEVATION LEGUMES

Broad bean (*Vicia faba*)

Alternative names are field bean, horse bean, fava bean.

General characteristics

The broad bean is an erect, annual herb, sometimes with a dwarf habit of growth and hollow stems, growing to 180cm. The alternate, divided leaves have between two and six leaflets which may grow to 7.5cm long. The fragrant flowers arise in small groups in the leaf axils and are white with purple markings. The large fleshy pods may reach 30cm in length and contain from two to six seeds per pod. The seeds are up to 2.5cm long.

Environmental factors

Fertile soils with reserves of calcium are preferable and clay soils are suitable if well drained. Most cultivars do not produce pods at low elevations, and altitudes from 1500−2000m are preferred since they normally provide a diurnal variation in temperature. Adequate soil moisture is required throughout the growing period.

Cultural techniques

Seeds are sown 5−8cm deep in rows 45−60cm apart with seeds 12−15cm apart. Irrigation may be necessary during dry periods and particularly at flowering time. A complete fertilizer should be applied before sowing and top dressings of phosphate and potash repeated at intervals up to flowering. Nitrogen is not normally required (see also Table 13 in the Appendix).

Immature pods may be harvested 100−150 days from sowing and yields of 6−6.5t/ha (0.6−0.7kg/m^2) of fresh pods may be obtained; dried bean yields are 600−800kg/ha (60−80g/m^2).

Pests

Aphid (*Aphis fabae*), Table 14 (45).

Diseases

Broad bean rust (*Uromyces fabae*), Table 15 (14).
Powdery mildew (*Erysiphe polygoni*), Table 15 (78).
Powdery mildew (*Leveillula taurica*), Table 15 (26).

French or Kidney bean (*Phaseolus vulgaris*)

Alternative names are common bean, bush bean, haricot bean, navy bean, snap bean, green bean, string bean.

General characteristics

The French bean is a bushy or twining annual herb with slender stems which are square in cross-section. The leaves are alternate, trifoliate, sometimes hairy and the leaflets are oval with pointed tips (Figures 14-8 and 14-9). The flowers are borne in axillary racemes and are 9–12mm in diameter. The pods are thin with from four to six seeds per pod, although the pods of some cultivars may contain up to 12 seeds. The seeds are variable in colour and may be mottled or striped. They are up to 1.6cm in length and are normally kidney-shaped.

Environmental factors

Many cultivars tolerate a range of soil conditions but optimum pH is 5.5–6.5. Climbing cultivars give economic yields in areas of high rainfall but dwarf types appear to be more sensitive to high moisture levels and are not generally successful when grown during the rainy season.

Nitrogenous fertilizers frequently produce an increase in growth rate since French beans are not as efficient as many other legumes in utilizing the nitrogen-fixing *Rhizobium* bacteria which inhabit the root system. Temperatures from 19–27°C are suitable for most cultivars.

European forms or cultivars are normally grown at elevations of over 600m, but a wide range of tolerance to climate exists and many cultivars will grow from sea level to 1000m.

Cultural techniques

Seeds are sown 1–3cm deep on beds or ridges. Dwarf cultivars are sown in rows 45–60cm apart with 15–20cm between plants. Climbing cultivars are grown in double rows 30–38cm apart with plants 20–30cm apart in the row and a spacing of 75cm between the double rows. Climbing cultivars require support by stakes or a trellis up to 2.5m in height. In some areas, the crop is interplanted with maize and the maize stalks provide support for the climbing beans.

Irrigation is mainly necessary just before and during flowering but earlier irrigation may be required during dry periods. A complete fertilizer should be applied before planting, followed by top dressings of potash or nitrogen (if necessary) to stimulate early growth (see also Table 13 in the Appendix).

Pods may be harvested from dwarf cultivars at 40–60 days from sowing and from the climbing cultivars at 70–90 days from sowing. Yields of fresh

Figure 14-8
French or kidney bean (*Phaseolus vulgaris*) dwarf variety

Figure 14-9
French or kidney bean, (*Phaseolus vulgaris*) climbing variety

beans from dwarf cultivars vary from 2−4 t/ha (0.2−0.4kg/m^2). Climbing cultivars yield 4−6 t/ha (0.4−0.6 kg/m^2).

Pests

Aphid (*Aphis fabae*), Table 14 (117).
Bean bruchid (*Acanthoscelides obtectus*), Table 14 (64).
Epilachna bettle (*Epilachna* spp.), Table 14 (143).
Red spider mite (*Tetranychus urticae, T. telarius*), Table 14 (46).

Diseases

Anthracnose (*Colletotrichum lindemuthianum*), Table 15 (133).
Bean yellow mosaic virus, common bean mosaic virus, Table 15 (72, 73).
Halo blight (*Pseudomonas phaseolicola*), Table 15 (74).
Powdery mildew (*Erysiphe polygoni*), Table 15 (78).
Root rot (*Fusarium oxysporum*), Table 15 (75).

Garden pea (*Pisum sativum*)

General characteristics

The pea is an annual herb with thin stems; both the climbing and dwarf (bush) forms are grown. The divided leaves have oval leaflets and a terminal tendril, and there are large stipules at the leaf bases. The flowers arise from the axils of the leaves in groups of two or three. The petals are normally white and up to 1.6cm long. The pods measure up to 15cm in length and have from two to 10 seeds per pod. The best results in tropical areas are produced by 'sugar pea' cultivars which are sometimes given varietal status as var. *macrocarpon* and are called Chinese pea pods, mange-tout or edible-podded peas (Figure 14−10).

Environmental factors and cultural techniques

Most cultivars require a cool humid climate with temperatures in the range 13−18°C. Most are therefore only suited to altitudes above 1200 m. High temperatures may affect fertilization. Fertile soils with a pH of 6.0−7.0 are required and acid or waterlogged soils are unsuitable. Some cultivars are adapted to short daylengths.

The pods and immature seeds may be harvested 60−80 days from sowing but seeds require up to 140 days to become fully mature. Yields of up to 7t/ha (0.7kg/m^2) have been recorded.

Pests

Cutworm (*Agrotis* spp.), Table 14 (29).
Pea weevil (*Bruchus pisorum*), Table 14 (68).
Root-knot nematodes (*Meloidogyne* spp.), Table 14 (119).

Figure 14-10
Garden pea (*Pisum sativum*)

Diseases

Ascochyta blight, leaf spot (*Ascochyta pisi*), Table 15 (76).
Blight, leaf or pod spot (*Mycosphaerella pinodes*), Table 15 (77).
Powdery mildew (*Erysiphe polygoni*), Table 15 (78).

Questions for study and review

1. What is the function of '*Rhizobium* bacteria'? Discuss the ways in which the presence of this bacterium contributes to the importance of leguminous vegetables in tropical areas.
2. Select six of the more important leguminous vegetables and compare their nutritional values from the data given in Table 10 in the Appendix.
3. Write a brief account of the importance of environmental factors on the growth of: Lima bean, winged bean, and French bean in tropical areas.

4. Compare the fresh pod yields given in the text for any four leguminous crops, preferably ones grown in your locality. How do these yields compare with those obtained by growers of these crops in your area?

5. Which fertilizers appear to be most essential in promoting the satisfactory growth of legumes? Attempt to identify the kinds of fertilizer used for legumes in your area and estimate the rates of application used.

6. What would you consider to be the three most serious pests and diseases of the legumes described in this chapter?

7. Many legumes have edible leaves in addition to the pods. Which do you consider to be the most important of these dual-purpose legumes?

8. Two legumes described in this chapter have tuberous roots in addition to edible pods. Identify and write brief notes on both of these crops.

9. Collect as many pods as possible of different legumes grown in your locality. Write descriptive notes on their appearance and indicate measurement of length and width. What do you estimate to be the average number of seeds per pod of each type of legume?

10. Write an account of the cultivation of the French bean, from soil preparation to harvesting. Indicate any ways in which you consider the cultural operations used in your locality could be improved.

CHAPTER 15

Miscellaneous Crops

Okra (*Hibiscus esculentus, Abelmoschus esculentus*)

Alternative names are okro, lady's finger, gumbo.

General characteristics

A widely grown crop in many tropical areas, okra is mainly produced for local domestic use. Many cultivars have been selected for local conditions but there are two main types: the long and the short duration. Pod characteristics vary widely, reflecting local preferences for okra as a food crop.

Okra is an annual herb growing to 2m tall (Figure 15-1). The alternate, three-lobed leaves may reach 30cm in length and are generally hairy. The flowers are borne singly in the leaf axils and are about 5cm in diameter. They are usually yellow with a dark yellow or purple base. The fruit is a capsule varying in colour from white through green to red; it is ridged and pointed at the apex. The seeds are dark green to grey-black and about 5mm in diameter.

Environmental factors

Seeds germinate only in warm soils and temperatures above 16°C are required. A wide range of soil types gives economic yields but well-drained, fertile soils with adequate organic material and reserves of the major elements are ideal. Some cultivars are sensitive to excessive soil moisture.

Most cultivars are adapted to high temperatures with little diurnal or seasonal fluctuation. An average temperature of 20−30°C is appropriate for growth, flowering, and pod development. Okra is tolerant of a wide variation in rainfall. Most selections are adapted to cultivation in the lowland humid tropics and up to 1000m.

Cultural techniques

Seeds should be sown in deeply-cultivated soil on ridges or beds in rows

Figure 15-1
Okra (*Hibiscus esculentus, Abelmoschus esculentus*)

50—70 cm apart with 30—40 cm between plants. More vigorous cultivars require wider spacing. The growing point is sometimes removed to encourage lateral branching. Irrigation may be required up to the fruiting period if rainfall is insufficient to maintain vigorous growth. A complete fertilizer should be applied before sowing followed by supplementary dressings if necessary (see also Table 13 in the Appendix).

Young pods may be harvested beginning 60—90 days from sowing; if pods remain on the plant for too long they become fibrous and tough. Yields of up to 500 kg/ha (500 g/10m²) of pods may be produced (equivalent to 0.5 kg per plant) over a harvesting period of 30—40 days.

Pests

Cotton leaf caterpillar, cotton leafworm (*Xanthodes graellsi*), Table 14 (98).

Cotton stainer (*Dysdercus superstitiosus*), Table 14 (99).
Leaf-eating beetle (*Lagria villosa*), Table 14 (115).
Pink bollworm (*Pectinophora gossypiella*), Table 14 (101).
Plant bug (*Mirperus jaculus*), Table 14 (102).
Spiny bollworm (*Earias insulana* and *E. biplaga*), Table 14 (123).

Diseases

Powdery mildew (*Erysiphe polygoni*), Table 15 (78).
Yellow vein mosaic virus, Table 15 (108).

Sweet corn (*Zea mays* var. *saccharinum*)

Alternative names are maize, mealies, corn, green maize, corn on the cob.

General characteristics

Although dried maize is a staple food in many warm climates, it is also widely consumed in the immature stage.

Sweet corn is an annual growing to 4 m tall. The root system is fibrous with additional prop roots on the lower stem which assist with anchorage. The leaves are alternate, straplike and up to 150 cm long. The flowers are unisexual with the male flowers in terminal panicles and the female flowers are axillary. Normally two or three female inflorescences are formed per plant. The mature cob may be up to 40 cm in length (Figure 15-2).

Sweet corn is a variety of the field maize which has been specifically selected and bred for use in the immature cob stage. It should therefore not be confused with field maize or mealies. Even the young cobs of maize are much more starchy and have a different texture and flavour from sweet corn, the seeds of which have a higher sugar content.

Environmental factors

Sweet corn grows well on most soils but medium, well-drained loams with a high organic content are preferred. Soil pH should be 5.5−6.8. Flowering and yield are adversely affected by heavy rainfall, and excessively wet soils will retard development. Rainfall of 1000−1500 mm per year is adequate.

Average temperatures of 16−24°C favour growth but the temperature should not exceed 35°C. Tillering is encouraged by diurnal temperature variation. Elevations of up to 1700 m are suitable for cultivars adapted to short days and high temperatures. Crops grown at higher elevations are often delayed in maturity and are liable to produce more tillers. Many cultivars grown in the tropics are sensitive to daylength variation.

Cultural techniques

Seeds are sown in rows 60−90 cm apart with 30−45 cm between plants; two seeds can be sown and thinned to one seedling. Seeds are usually sown at a depth of 2−3 cm, depending on the soil type. Since sweet corn has a

Figure 15-2
Sweet corn (*Zea mays* var. *saccharinum*)
(Photograph by Rex Parry)

relatively shallow root system, regular irrigation is essential, particularly from the stage just before the silks develop through to the fertilization stage. Sweet corn should not be grown within 100m of field maize because cross-pollination will ruin its quality, although planting corn cultivars with widely differing maturity dates can overcome this problem.

Soils can be supplied with compost or manure and an NPK dressing before sowing, followed by either nitrogen or NPK dressings at intervals until flowering (see also Table 13 in the Appendix).

The rate of development varies considerably due to the genetic composition of any particular cultivar and the environmental conditions. An approximate guide to the time of harvesting is about 20 days after 50 per cent of the crop has formed silks. A yield of $7-15$ t/ha (1 kg/m^2) depending on the planting density, cultivar, and environment is normal.

Pests

African army worm (*Spodoptera exempta = Laphygma exempta*),
Table 14 (130).
Corn earworm (*Heliothis armigera*), Table 14 (131).
Maize stalk borer (*Busseola fusca*), Table 14 (132).
Maize weevil (*Sitophilus oryzae*), Table 14 (133).
Other local pests occur.

Diseases

Corn smut (*Ustilago maydis*), Table 15 (124).
Dry rot of ear, stalk or cob (*Diplodia macrospora*), Table 15 (125).
Maize rust (*Puccinia polysora*), Table 15 (126).
Streak virus, Table 15 (127).

Questions for study and review

1. Collect mature fruits of all the types of okra grown in your locality and attempt to classify them on length and width of pod. State whether they are short- or long-duration cultivars.
2. Outline the main environmental factors which are important in growing okra, including fertilizer requirements.
3. Write brief notes on the various ways in which okra is used in your locality, including the dried pods.
4. Discuss the growing of sweet corn in your locality and list the various pests and diseases which growers have to deal with. How are these prevented or controlled? Include brief notes on bird damage and methods of control if this is a local problem.
5. What do you estimate to be the average yield of sweet corn grown in your locality? How does this compare with the figures given in this chapter?

Section III — Protected Environments

CHAPTER 16

Protected Environments

Reasons for protected environments

In many areas of the world which would otherwise be considered as having a warm climate, parts of the year are still too cold to allow the production of tropical crops, which are subject to chilling injury during the winter time. In addition some crops, such as the strawberry, while not being killed or severely damaged by the cold winter temperatures, may not produce economic yields during the cold parts of the year.

Temperatures in other areas may be suitable, however other features of the climate such as high winds, low relative humidities, and heavy rainfall may adversely affect yields. In these cases it is often desirable to utilize relatively inexpensive techniques to mitigate the environment. This is commonly done in the Middle East, North Africa, high elevations in southern Africa, and other areas of the world. Generally systems are used which will provide the desired change in environment with the least energy input. Thus, expensive glasshouses (such as are used in Europe and North America) are generally not used, as the period of the year in which they are useful is too short to justify their expense. The two most common modifications used are row-crop covers, which are used to cover individual rows or beds of plants, and plastic tunnel or quonset greenhouses.

Row-crop covers

Row-crop covers or cloches are used to modify the environment in the field by covering individual rows or beds with low covers, in effect by creating miniature greenhouses (Figure 16-1). The covers can be either supported by some type of frame (usually wire), or can be of the 'floating' type which are light in weight and are supported by the crops themselves.

Row-crop covers are used to protect the crop from frost and rain, to raise the air and soil temperatures around the crop, and in some cases to prevent entry by insects and birds. Their effectiveness in accomplishing these goals

Figure 16-1
Row-crop covers in the field

varies with the type of row-crop cover employed and the care with which they are used.

Materials

A number of different materials are used to manufacture crop covers and each material will have its advantages and disadvantages.

Polyester fabric This is a non-woven synthetic material which is light in weight (usually less than $35 g/m^2$) and porous; it can therefore be used without any supports and is usually applied just after seeding or transplanting. It is available in different weights and similar materials are made by several manufacturers. Since it is porous, no special irrigation is needed and air movement through the fabric decreases problems with excessive heat and humidity within the row. Thus it is not necessary to open and close the tunnels under most conditions. However, since water is not excluded, it is less effective than some of the other types of covers in preventing diseases (such as Botrytis in strawberries) which develop under wet conditions.

Polyester-alcohol based covers At the present time a Japanese company is manufacturing a line of crop covers with a polyester-alcohol base. These have many of the same advantages as the polyester covers but may conserve heat slightly more efficiently. A silver reflective cover is also available which reduces light transmission by 20 to 50 per cent, depending on type, and has some characteristics which repel insects.

Polyethylene Polyethylene has been the most commonly used material for row-crop covers and has some very definite advantages and disadvantages when compared with the previously mentioned materials. Polyethylene covers must be supported and cannot be permitted to lie directly on the plants, as is possible with the polyester fabric and the polyester-alcohol based covers.

Support for polyethylene covers is usually accomplished through the use of wire hoops which are bent in a half circle over the row. The polyethylene completely excludes water, so drip or furrow irrigation must be used in conjunction with it. At the same time, this exclusion of water is beneficial in reducing diseases which occur due to excess moisture on the crop. Air exchange is limited, so the warming effect of polyethylene is greater than that of other types of covers, and while this is beneficial on cool days, it necessitates the opening of the tunnels on warm sunny days (Figure 16-2). The problem of having to open and close the tunnels according to the weather is decreased by using 'slitted' tunnels. Slitted tunnels have slits cut in the side of the polyethylene so that some air exchange does occur. Some

Figure 16-2
Ventilation of strawberry crop being grown under polyethylene; note also the irrigation furrows between the rows

types are designed so that when the weather is warm and the polyethylene stretches a bit the slits will open, while tending to stay closed when the weather is cold. In areas where day temperatures may be quite warm while nights are cold, polyethylene with a white pigment may be used to reflect the heat. This gives a crop the normal night-time benefits provided by polyethylene, while not allowing the temperature to rise as high during the day. Pigmented polyethylene does exclude some light so should be used only where light intensities are relatively high.

Installation of covers

Hoop-supported types Row-crop covers requiring support are usually installed over 10-guage wire hoops spaced at 2 to 3 metre intervals. The cover is laid over the hoops, and then an additional hoop is placed over the cover and pushed into the ground, normally at every second lower hoop. The frequency of hoops is affected by the windiness of the area. Machines are available for installing both hoops and covers, although the operation is commonly done by hand. It is important that the covers are tightly stretched over the hoops to prevent them from sagging and then becoming more susceptible to damage from wind and rain. When ventilation is needed the covers can be raised at the sides to allow the movement of air across the crop.

Unsupported types Unsupported types of row-crop covers are laid loosely over the crop and the edges secured by a covering of soil or the use of 10-guage wire bent into a U-shape and pushed into the ground through the cover. Modified plastic mulch-laying equipment can be used to lay the crop covers or they can be laid by hand.

In many cases these covers can be reused if they have been kept reasonably clean and if they have not deteriorated too severely. To preserve them, roll the covers back onto a roll and store in a shady dry spot until they are again needed. With polyethylene it is usually best to use an inexpensive grade and to discard it when it is no longer needed. Polyethylene tends to discolour and light transmission is seriously reduced.

Aspects of the environment affected

Temperature and frost control The use of crop covers will generally raise night temperatures by 2−4°C but daytime temperatures may be raised by 5−40°C, depending on the type of cover and the weather. Unslit polyethylene will raise the temperatures the most, followed by slit or coloured polyethylene, while porous covers will raise daytime temperatures the least. Frost control usually does not differ significantly among the covers.

Relative humidity Relative humidity is increased the most in unventilated polyethylene covers and the least by the polyester-alcohol based covers, which have the ability to absorb some moisture.

Light intensity The effects on light intensity will vary from product to product. Most are designed to allow the maximum amount of light through, but all types of cover will decrease its intensity to some extent. Since most of the areas where these crop covers are used have relatively high light intensities, reductions are probably not important unless the covers have become dirty. Some products, such as tinted polyethylene, are designed to reduce light intensity. The covers are used where shade-requiring crops are being grown, and where the reduction in light intensity does not significantly hinder the growth of the crop, while the daytime cooling effect is desirable.

Wind The reduction of wind is a beneficial feature of all the crop covers — transpiration is reduced, as is possible mechanical damage to crops. Wind reduction also contributes to the warming and humidity-raising properties of the covers.

Hail All the covers are effective in reducing hail damage.

Pest exclusion An additional benefit of the porous crop covers is the exclusion of airborne pests such as insects and birds. Since these covers are placed directly over the crop and are not opened for any length of time, insects cannot enter. It has been demonstrated that large increases in yield can occur, particularly in crops which are subject to virus infections spread by white flies, aphids, and leaf hoppers. There may also be a saving in the use of pesticides. In polyethylene tunnels, pest problems may sometimes increase because of the warm moist environment. This is partially due to the late detection of pest attack — there is a tendency not to look closely at the crop when it is covered with plastic, the removal of which requires some effort.

Problems

Cost There are a number of disadvantages to the use of row-crop covers, the first of which is cost. The cost of the covering materials will vary from location to location, however it is usually economical to use them only for high value crops such as certain fruits and vegetables and at certain times of the year. The user should therefore make a careful economic analysis before deciding whether row crop-covers should be used.

Storage A problem which can sometimes occur is that of anchoring the covers firmly enough so that they do not blow away during periods of heavy wind. This is particularly a problem with polyethylene covers since they are not permeable to air. The polyester-type covers are less likely to blow away due to their partial permeability to air. To prevent problems, the covers should be firmly anchored and inspected periodically for signs of damage or instability.

Pests and diseases   Although some consideration has already been given to pests and diseases, it is worth adding a few comments here. The humidity within polyethylene tunnels can often cause increases in disease problems (such as *Botrytis*) if tunnels are not properly ventilated. During periods of rainy weather tunnels should be ventilated between storms, if temperatures allow, to decrease relative humidity within them. During periods of relatively dry weather, mite problems are sometimes encouraged by the high temperatures which occur inside polyethylene tunnels. Weed growth is often a serious problem inside tunnels and can be solved through the use of plastic mulches or herbicides. It is seldom economical to lift tunnels to handweed as weeds will regrow very rapidly.

Opening and closing   Another disadvantage of tunnels is the need to open and close them to allow pollination by insects, to care for, and in the case of plastic tunnels, to ventilate the plants they protect. Since temperatures and humidities can rise very quickly within unslit plastic tunnels, they must be opened when the weather allows and closed at night or during cloudy or cold weather.

Removal   Finally, tunnels will have to be removed when they are no longer needed. The supports, if used, are stored. The covers too are either rolled up and stored, if in good condition, or disposed of, if in poor condition.

Plastic tunnel greenhouses

Design

Frame   The plastic tunnel greenhouse is designed to take advantage of the quonset or semi-circular shape, which is strong and more wind-resistant than most other shapes and has excellent light penetration properties, since light tends to strike it at right angles regardless of latitude. It is well adapted, because of its relatively large open ends, to the use of tillage implements and to the need for good ventilation, which is achieved by raising the sides, opening the doors, or pulling the individual sheets of plastic back from each other. At the same time, the cost of materials for constructing this greenhouse is lower than for other types.

The frame of a tunnel is usually constructed of galvanized steel, although alumimun and even PVC pipe can be used. In order to take advantage of the inherent strength of the quonset shape, the tunnel should be a half-circle in cross-section. This means that it will be half as tall at the centre as it is wide. Though widths will vary, a common size is 2m tall by 4m wide. Figure 16−3 illustrates a simple design which was developed by the FAO for use in the Middle East region. The author has used this design in Lebanon and has found it to work quite well at a minimum cost. In this case the frame is constructed of 1.9cm diameter galvanized pipe which is bent to the

Figure 16-3
A plastic tunnel or quonset greenhouse; note that the polyethylene cladding is
secured by burying the edges
(FAO photograph)

appropriate shape. A disadvantage of this shape is that there is a considerable amount of wasted space at the sides of the house where overhead clearance is not sufficient to allow for the growth of a tall crop.

Doors The doors of the tunnel should be constructed so that the entire end of the house can be opened to allow a small tractor to pass straight through the greenhouse and perform necessary tillage operations and to allow better ventilation. Doors can be either hinged or can be of a roll-up type (Figure 16-4).

Polyethylene as a cladding material

After the tunnel frame has been constructed, it must be covered with a cladding material which possesses the right combination of properties to allow economical plant production. The most commonly used material for covering tunnel greenhouses is polyethylene, which combines relatively low cost with good light transmission and diffusion properties. Polyethylene is, however, subject to degradation from various environmental factors, and the least expensive types will seldom last for more that a few months before needing to be replaced.

Ultra-violet light The most important factor affecting the durability of polyethylene is the intensity of ultra-violet rays, since these rapidly degrade

Figure 16-4
Tunnel greenhouses with their roll-up plastic doors open

polyethylene. The rate of degradation can be decreased by utilizing polyethylene which contains effective ultra-violet inhibitors, and although such products are more expensive, they may last up to three years. The user will have to look carefully at the economics of using the better grades of polyethylene versus cheaper grades.

Temperature   Another factor affecting the degradation of cladding materials is temperature. Generally high temperatures with decrease the life of the materials. This is particularly true where the cladding is in contact with the frame. Frames constructed of smooth materials which do not abrade the plastic or become hot when the sun shines on them will prolong the life of the cladding material. Tubular frames of a light colour are preferable to solid frames of a dark colour because they tend to stay cooler.

Shading materials   The use of shading materials can also affect the longevity of the cladding material. As a general rule, shading will extend the life of the polethylene by decreasing the ultra-violet rays striking it, however, spray-on shading materials should not chemically react with the plastic. Lime-based shading compounds formulated for use on glasshouses are generally unsuitable for use on polyethylene as they extract certain additives from the polyethylene, thus weakening it.

Method of attachment   The method of attachment will also affect film life. The cladding should be attached during calm weather to minimize wind damage. Usually the plastic edges are buried in the ground and held in place by the weight of the soil. Where the cladding is attached to wood, as around doors in some designs, this should have been cured and any wood preservatives used should have dried thoroughly.

Pesticide splashing   Contact between plants and polythene film should be avoided and care taken to keep pesticide sprays off the film as much as possible as many pesticides are harmful to the plastic.

Alternative cladding materials

In addition to polyethylene, two other cladding materials are commonly used in warm climates. Both are more expensive than polyethylene but may have advantages which outweigh the cost for certain uses.

The first type of cladding material is made from one of the various ethylenevinyl acetate co-polymers (EVA). These last slightly longer than most polyethylenes and have better light transmission and diffusion properties. In addition, they are less transparent to infra-red energy than are most polyethylene materials (although special polyethylenes are available with similar properties) so that the tunnels stay warmer at night, and this may have important advantages to plant growth. An EVA material is usually somewhat more expensive than polyethylene but in most areas will last for three full years.

The second material is fiberglass. This is far more expensive than other materials, is less flexible, and requires more labour to attach, but will last for up to 15 years, depending on grade. Fiberglass is generally used in permanent and relatively expensive greenhouses for the production of very high value crops such as potted ornamental plants and some cut flowers.

Location factors

Soil   The tunnel greenhouse should be located where the soil is suitable for production of the desired crop. In cases where existing soil is not suitable, raised beds can be used (Figure 16−5) or existing soil amended, but this will increase production costs. The area should be well drained to allow for leaching.

Water   Since crops grown in tunnels will require irrigation, a source of sufficient good quality water must be available nearby. The water should be tested for soluble salt levels, pH, and the presence of toxic materials (such as excess boron) prior to a decision being made on the best site.

Orientation   The orientation of the tunnel greenhouse relative to the direction of the sun is unimportant at latitudes less than about 35°N or S. However, the direction of prevailing winds should be considered, as the open ends of the tunnel are sometimes aligned with the wind to allow for better ventilation and less wind resistance.

Figure 16-5
Tomato plants in carefully prepared raised greenhouse beds; the plants are
protected from the harmful organisms in the soil beneath by polyethylene

Exposure to sunlight   It is very important that greenhouses be located away
from possible sources of shade, such as tall buildings or trees. Shading is
most critical during the shorter days of winter when the sun is low on the
horizon, so the angle of the sun during mid-winter and the resulting shadow
lengths, should be considered.

Irrigation within the greenhouse

The most commonly used system for irrigation in tunnels is drip. Rows are
usually run along the length of the tunnel with a drip line at each row. For
the best results, a 'proportioner' which allows the injection of the desired
concentration of fertilizer directly into the irrigation line should be used.
Where proportioners are not available, a bypass tank can be utilized
whereby the desired quantity of soluble fertilizer is placed in the tank and
the irrigation water flushed through it during the last ten minutes of irrigation.

Growers should be aware that there will be a tendency for soluble salt levels to rise within the greenhouse due to high fertilization levels, and the use of irrigation with no opportunity (due to the cover) for rain to leach away the accumulated salts. The higher the level of salts in the water, the greater the problem of soil salinity will become. Soluble salt problems are reduced by using fertilizers with a low salt index, by not over fertilizing, and by leaching heavily using sprinklers between crops. Where visible salt accumulations occur on the soil surface, such as around the edges of areas wet by drippers, the top few millimetres of soil and salt can sometimes be scooped away with a shovel.

Supporting plants

The most economical way of utilizing tunnel space is to grow vining plants which utilize the vertical area of the greenhouse. This is why tomatoes and cucumbers are two of the most important crops grown in tunnels. Vining plants require support, and this is most often provided by strings which are attached to the greenhouse frame and to stakes in the ground. Plants are wound around these strings as they grow (Figure 16-6, see also Figure 16-5). The weight of these plants is substantial and one must be sure that the greenhouse frame is strong enough to support them.

Figure 16-6
Young cucumber plants growing with supporting strings and drip-line for irrigation

Climate control within the greenhouse

Temperature One of the most important aspects of climate control within the greenhouse is temperature control. In a greenhouse which is not ventilated, solar energy enters through the cladding material, is absorbed by objects within the greenhouse, and is remitted as long-wave or infra-red radiation, which is then unable to escape through most cladding materials. In addition, surfaces become warm and air in contact with them is also warmed. All of this contributes to a rapid rise in temperature within the greenhouse. This effect is called the 'greenhouse effect' and can be either beneficial or detrimental depending on the degree of warming which occurs.

Ventilation When the greenhouse becomes too warm, as will frequently occur on a sunny day, the most simple way to cool it is by the use of ventilation. That is, by allowing cooler outside air to displace the warm air inside. Ventilation also has the effect of lowering the relative humidity inside the greenhouse. Tunnels in warm areas must be designed to allow adequate venting, preferably through both ends and sides of the greenhouse.

Where natural ventilation is insufficient, extractor fans can be used to force warm air out of the greenhouse, and this air is then replaced by cooler outside air. Where fans are used, they should be of sufficient capacity to cause all the air to be replaced once each minute.

In areas where additional cooling is needed, and particularly in areas of low relative humidities, a pad-and-fan cooling system is used (Figure 16-7). In this system, porous pads of excelsior, aluminum, or other materials are placed on one wall of the tunnel and kept moist using a recirculating pump. Extractor fans are placed on the opposite walls and all vents are closed. The extractor fans will cause all air to be sucked in through the moist pads, and the evaporation of water from the pads will cool the air and humidify it at the same time. The capacity of the fans to draw in new air and the surface area of pads must be carefully calculated to ensure that the cooling capacity is sufficient to adequately cool the greenhouse.

Shading Another effective means of cooling tunnels is to apply a shading compound to the surface of the plastic to decrease the amount of solar energy which enters the structure. Shading does however have the disadvantage of reducing light intensity which, depending on the crop and the outside light intensity, can decrease crop growth. A simple shading compound which is quite effective is to dilute white latex (water-based) interior paint with eight parts of water and to spray it onto the outside of the cladding material. The degree of shading can be increased by spraying more heavily or by applying successive coats of the compound. Alternatively shade-cloths of varying densities can be hung on the outside of the greenhouse. Note that it

Figure 16-7
A pad-and-fan cooling system. An exhaust fan causes air at the opposite end of
the greenhouse to be drawn through wet pads — the air is cooled as the water
evaporates

is much more efficient to cool a greenhouse by having the shading on the
outside of the house, which then prevents the energy from entering it, than
to hang the shade inside the house which decreases the light striking the
plants but does not decrease the total amount of energy entering the
greenhouse.

Relative humidity Another important aspect of the greenhouse environment
is the relative humidity. High relative humidities encourage the development
of diseases, while very low relative humidities increase the incidence of mite
problems. One of the most important ways of controlling relative humidity
is by proper venting of the tunnel. As a rule, outside air will be less humid
than air inside the tunnel because of the moisture added to the air due to
transpiration. Thus, replacing the air in the tunnel with outside air generally
lowers relative humidity. The problem with this is that when the weather is
cold, especially at night, it is desirable to keep vents closed in order to
conserve heat. If vents are closed early in the evening or in the late afternoon,
the house will stay marginally warmer, however, as the air in the greenhouse
cools, the relative humidity will rise and often moisture will condense on the
cladding material and on the plants themselves. This can precipitate serious
disease problems. It is doubtful that the heat conserved by closing the
greenhouse early is substantial enough to compensate for the potential

disease hazard. Thus it is usually better to ventilate as much as possible and not to close the tunnel too early in the evening. When moisture condensation does occur, it usually occurs on the cladding material first and then drips onto the leaves of the plants. This problem can be reduced by using cladding materials whose surface properties cause condensation to run off into the ground along the wall of the greenhouse rather than to form large droplets which will then drip. Additives which will reduce dripping when applied to the inside of the cladding material are also available.

Tomato production in protected environments

Cultivars

The most important crop grown in tunnels is the tomato (see again 16-5). Since tomato production is discussed in Chapter 11, only those aspects which pertain specifically to tunnel production are discussed here. One of the most important decisions the grower will make is the cultivar to be grown. Cultivars which are recommended for field production are seldom suitable for tunnel production. The grower should therefore choose indeterminate cultivars which have fruit suitable for the proposed market and which possess resistance to any diseases which are likely to be a problem in the greenhouse. If nematodes are present, resistance to them will be important. Tremendous differences in yield exist between cultivars, so if recommendations are not available several cultivars should be evaluated.

Cultural techniques

Training In most cases, tomatoes are trained to a single stem which is twined around string hanging from the ceiling of the greenhouse. Since tomatoes fruit from the bottom up, the length of time that plants continue to produce can be increased by lowering the plants as they reach the top of the greenhouse. This is done by increasing the length of the string at the top of the greenhouse (extra string should previously have been coiled at the point of connection to the tunnel frame) and allowing the now bare stems at the base of the plants to coil on the ground.

Pest and disease control

Pest and disease control in the tunnel production of tomatoes must be very stringent because the environmental conditions favour rapid increases in pest problems. It is common practice to apply routine sprays of copper oxychloride or other protectant fungicide to decrease the incidence of fungal pathogens such as those causing early and late blights and grey leaf spot. Careful monitoring should occur to quickly locate and treat any pest outbreak before it spreads.

Bunchy top virus   Bunchy top virus and other viruses transmitted by white fly can be a particularly serious problem in a tunnel because of the rapid increase in vector populations, the high density of the tomatoes, and the enclosed nature of the tunnel which tends to concentrate vector feeding. Careful attention should be paid to controlling vectors, and virus-infected plants should be removed immediately to limit the spread of disease.

Orabonche   In the Middle East, orobanche is a serious parasitic weed which greatly decreases tomato yields. Care should be taken not to introduce the weed into the greenhouse, and if it should appear it should not be allowed to produce seed. By the time the above-ground portion of the plant appears, serious reductions in yield will have already occurred. There is some evidence that high rates of nitrogen fertilization just after transplanting tomatoes may decrease parasitism, however, where infections are heavy, consideration should be given to switching to non-host crops or to soil sterilization with methyl bromide.

Nematodes   These can be another serious pest problem to susceptible species, particularly in tunnels since these are usually intensively cropped. Crops should be monitored, and treatment with nematicides or fumigation with methyl bromide should occur when nematode populations have risen to damaging levels.

Cucumber production in protected environments

Cultivars

Cucumbers are second to tomatoes in the quantities which are produced in tunnels (Figure 16-6). Usually one of two types is grown, either the Beit Alpha type, which is popular in the Middle East and North Africa, or the Dutch or English type which is popular in Europe and is often grown elsewhere for export to Europe.

The Beit Alpha cucumber is short and cylindrical with a smooth medium-thick skin and rounded ends. It is usually three to three-and-a-half times as long as its diameter. Most cultivars are parthenocarpic and produce pre-dominately female flowers, so no pollination is necessary. The plants are high yielding and produce fruits of excellent quality, although they are relatively perishable.

The Dutch or English cucumber is also high yielding although it may be more demanding in its environmental requirements. The fruits are long and their skins thin and often ribbed. Because of their thin skins, they loose moisture rapidly and are often individually shrink-wrapped to extend their post-harvest life. Most are parthenocarpic and do not require pollination. Cultivars should be carefully selected in order that their environmental requirements coincide with those which can be provided at a particular time of the year. Thus a grower may choose to grow a different cultivar as a mid-winter crop to the one he would choose for a spring crop.

Cultural techniques

Most commonly, plants are planted in rows in the greenhouse and drip irrigation is utilized. Plants can either be started by direct seeding in place or, to save time and seed, they can be started in pots (peat pots are best) and transplanted into the greenhouse when they have between two and five true leaves. Seeds should have been treated with a fungicide, such as captan, and where angular leaf spot is a problem, with a bactericide such as streptomycin. Since cucumbers have a coarse root system and do not transplant easily, care should be taken to minimize root disturbance.

Soil   This should have been prepared by the addition of organic matter, the adjustment of soil pH to the 5.7−6.7 range, the leaching out of salts (if highly soluble salts occur), and the plowing in of phosphorus fertilizer according to soil test results.

Cucumbers are very sensitive to highly soluble salts, and since natural leaching by rainfall does not occur in tunnels, there is a tendency for salinity problems to occur. Periodic measurements of electrical conductivity should be made, and if irrigation water is of good quality, salts should be leached. Alternatively, the accumulated salts which are visible as a white crust at the soil surface (at the borders of the irrigated area) can be skimmed off and discarded periodically. In many locations in the Middle East, where salinity problems become severe, it is necessary to take the tunnel out of production for a rainy season, and by removing the cladding material allow the rainfall to leach out the salts. This would also be an excellent time to plant a cover crop of grass or a legume in the greenhouse, to be plowed down to increase the organic-matter content. Where highly soluble salts are a problem animal manure should not be used as a soil amendment, since it is high in soluble salts. Other sources of organic matter, such as carob pomace, straw, bark, or compost derived from plants are preferable.

Irrigation   After the soil has been prepared, the drip-lines are laid down and plants planted at each emitter (Figure 16-6). The spacing will vary between cultivars and locations, and either single or double rows may be utilized. Double rows have the advantage of wider aisles and thus allow for the easier picking and movement of the harvested fruits. In areas of high light intensity, a density of from three to four plants per square metre is common. This density is reduced where light intensity is low or when especially vigorous cultivars are grown. Initially water is applied at a rate of approximately one litre per plant per day and increased gradually to double that for large plants. The actual amount depends on the environment.

Training When the plants have reached about 15cm in height they will require support which is provided by strings attached to the frame of the tunnel (see again, Figure 16-6). Plants are often pruned or trimmed, although the extent to which this is done depends on the characteristics of the cultivar. In cultivars where fruit is borne very close to the ground and where branching is profuse, fruit produced at nodes less than 50cm from the ground is removed and side branches trimmed at the second node. Once plants have reached the top of the greenhouse one of two practices can be used; the top can be pinched out to allow side branches to develop and produce the late crop, or the stem can be bent over and allowed to grow back down the wire. At this stage all side branches are allowed to develop.

Fertilizer This is best applied through the irrigation water, either by supplying it in small amounts with every watering (constant feed), utilizing a fertilizer injector, or by applying it on a weekly basis using a higher amount. Though fertilizer requirements will vary with the soil, approximately 500kg/ha each of N and K will be required for a ten week cucumber crop. If superphosphate has been added during soil preparation, additional phosphorus should not be needed.

Pest and disease control

Greenhouse cucumbers are susceptible to a wide variety of pests which must be controlled through proper cultural practices and spray programmes. Sanitation in and around the greenhouse is essential. A scrupulously clean area should be maintained around the greenhouse — this should preferably be gravelled but at least be maintained free of any plant material, either living or dead. Likewise, the interior of the greenhouse should be maintained free of weeds and debris. Mulching will help with weed and disease control, as well as conserving moisture, adding carbon dioxide to the air (which can be a limiting factor in a closed greenhouse), reducing nematodes, and improving the soil. Where susceptible crops are grown repeatedly (such as cucumber and tomatoes), periodic soil fumigation with methyl bromide is recommended to control nematodes.

Powdery mildew This causes round spots covered with a white powder, mostly on the upper leaf surface. It is the single most serious disease of cucumbers and can be minimized by growing resistant cultivars. Where it is a problem, plants should be sprayed with protectant fungicides on a regular basis, and when outbreaks occur chemotherapeutant fungicides, such as imazalil or fernarimol, can be used. In many instances the productive life of a cucumber planting is ended when powdery mildew becomes so severe that it cannot be controlled.

Downy mildew This causes irregular vein-bounded spots which are translucent at first and then turn yellow. The disease can be a serious problem, especially where air circulation in the greenhouse is poor, or where vents are closed early in the day allowing the humid air in the greenhouse to cool and water to condense on the leaves. If possible, resistant cultivars should be grown. When conditions for downy mildew development occur, protectant fungicide sprays (such as copper or maneb) can be used; existing infections can be cured using chemotherapeutants (such as propamocarb or metalaxyl).

Virus infections   These can also be a problem and are controlled by the following means: growing resistant cultivars (where available); controlling the aphid, white fly and leaf hopper vectors; and removing infected plants in and around the greenhouse.

Other disease problems   These include scab and angular leaf spot, which are controlled by resistant cultivars, copper-cotaining sprays, and seed treatment.

White fly   This is the most serious insect pest of greenhouse cucumbers although aphids can also be a problem. Both are controlled by insecticide sprays, with particular attention being paid to the thorough coverage of the underside of the leaves. Mites can also be serious especially under hot, dry conditions and are best controlled by applications of acaracides applied to the undersides of the leaves. Since pest populations are confined in greenhouses, the rotation of pesticides between different chemical groups is especially important if resistance is to be avoided.

Questions for study and review

1. Describe the advantages and disadvantages of the two main types of row-crop covers.
2. Discuss the types of covers available for quonset type greenhouses. What are the advantages and disadvantages of each? Which is most commonly used in your country?
3. Describe and explain the 'greenhouse effect' in one paragraph. A diagram may help.
4. How does shading help to cool a greenhouse? What is the disadvantage of using shading as a cooling technique?
5. Describe a pad-and-fan greenhouse cooling system.
6. Describe how the proper timing of ventilation affects temperature and relative humidity in a greenhouse.
7. What is the vector of bunchy top virus in greenhouse tomatoes?
8. How is orabanche controlled in greenhouse tomatoes?
9. What are the most serious pest problems in greenhouse cucumbers?
10. Distinguish between the Beit Alpha and the Dutch type of cucumber.

Appendix

Table 1
Chilling and pollination requirements of selected temperate fruit and nut cultivars adapted to the upland tropics

Cultivar	Pollination	Chilling requirement in hours
Apple		
Alexander	–	–
Anna	Dorsett Golden, 28, Ein Shemer	Low
Beverly Hills	Pollinator needed	Low
Blenheim Orange	–	–
Dorsett Golden	–	Med
Early Harvest	–	–
Emilia	–	–
Golden Delicious	Starking Delicious Granny Smith	Med
Granny Smith	Golden Delicious, Rome Beauty	Low
Hume	–	–
Hyslop	–	–
Rome Beauty	Partly self-fruitful	High
Starking Delicious	Granny Smith	Med
Transcendent	–	–
Tropical Beauty	–	Low
White Permain	Blackjon, Commerce Delicious, Winter Banana	Low
Wickhoff	–	–
Winter Banana	White Permain	Low
Wolf River	–	–
Yellow Siberian	–	–
28	Anna	Low
Pecan		
Curtis	–	Low
Desirable	–	Low
Elliot	–	Med
Mataffin	–	Low
Moreland	–	Low
Stuart	Self-fertile	Med
Pear		
Ayres	Pollinator needed	Med
Baldwin	Pollinator needed	Med
Beurre Bosc	Beurre Hardy, Doyennede Comice, Josephine	Med
Ceres	Starkrimson or Clapp's favourite	250–400
Eldorado	Forelle	Low
Hood	Pollinator needed	Low
Kieffer	LeConte	Med
LeConte	–	250–500
Malines	–	Low

Table 1 (*Contd.*)

Cultivar	Pollination	Chilling requirement in hours
Orient	Self-fertile	Med
Packhams Triumph	Louie Bon Clapp's favorite	400
Pineapple	Pollinator needed	250
Starkrimson	Packham's Triumph	Low
Wilder	–	500
Peach and Nectarine		
Albatros	Self-fertile	300–500
Angel	Self-fertile	Low
Armking nectarine	Self-fertile	Low
August Pride (Genetic Dwarf)	Self-fertile	250–300
Babcock	Self-fertile	200
Boland	Self-fertile	Low
Clocolan	Self-fertile	Low
Culemborg	Self-fertile	100
Desert Dawn (Genetic Dwarf)	Self-fertile	250
De Wet	Self-fertile	Low
Desert Gold	Self-fertile	350
Desert Pearl	Self-fertile	Low
Don Elite	Self-fertile	Low
Donnarine nectarine	Self-fertile	Low
Dorothy Ann	Self-fertile	Med
Du Plessis	Self-fertile	Low
Earlibelle	Self-fertile	Low
EarliGrande	Self-fertile	200
Early Amber	Self-fertile	350
Estella	Self-fertile	Low
Flordabeauty or Flordalis	Self-fertile	150
Flordabelle	Self-fertile	150
Flordagem	Self-fertile	250
Flordagold	Self-fertile	325
Flordahome	Self-fertile	400
Flordabing	Self-fertile	400
Flordaprince	Self-fertile	150
Flordared	Self-fertile	100
Flordasun	Self-fertile	300
Flordawon	Self-fertile	150
Floridagem	Self-fertile	250
Golden Amber	Self-fertile	Low
Imperuni	Self-fertile	Med-Low
Ingwe	Self-fertile	Low
Jewel	Self-fertile	300
Kakamas	Self-fertile	Low
Keimoes	Self-fertile	Low
Killiekrankie	Self-fertile	Very Low
Kol van Dyke	Self-fertile	Low

Table 1 (*Contd.*)

Cultivar	Pollination	Chilling requirement in hours
Maluti	Self-fertile	Low
Maravilha	Self-fertile	150
Mayglo nectarine	Self-fertile	Low
McRed	Self-fertile	200
Mid-Pride (Genetic Dwarf)	Self-fertile	300−400
Moreira Cling	Self-fertile	Low
Moreira Jubilee	Self-fertile	Low
Nectared nectarine	Self-fertile	Low
Okinawa	Self-fertile	100
Oom Sarel	Self-fertile	Low
Orion	Self-fertile	300−500
Pallas	Self-fertile	Low
Peento (Salicer)	Self-fertile	Low
Precident	Self-fertile	Low
Prof. Black	Self-fertile	Low
Prof. Malherbe	Self-fertile	Low
Prof. Neethling	Self-fertile	Low
Red Ceylon	Self-fertile	50
Rhodes	Self-fertile	Low
Rio Grande	Self-fertile	450
Rochon	Self-fertile	Low
Sabella	Self-fertile	Low
San Pedro	Self-fertile	Low
St. Helena	Self-fertile	Low
Safari	Self-fertile	Low
C. O. Smith	Self-fertile	Low
Sunlite nectarine	Self-fertile	450
Sunray nectarine	Self-fertile	Low
Sunred nectarine	Self-fertile	250
Sunripe nectarine	Self-fertile	375
Tejon	Self-fertile	400
Tokane	Self-fertile	Low
Transvaal	Self-fertile	250−300
Van Riebeeck	Self-fertile	Low
White Knight	Self-fertile	300
Woltemade	Self-fertile	Low
Apricot		
Alpha	Self-fertile	Low
Amor leach	Self-fertile	250−300
Bulida	Self-fertile	Low
Canino	Self-fertile	250−300
Early Cape	Self-fertile	Low
Late Cape	Self-fertile	Low
Malan Royal	Self-fertile	Low
Old Cape	Self-fertile	Low
Palsteyne	Self-fertile	Low
Peeka	Self-fertile	Low

Table 1 (*Contd.*)

Cultivar	Pollination	Chilling requirement in hours
Piet cillie	Self-fertile	Low
Royal	Self-fertile	Low
Soldonne	Self-fertile	Low
Supergold	Self-fertile	Low
Plum		
Apple plum	Santa Rosa	200−600
Eclipse	Santa Rosa	Low
Eldorado	Santa Rosa	Med
Harry Pickstone	Mostly self-fertile	Low
Kelsey	Methley, Santa Rosa, Satsuma	800
Mariposa	Methley, Santa Rosa	Low
Methley	Self-fertile	200−250
Nubiana	Santa Rosa	Low
President	Pride of England	Med
Prune d'Agen	Self-fertile	800−900
(Van der Merwe)		Low
Red Ace	Santa Rosa	Med
Ruby Nel	Santa Rosa	Low
Santa Rosa	Partially self-fertile but use Methley, Kelsey or Wickson	500
Satsuma	Santa Rosa, Kelsey	Low
Sungold	Santa Rosa, Eldorado	Low
Wickson	Santa Rosa	500
Red Beaut	Santa Rosa, Casselmann	Low
Ruby Nel	Santa Rosa	Low
Oriental Persimmon		
Fuyu (Fuyugaki)	Self-fertile	Low
Hachiya	Self-fertile	Low
Hana fuyu	Self-fertile	Low
Hayakume	Self-fertile	Low
O'Gosho	Self-fertile	Low
Saijo	Self-fertile	Low
Tamopan	Self-fertile	Low
Tanenashi	Self-fertile	Low
Almond		
Ayerbe 1	Pollinator required	450
Ayerbe 5	Pollinator required	300
Basilia 1	Pollinator required	90−310
Desmayo	Pollinator required	175−300
Non Pareil	Jordanolo, Kapareil	Med
Texas	Nonpareil, Hall	300

Table 2

Representative nutritional values of fruits consumed in tropical countries (100g edible portions)

Fruit	Water (%)	Cal. (K Cal)	Protein (g)	Fat (g)	Carbo- hydrate (g)	Fibre (g)	Ca (mg)	Iron (mg)	Vit. A (IU)	Thia- mine (mg)	Ribo- flavin (mg)	Nicotin- amide (mg)	Vit. C (mg)
Acerola	82.7	32.0	1.7	0.1	13.5	1.0	21.4	1.0	101.7	0.01	0.04	0.5	15.3
Akee	69.2	196	5.0	20	4.6	1.6	40	2.7	–	0.13	0.14	1.4	26
Almond	5.0	657	20.0	59.0	12.0	1.7	150	3.5	0	0.30	0.60	4.5	0
Annona spp.	75.0	93	1.0	–	22.0	1.0	25	0.5	0	0.10	0.08	0.8	30
Apple (malus)	84	61	0.3	0.4	14.0	1.0	4	0.3	20	0.04	0.02	0.2	5.0
Apricot	90	36	1.0	0	8.0	0.4	15	1.0	2000	0.03	0.05	0.5	5.0
Avocado	75	165	1.5	15.0	6.0	1.5	10	1.0	200	0.07	0.15	1.0	15.0
Banana	70	116	1.0	0.3	27.0	0.3	7	0.5	100	0.05	0.05	0.7	10.0
Breadfruit	70	113	1.5	0.4	26.0	1.3	25	1.0	0.100	0.10	0.06	1.2	20
Breadnut	4.2	391	11.3	4.9	76.3	2.8	110	–	–	–	–	–	–
Canefruit	80	63	1.0	1.0	0.13	5.0	30	1.5	20	0.03	0.03	0.4	15
Carambola	87.4	45	0.4	0.3	11.5	0.8	6	0.9	–	0.03	0.02	0.4	35
Cashew (apple)	85.0	56	0.7	0	13.0	0.6	2	0.5	150	0.02	0.02	0.5	250
Cashew (nut)	5.0	590	20.0	45.0	26.0	1.3	50	5.0	0	0.60	0.20	2.1	0
Cherimoya — *see* annona													
Coconut (kernel, fresh)	45	375	4.0	35.0	11	4.0	10	2.0	0	0.05	0.02	0.6	0
Coconut milk	96	14	0.2	0.2	3	0	15	0	0	0	0	0	1
Date	22.5	274	2.2	0.5	72.9	2.3	59	3.0	50	0.09	0.10	2.2	0
Durian	59.9	147	2.0	1.2	36.1	1.9	18	1.1	Tr	0.32	0.28	1.1	44
Feijoa	84.9	–	0.8	0.2	4.2	–	36	–	–	–	–	–	–
Fig	85	49	1.3	0	11	2.0	50	1.0	80	0.05	0.05	0.4	2
Granadilla (Purple)	75	92	2.3	2.0	16	3.5	10	1.0	20	0	0.1	1.5	20
Grape (V. vinifera)	80	76	1.0	0	18	0.5	20	0.3	50	0.04	0.02	0.3	5
Grapefruit	90	37	0.5	0	9	0.3	20	0.5	0	0.04	0.01	0.2	40
Guava	80	58	1.0	0.4	13	5.5	15	1.0	200	0.05	0.04	1.0	200

Table 2 (*Contd.*)

Fruit	Water (%)	Cal. (K Cal)	Protein (g)	Fat (g)	Carbo-hydrate (g)	Fibre (g)	Ca (mg)	Iron (mg)	Vit. A (IU)	Thia-mine (mg)	Ribo-flavin (mg)	Nicotin-amide (mg)	Vit. C (mg)
Jackfruit	72.9	94	1.7	0.3	23.7	0.9	27	0.6	39	0.09	0.11	–	9
Kiwi	84.0	40.0	1.1	0.4	10.5	1.1	40	0.4	56	0.02	0.05	0.5	98
Lemon	90	36	0.7	0	8	0.5	22	0.5	0	0.05	0	0.2	40
Lime — *see* lemon													
Litchi	82	71	0.9	0.5	16	0.3	5	0.5	–	0.04	0.04	0.3	50
Longan	82.8	60	1.3	0.1	15.1	0.4	1	0.1	–	0.3	0.14	0.3	84
Loquat	88.6	40	0.5	0.2	10.2	0.5	18	0.2	1291	0.2	0.04	0.2	4.0
Macadamia	1.2	–	9.2	78	10	1.8	53	2	–	0.2	0.12	1.6	
Mandarin	86	53	0.8	0	13	0.3	30	0.5	30	0.08	0.03	0.2	30
Mango	83	63	0.5	0	15	0.8	10	0.5	600–2000	0.03	0.04	0.3	30
Mangosteen	80	76	0.7	0.8	18.6	1.3	18	0.3	0	0.06	0.01	0.04	2
Mulberry	85	60	1.5	0	14	0.7	1.5	3.0	26	0.04	0.08	10	39
Olives (green)	75.2	114	1.5	13.5	4.4	–	90	2.0	–	0.02	0.02	0.1	0
Orange	86	53	0.8	0	13	0.3	30	0.5	30	0.08	0.04	0.4	43
Papaya	89	39	0.6	0	9	0.7	20	0.5	1000	0.03	0.03	0.2	50
Passion fruit — *see* Granadilla													
Peach	85	56	0.8	0	13	0.5	8	0.5	300	0.02	0.03	0.3	10
Pear (peeled and cored)	84	59	0.3	0	15	0.9	7	0.4	0	0.02	0.02	0.2	4
Pineapple	85	57	0.4	0	14	0.5	20	0.5	100	0.08	0.03	0.1	30
Pitanga Cherry	89	38	0.5	0.1	10.2	0.3	7	0.1	1866	0.02	0.05	0.2	38
Plum	88	45	0.7	0	11	0.4	10	0.4	30	0.02	0.03	0.3	5
Pomegranate (pulp)	80	77	1.0	0	18	0.2	3	0.7	0	0.02	0.03	0.2	8
Quince	83	60	0.4	0	15	1.5	5	0.6	0	0.02	0.02	0.2	15
Rambutan	82	64	1.0	0.1	16.5	1.1	20	1.9	–	0.01	0.06	0.4	40
Sapodilla	75	97	0.4	1.0	22	1.5	22	0.8	15	0	0.03	0.2	15
Soursop — *see* Annona													

Table 2 (*Contd.*)

Fruit	Water (%)	Cal. (K Cal)	Protein (g)	Fat (g)	Carbo-hydrate (g)	Fibre (g)	Ca (mg)	Iron (mg)	Vit. A (IU)	Thia-mine (mg)	Ribo-flavin (mg)	Nicotin-amide (mg)	Vit. C (mg)
Sweetsop — *see* Annona													
Strawberry	90	34	0.7	0.1	8	1.3	30	1.0	30	0.03	0.03	0.3	60
Tree Tomato	86	46	1.5	0.5	9	2.2	50	0.2	150	0.04	0.03	1.0	25

Source: Platt, B.S. (1975). *Tables of Representative Values of Foods Commonly Used in Tropical Countries*, Med. Res. Counc. Sp. Rept. Ser. No. 302 (London: Her Majesty's Stationery Office: and FAO and U.S. Dept. Welfare (Nutrition Program) Bethsda. Maryland) Nagy, S. and Shaw, P.E. (1980). *Tropical and Subtropical Fruits*. (AVI Publishing, Westport, Connecticut, USA)

Table 3
Chemical weed control in fruits and nuts

Fruit/Chemical	Rate (kg/ha active ingredient)	Weeds controlled	Notes
General Use as Directed Sprays			
Fluazifop-butyl	0.125–2.0+non-ionic surfactant	Annual and perennial grass	Use higher rates on perennials. Pre-or post-emergence.
Glyphosate	1.5–3.75 150m/18l spot spray	Annual and perennial	Post-emergence most effective Directed spray to actively growing weeds with maximum leaf area. Post-emergence
Paraquat	0.5–1.0+non-ionic surfactant	Top kill of all weeds	Apply when weeds are up to 15 cm tall. Extremely toxic to humans and animals. Post-emergence
Sethoxydim	0.25–1.0+oil	annual and perennial grasses	Same as Fluoazifop-butyl
Akee (Blighia sapida) *See General Use as Directed Sprays*			
Acerola (Malpighia glabra)			
Alachlor	2.0	Annual weeds	Pre-emergence
Dalapon	2–11 (product)	Grasses	Post-emergence
Napropamide	2–6	Annual weeds	Pre-emergence
Almond (Prunus amygdalus)			
Dichlobenil	4–6	Many annuals and perennials	Apply to established trees only. Follow with 2–3 cm of irrigation or incorporate. Pre-emergence

Table 3 (*Contd.*)

Fruit/Chemical	Rate (kg/ha active ingredient)	Weeds controlled	Notes
EPTC	3	Annual grasses, nutsedge	Incorporate after application; pre-emergence
MSMA	2	Annual grasses, nutsedge suppression	Apply when weeds are small and actively growing. Do not contact tree. Up to 3 applications per year may be made. Pre-and Post-emergence
Napropamide	4	Annual grasses and broadleaf weeds	Apply just prior to beginning of rains. Pre-emergence
Oryzalin	2–4	Annual grasses and broadleaf weeds	Apply once yearly to established trees. Pre-emergence
Simazine	1–2	Annual grasses, broadleaf weeds	Apply once yearly. Do not apply to mission variety. Pre-emergence.
Trifuralin	New plantings: 0.5–1 established 1–2	Annual grasses and broadleaf weeds	New plantings: incorporate before planting. Established: directed spray followed by irrigation. Pre-emergence

Ambarella (Spondias cytherea)
See General Use as Directed Sprays

Annona (Annona spp.)
See General Use as Directed Sprays

Table 3 (*Contd.*)

Fruit/Chemical	Rate (kg/ha active ingredient)	Weeds controlled	Notes
Avocado (Persea americana)			
Dichlobenil	4–6	Annuals and perennials	Apply to established trees only. Incorporate or irrigate. Pre-emergence.
Monuron	4	Annuals	Directed spray once yearly. pre-and post-emergence.
Napropamide	8	Annuals	Pre-emergence, may be applied twice yearly
Simizine	2–4	Annuals	Directed spray, pre-emergence
Banana (Musa spp.)			
Dalapon	5–10	Annual and perennial grasses	Apply in 780–950 l water per hectare. Repeat up to 3 times per year. Do not contact banana plants. Pre-and post-emergence
Ametryn	4–8	Annual and perennial grasses and broadleaf weeds	Apply pre-or post-emergence (directed spray). Repeat every 3–4 months but do not exceed 26.8 kg/ha per year.
Diuron	2–4	Annual broadleaf weeds and grasses	Directed spray on established plants. Keep away from base of bananas. Pre-emergence

Table 3 (*Contd.*)

Fruit/Chemical	Rate (kg/ha active ingredient)	Weeds controlled	Notes
Breadfruit/Breadnut. (*Artocarpus altilis*) See *General Use as Directed Sprays*			
Cane fruits (*Rubus* spp.) Dichlobenil	4–6	Annuals and perennials	Apply to established plantings only. Water-in. Pre-emergence
Diuron	1.6–3.2	Annuals	Directed spray to plantings at least one year old. Pre-emergence
Diphenamid	4	Annuals	Directed spray, pre-emergence
Chlorpropham	4–6	Annuals	Directed spray, pre-emergence
Napropamide	8	Annuals	Pre-emergence
Simizine	2–4	Annuals	Pre-emergence
Norflurazon	3–6	Annuals	Pre-emergence
Pronamide	1–2	Annuals	Pre-emergence
Carambola (*Averrhoa carambola*) See *General Use as Directed Sprays*			
Cashew (*Anacardium occidentale*) See *General Use as Directed Sprays*			

Table 3 (*Contd.*)

Fruit/Chemical	Rate (kg/ha active ingredient)	Weeds controlled	Notes
Citrus (*Citrus* spp.)			
Dalapon	2–4 product	Annual and perennial grasses	Directed spray. Repeat up to 3 times per year. Post-emergence
MSMA	2	Annual grasses, nutsedge suppression	Directed spray. Repeat up to 3 times per year. Post-emergence
Dichlobenil	3–6	Annual and perennial grasses and broadleaf weeds	Directed spray. Incorporate after application. Use only on plants more than 1 year old. Pre-emergence
Napropamide	4	Annuals	Apply before commencement of rains. Pre-emergence
Diuron	1.6–3.2	Annuals	Directed spray. Pre-emergence
Diphenamid	4	Annuals	Directed spray. Pre-emergence
EPTC	3–6	Annuals and nutsedge	Directed spray. Incorporate mechanically or by irrigation. Pre-emergence
Ametryn	4–6.4	Annuals, *Cynodon* spp	Directed spray before weeds are 10 cm tall. Trees established more than 2 years only. Pre- and post-emergence
Pendimethalin	1.0–2.0	Annuals	Pre-emergence or early post-emergence (1–2 leaf stage)

Table 3 (*Contd.*)

Fruit/Chemical	Rate (kg/ha active ingredient)	Weeds controlled	Notes
Bromacil	1.6–6.4	Annual and Perennial weeds	Directed spray to orchards established more than 4 years. Pre-emergence
Simazine	4–9	Annuals	Directed spray to trees established more than 1 year. Pre-emergence
Terbacil	1.6–6.4	Annuals and some perennial grasses	Directed spray to orchards established more than 2 years. Pre-and post-emergence
Oryzalin	2–4	Annuals	Directed sprays once yearly to established trees. Pre-emergence
Trifluralin	0.5–1 (new planting) (established)	Annuals	Directed sprays. Incorporate mechanically or by irrigation. Pre-emergence
Coconut (Cocos nucifera)			
Atrazine	2–4	Annuals	Directed spray in established plantings. Pre-emergence
Diuron	1.6–3.2	Annuals	Directed spray in established plantings. Pre-emergence
2,4–D amine	1 acid equivalent	Broadleafs	Directed spray. Pre- and post-emergence
Dalapon	2–11 product	Grasses	Directed spray, repeat up to 3 times yearly. pre- and post-emergence
Ametryn	1.6–3.2	Annuals	Pre-emergence
Asulam	2.8–4.8	Grasses	Post-emergence

Table 3 (*Contd.*)

Fruit/Chemical	Rate (kg/ha active ingredient)	Weeds controlled	Notes
MSMA	2.5	Grasses	Post-emergence
Date (*Phoenix dactylifera*)			
oxyfluorfen	1–2.2	Annual weeds	Pre-emergence and early post-emergence
sethoxydim	0.2–0.4	Annual and perennial grasses	Post-emergence (add 2 1 oil-ha)
Durian (Durio zibethinus) See *General Use as Directed Sprays*			
Feijoa (Feijoa sellowiana)			
Dichlobenil	4–6	Annuals and perennials	Directed spray. Incorporate mechanically or by irrigation.
Napropamide	8	Annuals	Pre-emergence
Oryzalin	2–4	Annuals	Pre-emergence
Oxyfluorfen	1–2.2	Annuals	Pre-emergence
Pronamid	1–2	Annual and perennial grasses	Pre- and early post-emergence
			Directed spray.
Simizine	2–4	Annuals	Pre-emergence
			Directed spray.
Trifluralin	0.6–1.12	annuals	Pre-emergence
			Pre-emergence. Incorporate
Fig (Ficus caria)			
Asulam	2.8	Grasses	Post-emergence
Bromacil	2	Annuals	Pre-emergence
Dalapon	5–10 Product	Grasses	Post-emergence

Table 3 (*Contd.*)

Fruit/Chemical	Rate (kg/ha active ingredient)	Weeds controlled	Notes
Dichlobenil	4–6	Annuals and perennials	Directed spray. Incorporate. Pre-emergence
Dinoseb	1.25–1.9	Annuals	Directed spray. Pre-and early post-emergence, very toxic
Granadilla (Passiflora edulis) See *General Use as Directed Sprays*			
Grape (Vitis spp.)			
Dalapon	10 (product)	Annual and perennial grasses	Directed spray. Post-emergence
Dichlobenil	4–6	Annuals and perennials	Directed sprays in established plantings. Incorporate. Pre-emergence
Napropamide	4	Annuals	Directed sprays. Pre-emergence
Diuron	1.6–4.8	Annuals	Directed sprays. Pre-emergence
EPTC	3	Annuals and nutsedge	Directed sprays incorporate. Pre-emergence
Simazine	2–4.8	Annuals	Directed sprays in plantings older than 3 years. Pre-emergence
Trifluralin	0.5–2	Annuals	Directed spray. Do not use on newly transplanted vines. Pre-emergence
MSMA	2	Grasses	Directed spray, incorporate. Pre-emergence. Directed spray. Post-emergence.

Table 3 (*Contd.*)

Fruit/Chemical	Rate (kg/ha active ingredient)	Weeds controlled	Notes
Guava (Psidium guajava)			
Oryzalin	2–4	Annuals	Directed sprays once yearly to established plants.
Napropamide	8	Annuals	Pre-emergence
Prodiamine	4	Annuals	Pre-emergence
			Pre-emergence
Jackfruit (Artocarpus heterophyllus) See *General Use as Directed Sprays*			
Jambolan (Syzgium cumini) See *General Use as Directed Sprays*			
Kiwi fruit (Actinidia chinensis)			
Oryzalin	2–4	Annuals	Directed sprays once yearly to established vines.
Napropamide	8	Annuals	Pre-emergence
			Pre-emergence
Langsat (Lansium domesticum) See *General Use as Directed Sprays*			
Litchi (Litchi chinensis) See *General Use as Directed sprays*			

Table 3 (*Contd.*)

Fruit/Chemical	Rate (kg/ha active ingredient)	Weeds controlled	Notes
Longan (*Euphoria longan*) See *General Use as Directed Sprays*			
Loquat (*Eriobotrya japonica*) See *General Use as Directed Sprays*			
Macadamia (*Macadamia integrifolia*)			
Atrazine	2–4	Annuals	Directed spray. Do not exceed 9 kg/ha per year. Pre emergence
Dalapon	5–10 (product)	Annual and perennial grasses	Directed spray. Do not exceed 22 kg/ha per year. Post-emergence
Diuron	2–4	Annuals	Directed spray. Plantings over 1 over year old only. Pre-emergence
Oryzalin	2–6	Annuals	Do not exceed 11 kg/ha per year. Pre-emergence
Simazine	2–4	Annuals	Directed spray just before rains. Do not exceed 11 kg/ha per year. Pre-emergence
Mamey (*Mammea americana*) See *General Use as Directed Sprays*			

Table 3 (*Contd.*)

Fruit/Chemical	Rate (kg/ha active ingredient)	Weeds controlled	Notes
Mamey sapote (*Pouteria sapota*) See *General Use as Directed Sprays*			
Mango (*Mangifera indica*) Dichlobenil	4–6	Annuals and perennials	Directed spray in established plantings. Incorporate. Pre-emergence
Mangosteen (*Garcinia mangostana*) See *General use as Directed Sprays*			
Mulberry (*Morus* spp.) Benfluralin	1–1.5	Annuals	Pre-emergence
Naranjilla (*Solanum quitoense*) See *General Use as Directed Sprays*			
Olive (*Olea europaea*) Diuron	2–4	Annual broadleaf weeds and grasses	Pre-emergence
Simazine	2–4	Annual broadleaf weeds and grasses	Pre-emergence
Chlorthiamid	6–7.5	Annual weeds	Pre-emergence
Napropamide	8.0	Annual weeds	Pre-emergence

Table 3 (*Contd.*)

Fruit/Chemical	Rate (kg/ha active ingredient)	Weeds controlled	Notes
Papaya (Carica papaya)			
Ametryn	2	Annuals	Pre-emergence
Asulam	2.8–4.8	Grasses	Post-emergence
Diuron	2–4	Annuals	Directed sprays on established plants. Pre-emergence
Dalapon	5–10	Annual and perennial grasses	Directed spray. Post-emergence
Napropamide	8	Annuals	Pre-emergence
Oryzalin	2–4	Annuals	Pre-emergence
Passionfruit-see granadilla			
Persimmon (Diospyros kaki)			
Napropamide	5	Annuals	Pre-emergence
Pineapple (Ananas comosus)			
Ametryne	2	Annuals	Directed spray. Pre-emergence
Atrazine	2–4	Annuals	Apply immediately after planting. Pre-emergence
Bromacil	5	Annuals	Apply immediately after planting. Pre-emergence
Dalapon	4 (product)	Annual and perennial grasses	Pre-plant application before land preparation
Diuron	2–4	Annuals	Apply immediately after planting. Pre-emergence
Simazine	2–4	Annuals	Apply immediately after planting. Pre-emergence
Terbacil	5	Annuals and some perennials	Apply immediately after planting. Pre-emergence

Table 3 (*Contd.*)

Fruit/Chemical	Rate (kg/ha active ingredient)	Weeds controlled	Notes
Pomegranate (Punica granatum)			
Simizine	2–4	Annuals	Pre-emergence
Oryzalin	2–4	Annuals	Pre-emergence
Sapodilla (Manilkara achras)			
See General Use as Directed Sprays			
Strawberry (Fragaria × ananassa)			
DCPA	9	Annual grasses and a few annual broadleaf weeds	Broadcast over plants before flowering or after harvest. Up to 2 applications per year. Pre-emergence
Diphenamid	4–6	Annuals	Broadcast over plants. Do not apply within two months of harvest. Do not use on Shasta cultivar. Pre-emergence
Sesone	3–5.5	Annuals	Do not apply during harvest season. Varietal tolerance varies. Pre-emergence
Lenacil	1.6–2	Annuals	Broadcast in established plantings. Pre-emergence
Napropamide	8	Annuals	Pre-emergence
Simizine	1.1	Annuals	Apply only once per year and at least 4 months after transplanting. Irrigate after application.

Table 3 (*Contd.*)

Fruit/Chemical	Rate (kg/ha active ingredient)	Weeds controlled	Notes
Temperate Tree Fruits			
Dalapon	2–6 product	Annual and perennial grasses	For use as a directed spray when trees are well established. Repeat as needed. Post-emergence
MSMA	2	Annual grasses, nutsedge suppression	Apply as directed spray when weeds are small and actively growing. Reapply up to 3 times per year. Post-emergence
Dichlobenil	4–6	Annual and perennial grasses and broadleaf weeds	Apply to surface and water in. Established trees only. Pre-emergence
Napropamide	4	Annual grasses and broadleaf weeds	Apply before rains commence. Pre-emergence
Simazine	1.6–4	Annual broadleaf weeds and grasses	Apply once yearly as directed to established trees. Pre-emergence
Oryzalin	2–4	Annual broadleaf weeds and grasses	Directed spray around established trees. Pre-emergence
Trifluralin	0.5–1 (new plantings) 1–2 (established)	Annual broadleaf weeds and grasses	Incorporate prior to planting, directed spray followed by irrigation. Pre-emergence
Oxyfluorfen	0.25–20	Annual broadleaf weeds and grasses, striga, especially effective on *Malva* spp.	Apply as a directed spray when trees are dormant. Pre- and post-emergence
Tree Tomato (Cyphomandra betacea)			
Prometryn	0.5–1	Annuals	Pre-emergence

Appendix

Table 4
Recommended techniques for fruit tree propagation

Fruit species	Cuttings	Grafting	Layering	Division	Seed
Acerola (*Malpighia glabra*)	Leafy hardwood, use IBA	Side veneer graft, cleft graft, shield bud	Roots in 6—8 weeks	—	Variable
Akee (*Blighia sapida*)		—	—	—	Normal seed
Almond (*Prunus amygdalus*)	Hardwood (difficult and uncommon)	T-bud		Mound layering (occasional)	Normal seeds. Stratify for 20—30 days at 7°C (for rootstocks)
Ambarella (*Spondias cytherea*)	Roots easily from hardwood cuttings	Veneer graft, shield bud	Successful	—	Variable
Annona (*Annona* spp.)		Cleft graft or T-bud	—	—	Normal seeds although some may come nearly true from seed
Apple (*Malus domestica*)	Hardwood (occasional), softwood plus IBA under mist (occasional), root cuttings for some rootstocks	T-bud	Mound or simple layering	—	Mostly normal although some Asian species are apomictic. Stratify 60—90 days at 2—7°C (for rootstocks)
Apricot (*Prunus armeniaca*)		T-bud	—	—	Normal seeds. Stratify 20—30 days at 5°C (for rootstocks)
Avocado (*Persea americana*)	Semi-hardwood with etiolated bases—uncommon	T-bud, whip or splice graft	—	—	Normal should not be allowed to dry out

Table 4 (*Contd.*)

Fruit species	Cuttings	Grafting	Layering	Division	Seed
Banana (*Musa* spp.)	—	—	—	Division of rhizome or corm, separate offshots	—
Breadfruit (*Artocarpus altilis*)	Root cuttings, root suckers	—	—	—	—
Breadnut (*Artocarpus altilis*)	Root cuttings, root suckers	—	—	—	Normal
Cane fruit (*Rubus* spp.) (blackberries)	Root cuttings (thornless types will revert to thorny); softwood, or leaf-bud cuttings using IBA in high humidity	—	Tip layering	—	—
Carambola (*Averrhoa carambola*)	—	T-bud	Air layer	—	Normal seed
Cashew (*Anacardium occidentale*)	Softwood under mist–difficult	Approach graft; T-bud, patch bud uncommon	Air layer	—	Normal seed
Cherimoya — *see* Annona					
Citrus (*Citrus* spp.)	Semi-hardwood, leaf-bud	T-bud, cutting-graft	Air layering	—	Normal or apomictic seeds, should not be allowed to dry out
Coconut (*Cocos nucifera*)	—	—	—	—	Normal seed
Date (*Phoenix dactylifera*)	—	—	—	Remove offshoots	Variable offspring

Table 4 (*Contd.*)

Fruit species	Cuttings	Grafting	Layering	Division	Seed
Durian (*Durio zibethinus*)	Use bottom heat; possible but not common	Patch bud, approach graft	Successful	—	Variable, plant immediately after removal from fruit
Feijoa (*Feijoa sellowiana*)	Softwood with IBA and high humidity—slow and low rooting percentage	Approach	—	—	Normal seeds
Fig (*Ficus carica*)	Hardwood, 1–1.3 m cuttings may be inserted directly in the field during the rains	T-bud, patch bud	Air layering	—	—
Granadilla (*Passiflora edulis*)	Semi-hardwood	T-bud	tip layering	—	Normal seeds
Grape (*Vitis* spp.)	Dormant hardwood, leafy softwood, under mist	Greenwood splice, graft, chip budding	—	—	Normal seed for breeding stratify at 0.5–4°C for 100 days.
Grapefruit — *see* Citrus					
Guava (*Psidium guajava*)	Softwood under mist plus 200 ppm IBA root suckers	Chip budding	Simple or mound air layer (500 ppm IBA in lanolin)	—	Normal seed
Jackfruit (*Artocarpus heterophyllus*)	—	Approach	—	—	Normal seed
Jambolan (*Syzygium cumini*)	Successful with high IBA concentrations	T-bud modified Forkert bud, approach graft	Successful	—	Fresh seed only

Table 4 (*Contd.*)

Fruit species	Cuttings	Grafting	Layering	Division	Seed
Kiwi fruit Chinese gooseberry (*Actinidia chinensis*)	Softwood tip with 0.8% IBA in talc	Whip or tongue . using dormant scion wood, T-bud	—	—	Normal seed, dry for 2 weeks, then fluctuate temperature—10°C night, 20°C day for 20–30 days.
Langsat (*Lansium domesticum*)	Hardwood and semi-hardwood	Cleft or side graft, modified Forkert budding	Roots in 8 weeks	—	Variable
Lemon — *see* Citrus					
Lime — *see* Citrus					
Litchi (*Litchi chinensis*)	Softwood under mist in full sun, hardwood	—	Air layering (most common)	—	Normal. Seed planted immediately upon removal from the fruit, long juvenile period
Longan (*Euphoria longan*)	—	Veneer graft	Roots in 10–12 weeks	—	Variable
Loquat (*Eriobotrya japonica*)	—	T-bud, side graft, cleft graft	Air layering with 250 ppm IBA	—	Normal seed
Macadamia (*Macadamia integrifolia*)	Semi-hardwood using IBA at 8000–10,000 ppm in closed propagating frame or under intermittent mist	Side graft, approach graft (ring scion several weeks before grafting) approach graft	Air layering	—	Normal seed (do not crack) soak in hot water
Malay rose apple (*Syzygium malaccense*)	Successful	Bud	Successful	—	Variable
Mammey (*Mammaea americana*)	—	Approach graft, shield bud	—	—	Variable

Table 4 (Contd.)

Fruit species	Cuttings	Grafting	Layering	Division	Seed
Mamey sapote (*Pouteria sapota*)	—	Approach graft, veneer graft, chip bud	—	—	Variable
Mandarin — *see* Citrus					
Mango (*Mangifera indica*)	Softwood under mist plus IBA (difficult)	Veneer graft or chip bud, T-bud approach graft possible but uncommon	Air layering (IBA at 10,000 ppm)	—	Normal and apomictic seedlings
Mangosteen (*Garcinia mangostana*)	—		—	—	Apomictic seeds (must be planted 3–5 days from harvest)
Mulberry (*Morus alba, Morus nigra*)	Hardwood 25–30 cm long	—			—
Naranjilla (*Solanum quitoense*)	Roots easily	Possible	—		Most common
Olive (*Olea europea*)	Leafless hardwood cuttings; leafy semi-hardwood	T-bud, patch bud, whip graft, side-tongue graft	Possible	Dig up suckers	Variable
Orange — *see* Citrus					
Papaya (*Carica papaya*)	Entire branches with basal swelling (bottom heat)	Patch budding side graft	—		Normal seeds
Passion fruit — *see* granadilla					
Peach (*Prunus persica*)	Softwood + IBA under mist	T-bud, cleft graft	—	—	Normal seeds — stratify for 90–120 days at 4°C (for rootstock)

Table 4 (*Contd.*)

Fruit species	Cuttings	Grafting	Layering	Division	Seed
Pear (*Pyrus communis*)	Hardwood; softwood plus IBA under mist	T-bud, whip or tongue root graft	—	—	Normal seed — stratify for 60–100 days at 4°C (for rootstock)
Pineapple (*Ananas comosus*)	Suckers, slips, or crowns planted directly in the field	—	—	—	—
Pitanga cherry (*Eugenia uniflora*)	Semi-hardwood	—	Air layering, simple layering	—	Normal seed
Plum (*Prunus* spp.)	Hardwood, softwood under mist	T-bud	—	—	Normal seed — stratify at 2–4°C for 100 days (for rootstock)
Pomegranate (*Punica granatum*)	Hardwood, softwood under high humidity, root suckers	—	Simple layering, air layering	—	Normal seed
Quince (*Cydonia oblonga*	Hardwood	T-budding approach. grafting, shield budding, cleft grafting (ring scions 6 weeks before grafting)	Mound or air	—	—
Rambutan (*Nephelium lappaceum*)	—	Patch bud, modified Forkert	Successful	—	Plant immediately after removal from fruit
Rose apple (*Syzygium jambos*)	Successful under mist	Shield budding, modified Forkert budding	Successful	—	Variable

Table 4 (*Contd.*)

Fruit species	Cuttings	Grafting	Layering	Division	Seed
Sapodilla (*Manilkara achras*)	—	Approach grafting, shield budding, cleft grafting, (ring scions 6 weeks before grafting)	Air layering	—	Normal seed
Soursop — *see* Annona					
Sweetsop — *see* Annona					
Strawberry (*Fragaria* spp.)	Runners	—	—	Crown division European Strawberry (*F. vesca*) only	Cultivar 'Sweetheart' from seed, *Fragaria vesca* from seed.
Tree tomato (*Cyphomandra betacea*)	Softwood (uncommon)	—	—	—	Normal seed
Watery rose apple (*Syzygium aqueum*)	—	Successful	Air or mound	—	Variable

Table 5

Symptoms of nutrient deficiencies in selected crops

Nutrient	Vegetables	Peach	Mango	Banana	Pineapple
Boron	Distorted leaves, death of growing points, internal browning of root crops	Water-soaked cankers below growing points, excessive branching, corky bark lesions, premature defoliation	Rare	Stunting, chlorotic streaks perpendicular to veins, corky areas in fruit	No leaf symptoms, fruit small, cracked between fruitlets, poor slip and sucker production
Calcium	Death of growing point of stem and roots, blossom end rot of tomatoes, tip burn of lettuce	Decreased growth, short internodes, inward rolling of leaves	Smaller leaves, paler green than healthy leaves	Stunting, reduced leaf size, uncommon in field	Not common, pale green leaves necrotic leaf tips, internal discolouration of fruit
Copper	Yellow leaves, soft onion bulbs	Rare	Dieback of new growth	Drooping leaves, short leaves, chlorosis	Curling and twisting of heart leaves
Iron	Interveinal chlorosis of young leaves	Interveinal to complete chlorosis	Rare	Interveinal or general chlorosis	Leaves red-tinged, light green often similar to N deficiency
Magnesium	Interveinal chlorosis of old leaves first	Interveinal chlorosis producing net-like effect	Growth reduction, premature defoliation, bronze leaf margins	Chlorosis, blue petiole, purple blotches	Chlorosis of older leaves, yellow mottling coalescing to become a chlorotic stripe along leaf margins
Manganese	Mottled interveinal chlorosis of young leaves. Yellow stripes on onion leaves	Dull, light green colour becoming darker near veins	Growth reduction, interveinal chlorosis	Marginal interveinal chlorosis followed by dark spots near veins	Slight chlorosis, not common

Table 5 (*Contd.*)

Nutrient	Vegetables	Peach	Mango	Banana	Pineapple
Nitrogen	Pale green leaves, reduced plant size, slow growth	Pale yellow leaves, slow growth, reddish tints on leaves	Small yellow leaves, retardation of growth	Decrease in leaf size (some cultivars), yellowing of entire plant, slow growth	Yellowing of new leaves, reddish margins
Phosphorus	Thin, shortened leaves, purple colour, delayed maturity	Early dormancy, tough leathery texture	Tip dieback, growth retardation, dropping of older leaves, purple, pigment along margins (some cultivars). Sulphur deficiency causes similar symptoms	Slower leaf production, chlorosis of leaf margins in older leaves, shortened internodes	Narrow, purple leaves, chlorotic mottling in older leaves, decreased production of slips, fruits
Potassium	Scattered chlorosis, greying followed by necrosis of leaf margins	Long internodes, puckering of young leaves, leaf scorch, marginal necrosis	Chlorotic spots on old leaves, small, thin leaves	Slow growth, yellowing of old leaves	Necrotic leaf tips and spots on older leaves first, new leaves reddish brown, stunted growth, small fruits which do not ripen on top
Sulphur	Yellowing of lower leaves, hard, thin stems	Pale green new leaves, rosettes of lateral branches below terminal buds which cease growth early	Similar to phosphorus deficiency, scorch along leaf margins	Chlorosis of new leaves	Blistering on older leaves, deformed fruits with hollow core
Zinc	Red spots on bean cotyledons, interveinal chlorosis of beets	Rare	Thickened, small leaf blade, yellow veins, rosetting	Stunting, narrow chlorotic leaves, bronzing, twisted small fruit	Chlorosis, yellow spots becoming necrotic

Table 6
Composition of fertilizers and manures

Source	N %	P_2O_5 %	Potash (K_2O)
Ammonium nitrate	33−33.5	—	—
Ammonium phosphate	11.0	48.8	—
Ammonium sulphate	20.5	—	—
Calcium ammonium nitrate	20.5	—	—
Calcium cyanamide	21.0	—	—
Calcium nitrate	15.5	—	—
Fresh manure with some bedding or litter:			
Chicken (73% H_2O)	1.10	0.90	0.05
Cow (86% H_2O)	0.55	0.15	0.05
Pig (87% H_2O)	0.55	0.30	0.045
Sheep (68% H_2O)	0.10	0.75	0.40
Muriate of potash	–	—	48−62
Nitrate of potash (potassium nitrate)	13.0	—	44.0
Nitrate of soda (sodium nitrate)	16.0	—	–
Potassium ammonium nitrate	16.0	—	27.0
Sulphate of potash (potassium sulphate)	–	–	48−52
Sulphate of potash magnesia	–	–	21−22
Superphosphate	–	16−20	–
Double superphosphate	–	32	–
Treble superphosphate	–	40−47	–
Urea	42−46	—	–

Table 7
Diseases of fruits and suggested control measures

Crop	Disease	Pathogen	Symptoms	Control
Annona. spp.	Anthracnose	*Colletotrichum gloeosporioides*	Leaf spots, black cankers on twigs which may die causing witches broom	Bordeaux, fermate, phygon, zerlate
Apple	Apple mildew	*Podosphaera leucotricha*	White powdery fungus on leaves and young shoots causing leaf roll and shoot dieback	Benomyl, binapacryl, chinomethionat, carbendazim, copper oxychloride, dinocap, sulphur, sulphur/mancozeb, triforine
	Bitter rot	*Glomerella cingulata*	Brown or black sunken lesions on fruit with spore pustules in concentric rings	Captan, mancozeb, sulphur/mancozeb
	Scab	*Venturia inaequalis*	Irregular black spots on leaves and fruit, cracking and distoration of fruit	Benomyl, captan, carbendazim, copper oxychloride, dodine, mancozeb, metiram, sulphur, sulphur/mancozeb, thiram, triforine, zineb
	Sooty blotch	*Gloeodes pomigena*	Sooty mould covering fruit	Benomyl, captan mancozeb, sulphur/mancozeb
Avocado	Anthracnose	*Colletotrichum gloeosporioides*	Rust-brown necrosis beginning at leaf margins, flowers brown and fall, brown spots on fruits	Sanitation, Bordeaux, cuprous oxide, copper oxychloride
	Cercospora Spot blotch	*Cercospora purpurea*	Brown, slightly sunken lesions on surface of fruit	Benomyl, copper

Table 7 (*Contd.*)

Crop	Disease	Pathogen	Symptoms	Control
	Physalospora canker	*Physalospora persea*	Sunken discoloured cankers on branches and trunks, fruit becomes irregular, rough, and corky	prune infected limbs
	Scab	*Spaceloma persea*	Circular brown-to-black lesions on upper surfaces of young leaves and on stems and fruits	Benomyl, copper
	Phytophthora root rot and stem canker	*Phytophthora cinnamomi*	Small roots become necrotic and brittle, progressive decline and death of shoot, discoloured bark near ground level preceding general decline and death in the case of stem canker	Provide good drainage, ethazol, fosetyl-a1, metalaxyl, mulch root zone
Banana	Bunchy top virus	Virus transmitted by the banana aphid *Pentalonia nigronervosa*	Dark green discontinuous streaks along main veins. Leaves later form a rosette at the top of the plant.	Control vector and plant disease free propagules; rogue infected plants
	Cigar end disease	*Trachysphaera fructigena* *Verticillium theobromae*	Shrinking dry rot progresses slowly from the perianth end of fruits	Remove floral remains, copper fungicides, benzimidazole
	Panama disease	*Fusarium oxysporum* *F. cubense*	Purpling of rhizome followed by leaf yellowing and death	Soil-borne. Plant infected soil to resistant bananas of the Cavendish subgroup, inject 2% carbendazim into corm

Appendix

Table 7 (*Contd.*)

Crop	Disease	Pathogen	Symptoms	Control
Banana (*contd.*)	Sigatoka disease (leaf spot)	*Mycosphaerella musicola*	Yellow spots on leaves which coalesce with necrotic centres parallel to lateral veins. Necrosis of leaf margins	15–20 l/ha mineral oil or mineral oil mixed with copper or maneb, imazalil, benomyl, throphanate-methyl, tridemorph, propiconazole, chlorothalonil, carbendazim
Breadfruit	Black root rot	*Rosellinia pepo, R. bunodes*	Wilting and death of all or part of the host, a mycelial fan encircles stem at ground level	Remove infected debris from soil
	Leaf spot	*Cercospora artocarpi*	Black spots on leaves	Copper
	Fruit rot	*Phytophthora palmivora*	Water-soaked lesions with light brown centres spreading to rot the fruit which then mummifies on the tree	Bordeaux, copper, fentinhydroxide, fentinacetate, difolatan fosetyl-al
	Soft fruit rot	*Rhizopus stolonifer R. artocarpi*	Soft rot of fruit	Sanitation
Breadnut	*See* Breadfruit			
Carambola				
Cashew	Sudden death	*Valsa eugeniae*	Rapid yellowing of leaves within 10–14 days	Destroy infected plants by burning
	Anthracnose	*Colletotrichum gloeoporioides*	Red-brown water soaked lesions on leaves, stems, flowers, nuts followed by death of infected parts	Bordeaux, lime sulphur

Table 7 (*Contd.*)

Crop	Disease	Pathogen	Symptoms	Control
Cherimoya	*See* Annona			
Citrus	Black spot	*Guignardia citricarpa*	Tan sunken spots on ripe fruit; fringed black spots on older leaves which later yellow and fall off	Copper oxychloride
	Gummosis	*Botrytis cinerea* *Sclerotinia sclerotiorum/Diaporthe citri* *Glomerella* spp. *Phytophthora citrophthora*	Profuse gumming on surface of infected bark near soil level; eventual death of the tree	Improve drainage, use sweet or sour orange rootstocks budded at least 50cm above ground. Fumigate soil before replanting. Excise lesions and paint with copper paste.
	Citrus greening	Mycoplasma-like organism transmitted during propagation or by the citrus psyllid	Dark green leaf colour, twiggy upright trees, chlorosis resembling magnesium deficiency, failure of fruit to colour, sour fruit	Plant disease-free stock, control citrus psyllid, rogue infected trees, pressure trunk injections of tetracycline hydrochloride
	Psorosis (scaly bark)	Virus transmitted during propagation	Bark lesions resembling pimples at first and then enlarging until bark scales and falls off	Excise lesions and scrape away all infected tissue. Propagate only from trees known to be free of the disease
	Tristeza (quick decline)	Virus	Stem pitting on grapefruit, limes and citrons accompanied by poor vigour, and small fruit. Quick decline occurs when sour orange is used as rootstock for grape-fruit, mandarin, or orange	Control aphids; do not use sour orange rootstocks

Table 7 (*Contd.*)

Crop	Disease	Pathogen	Symptoms	Control
Coconut	Bud rot	*Phytophthora palmivora*	Chlorosis and death of leaves beginning with the youngest and proceeding to the oldest	Apply Bordeaux paste to the bud
	Lethal yellowing (Cape St. Paul Wilt) (Awka Wilt) ('Maladie de Kaincope')	Believed to be caused by a mycoplasma spread by a leafhopper	Fruit drop, leaf chlorosis, death	Rogue infected plants and replant with resistant dwarf cultivars; pressure inject tetracycline or oxytetracycline
	Stem bleeding	*Ceratocystis paradoxa*	Brown sap oozes from trunk 1.5–2 m above ground	Excise infected tissue and treat with copper paste
Date	Bayoud	*Fusarium oxysporum*	Spines or leaflets near base of a recently matured leaf whiten. Whitening progresses up one side of the rachis to apex and down the other side. Brown stain on upper side of rachis progresses to other leaves and then kills the apical bud. Tree dies	Sanitation to prevent the spread of soil/seed-borne pathogen; resistant cultivars; strict quarantine
	Fruit rots (various)		Fruit rots	Cover bunches with paper to exclude rain and moisture; dust with a mixture of 50% sulphur, 5% ferbam, 5% malathion, and 40% inert carrier
	Graphiola leaf spot	*Graphiola phoenicis*	Small spots on both sides of pinnae during humid weather	Remove infected leaves spray with Bordeaux resistant cultivars

Table 7 (*Contd.*)

Crop	Disease	Pathogen	Symptoms	Control
Durian	Patch canker	*Phytophthora palmivora*	Dieback, bark decay, gummosis; reddish, brown cankers which develop slowly during the rains	Bordeaux or other copper fungicide, perecol (60g/1), prune and paint wounds with difolatan; fentinhydroxide; fentinacetate; orthodifolatan
	Anthracnose	*Colletotrichum* spp.	Dark spots on leaves, fruit	Copper oxychloride, mancozeb, benomyl
	Claret canker	*Pythium vexans*	Cankers on lateral roots just below soil; brown dry, cortical lesions	Sanitation in nursery
Feijoa	Charcoal rot	*Macrophomina phaseoli*	Root and lower stem rot	Captan or zineb before flowering, after fruit set, and 20–25 days before harvest
	Botrytis	*Botrytis cinerea*	Rot of leaves, fruits, obvious grey mould present	As for *Botrytis cinerea*
	Leaf spots	*Pestalotiopsis guepini* *Colletotrichum gloeosporoides* *Phyllosticta feijoae* *Phoma feijoae*	Spots on leaves and fruit	Mancozeb, benomyl, copper
Fig	Mosaic	Virus spread by vegetative propagation and the eriophyid mite *Aceria ficus*	Light green spots with indefinite margins on leaves and fruit	Rogue infected plants, use clean propagating stock, control vector
	Rust	*Cerotelium fici*, *Uredo ficina*	Angular brownish spots on upper leaf surfaces	Zineb, maneb, daconil

Table 7 (*Contd.*)

Crop	Disease	Pathogen	Symptoms	Control
Granadilla	Brown spot	*Alternaria passiflorae* or *Septoria passiflorae*	Dark brown spots which enlarge and have light brown centres; stem lesions and defoliation may also occur	Thin vines to allow better air movement. Spray with Bordeaux, maneb, zineb, danconil
	Wilt	*Fusarium oxysporum, F. passiflorae*	Sudden and general wilt of plants, vascular discolouration of roots	Use rootstocks of *Passiflora aurantia, P. herbertiana. P. coerulea. P. incarnata* or a select strain of *P. edulis f. flavicarpa*
	Woodiness (bullet mosaic)	Virus complex transmitted by aphids and vegetative propagation	Yellow vein net, puckered leaves, misshapen shrivelled fruit	Rogue infected plants
Grape	Downy mildew	*Plasmopara viticola*	Yellow spots on leaves, white mycelium on undersides, eventual necrosis and defoliation	Sulphur dust early in the season, later after disease develops, benomyl, maneb, zineb, daconil. Copper may be used late in the season
	Powdery mildew	*Uncinula necator*	White powdery growth on upper leaf surface	Same as downy mildew
Grapefuit	*See* Citrus			
Guava	Fruit rot	*Glomerella cingulata*	Mummifies green fruit and rots ripe fruit	Captan or copper
Jackfruit	*see* Breadfruit			
Jambolan	Root rot	*Phytophthora* spp.	Yellowing of leaves general plant decline	Mulch heavily; improve drainage; fosetyl-Al, metalaxyl
Kiwi	Botrytis rot	*Botrytis cinerea*	Flower petals become infected and fail to drop. The disease then spread to fruit causing fruit drop	Thiram at petal fall

Table 7 (*Contd.*)

Crop	Disease	Pathogen	Symptoms	Control
	Leaf spot	*Glomerella* spp., *Phoma* spp., *Myxosporium* spp.	Spots on leaves and fruit which may lead to premature defoliation	Bordeaux applied when plants are dormant
Lemon	*See* Citrus			
Lime	*See* Citrus			
Litchi	Leaf spot	*Botryodiplodia theobromae, Colletotrichum gloesporioides, Pestalotia pauciseta*	Brown lesions near leaf tips spreading to leaf margins. Defoliation may occur	Lime sulphur, Bordeaux, copper oxychloride
Longan	Anthracnose	*Colletotricum gloeosporioides*	Leaf and fruit spots	Mancozeb, benomyl, copper
Loquat	Anthracnose	*Colletotrichum gloesporioides*	Dark spots on leaves which may cause defoliation and dieback of twigs	Copper oxychloride, maneb, daconil
Macadamia	Stick-tight nuts (Anthracnose)	*Colletotrichum* spp.	Husks of nuts are killed causing nuts to hang on the tree after maturity	Plant resistant varieties such as 'Keauhou'
	Flower blight	*Botrytis* spp., *Phytophthora* spp.	Flowers killed	Not usually necessary
Mandarin	*See* Citrus			
Mango	Anthracnose	*Colletotrichum gloeosporioides*	Leaf spot, flower blight, fruit spot and rotting	Benomyl, copper, daconil, maneb
	Bacterial black spot	*Pseudomonas mangifera-indicae*	Small angular water-soaked areas on leaves and fruit which may crack and exude gum	Copper oxychloride 3 weeks after blossom and again when fruit is developed
	Powdery mildew	*Oidium mangiferae*	Grey mycelium primarily on inflorescences	Sulphur dust, copper oxychloride, maneb, imazalil

Table 7 (*Contd.*)

Crop	Disease	Pathogen	Symptoms	Control
Mangosteen	No Serious pests reported			
Mulberry	Mildew	*Phyllactinia corylea Uncinula geniculata*	White powdery coating on lower leaf surface	Sulphur dust, copper oxychloride, karathane, imazalil
Naranjilla	Various viruses			Do not handle plants after smoking; control aphid and leafhopper vectors
Olive	Peacock spot	*Cycloconium oleaginum*	3–9 mm dark green to black spots on upper leaf surface sometimes with a faint yellow halo	Improve air circulation by pruning; Copper sprays during rains
	Olive knot	*Bacterium sevastanoi*	Roundish galls on twigs, branches, roots, trunks, or petioles, especially after cold winters.	Propagate from clean plants, remove infected branches, sterilize secateurs
	Verticillium wilt	*Verticillium alboatrum*	Sudden collapse of one or more branches; leaves remain on the dead branches.	Do not plant in soils previously planted to cotton, tomatoes, or potatoes; grow a grass cover crop between rows; use resistant rootstocks
Orange	*See* Citrus			
Papaya	Anthracnose	*Colletotrichum gloeosporioides*	Spots on leaves, water-soaked spots on fruit which develop into sunken lesions up to 5 cm in diameter	Copper oxychloride, dithane M-45, maneb
	Cercospora leaf spot	*Cercospora papayae*	Round, grey-white lesions on leaves, black sunken lesions on fruit	Copper oxychloride, captan, maneb, zineb, ziram

Table 7 (*Contd.*)

Crop	Disease	Pathogen	Symptoms	Control
	Mildews	*Oidium caricae* *Ovulariopsis papayae* *Sphaerotheca* spp.	Mycelial growth primarily on upper side of leaves	Benomyl, copper oxychloride, sulphur, dithane, karathane
	Ringspot	Virus transmitted by aphids	Chlorosis in youngest leaves, vein clearing, mottling of laminae, stunting, shortened petioles	Rogue infected plants
Passionfruit-See Granadilla				
Peach	Brown rot	*Monilinia fructicola*	Fruit rots on tree and often remains mummified on the tree	Benomyl, captan, carbendazim, dicloran, sulphur/mancozeb
	Freckle	*Fusicladium carpophilum*	Small greenish-brown spots on fruit, possible fruit splitting, shot-hole of leaves	Captan, mancozeb, sulphur/mancozeb
	Fruit rot	*Rhizopus* spp.	Fruit rots on tree	Benomyl, captan, carbendazim, dicloran, sulphur/mancozeb
	Gum spot	*Coryneum beijerinckii*	Small purple spots on shoots, scorch of buds	Mancozeb
	Leaf curl	*Taphrina deformans*	Malformed, curled leaves with blister, often thickened with red or yellow colour	Copper oxychloride, lime-sulphur, metiram, sulphur, thiram
	Powdery mildew	*Sphaerotheca pannosa*	White powdery growth on young shoots and leaves	Benomyl, chinomethionat, copper oxychloride, lime-sulphur, sulphur/mancozeb

Table 7 (*Contd.*)

Crop	Disease	Pathogen	Symptoms	Control
Peach (*contd.*)	Rust	*Tranzschelia prunispinosae*	Brown pustules on lower leaf surfaces	Chinomethionat, mancozeb, sulphur, sulphur/mancozeb, zineb
	Scab	*Fusicladium carpophilum*	Fruit scabs and leaf shot-hole	Mancozeb
Pineapple	Black spot	*Penicillium funiculosum, Fusarium moniliforme*	Tissue between the seed cavities turns brown and then black and corky	Control mealybugs
	Mealybug wilt	Toxin produced by mealybugs	Yellow spots on leaves, followed by general yellowing and then reddening	Control mealybugs
Pitanga cherry	No serious pest			
Plum	Bacterial canker	*Pseudomonas pruni*	Dark cankers on twigs which open to exude sap and may cause twig dieback	Copper sulphate, zinc sulphate lime-sulphur
	Bacterial spot	*Pseudomonas pruni*	Small dark spots on leaves and fruit followed by cracking in fruit	Copper sulphate, zinc sulphate, lime-sulphur
	Brown rot	*Monolinia fructicola*	Fruits rot and remain mummified on tree	Benomyl, captan carbendazim
	Leaf curl	*Taphrina deformans*	New leaves malformed with blisters, yellow colour	Copper oxychloride
	Powdery mildew	*Sphaerotheca pannosa*	White growth on upper leaf surface	Benomyl, copper oxychloride, dinocap, sulphur, bayleton, imazalil.
	Rust	*Tanzschelia prunispinosae*	Brown pustules on lower leaf surfaces	Mancozeb, sulphur/mancozeb zineb, bayleton

Table 7 (*Contd.*)

Crop	Disease	Pathogen	Symptoms	Control
Pomegranate	Anthracnose	*Sphaceloma punicae*	Elliptical lesions on leaves and fruit	Copper oxychloride
Quince	*See* Apple			
Sapodilla	Phytopthora fruit rot	*Phytopthora palmivora*	Rotting of fruit especially lower fruits on each tree	Sanitation
Soursop	*See* Annona			
Strawberry	Botrytis (grey mould)	*Botrytis cinerea*	Grey mould on fruit	Benomyl, captan, carbendazim, dicloran
	Leaf spot	*Mycosphaerella fragariae*	Red spots on leaves and ultimate defoliation	Captan, copper oxychloride
	Powdery mildew	*Sphaerotheca humuli*	White mould on upper leaf surface	Benomyl, carbendazim, copper oxychloride, pyrazophos
	Rhizopus rot	*Rhizopus stolonifer*	Fungal growth on fruit surface	Dicloran
Sweetstop	*See* Annona			
Tangelo	*See* Citrus			
Tangerine	*See* Citrus			
Tree Tomato	Cucumber mosaic	Virus transmitted by aphids	Stunted plants with mottled leaves	Rogue infected plants-control vector
	Potato virus Y	Virus transmitted by aphids and mechanically	Leaf mottling along veins first and then spreading; fruits develop irregular dark blemishes and yield is reduced.	Rogue infected plants

Table 8
Pests of fruits and suggested control measures

Crop	Insect name Common	Scientific	Damage	Control
Acerola				
Akee				
Amberella	Mexican fruit fly	*Anastrepha* spp.	Necrotic lesions on peel at places of oviposition, then rotting at these sites	Traps of borax plus cotton-seed hydrolysate; collect and destroy dropped fruit several times weekly; repeated bait sprays of protein hydrolysate plus malathion, trichlorophon, diazinon or fenthion, plus water
	Mediterranean fruit fly	*Ceratitis capitata*	As above	As for Mexican fruit fly
	Nematode	*Rotylenchus reniformis*	Poorly developed root system lacking in fibrous roots	Plant clean stock in non-infested soil
Annona	Citrus black fly (aphid)	*Aleurocanthus woglumi*	Sucking insect on leaf undersides, abundant honey dew and sooty mould	Biological control possible in many instances, malathion, dimethoate, acephate, others
	Spirea aphid	*Aphis spiraecola*	Sucking insect producing abundant honeydew and sooty mould	Many natural predators, malathion, dimethoate, acephate, others
	Florida red scale	*Chrysomphalus ficus*	Yellow chlorotic spots	Azinphos-methyl, diazinon, parathion, malathion, others

Table 8 (*Contd.*)

Crop	Insect name Common	Scientific	Damage	Control
	Citrus mealybug	*Planococcus citri* or *Planococcus lilacinus*	Sucking insect, honeydew and sooty mould	Treat base of tree against ants with lindane, chlordane, or methoxychlor. Spray foliage with malathion plus a light oil, dimethoate or azinphos
	Hemispherical scale Black citrus aphid	*Saissetia coffeae* *Toxoptera aurantii*	See *P. citri* Rolled leaves at growing points	Treat base of tree against ants with lindane, chlordane, or methoxychlor. Spray foliage with malathion, dimethoate, endosulfan, acephate, others
Apple	Aphids	*Aphis* spp.	Sucking insects whose feeding causes distortion of new growth	Malathion, acephate, demeton-s-methyl, dimethoate, endosulfan, cypermethrin
	Ants	Many	Attend mealybugs, scale, aphids	Brush or spray trunk with lindane, methoxychlor, or chlordane
	Australian bug	*Icerya purchasi*	Sucks sap and excretes honeydew	dimethoate,

Table 8 (*Contd.*)

Crop	Insect name Common	Scientific	Damage	Control
Apple (*contd.*)	Bryobia mite	*Bryobia proetiosa*	Spotted discolouration of leaves	Mineral oil, Azinphos methyl, demeton-s-methyl diazinon (not on green or yellow cultivars) dimethoate, fenthion, lime-sulphur, malathion, oxydemeton-methyl, parathion
	Grasshopper	*Zonocerus elegans*	Chews leaves	Carbaryl (not within 1 month of flowering)
	Fruitfly	*Ceratitis capitata* *Pterandrus rosa*	Larvae feed on interior of fruit	Dimethoate bait (150ml (40 e.c.) + 5kg fishmeal + 100l water); malathion bait (100g (50 WP) + 7.5kg sugar + 100l water; trichlorfon bait (50g (95sp) + 6kg sugar + 100l water); fenthion spray
	Grey scale	*Diaspidiotus africanus*	Sucking insect causing discoloration of surrounding tissue	Mineral oil, malathion, parathion
	Leaf rollers	Tortridiaceace	Small caterpillars which feed on new growth and bind leaves together	Carbaryl (not within 1 month of flowering) azinphos-methyl
	Mealybug	*Planococcus* spp. *Pseudococcus* spp.	Flat sucking insect with waxy white covering, secretes honey and is tended by ants	Dimethoate, limesulphur, malathion mineral oil

Table 8 (*Contd.*)

Crop	Insect name Common	Scientific	Damage	Control
	Pear bud mite	*Eriophyes pyri*	Deforms buds and causes blisters on young leaves and fruit	Carbaryl (not within 1 month of flowering)
	Pernicious scale	*Quadraspidiotus perniciosus*	Sucking insect causing fruit discoloration and tree dieback	DNOC, lime-sulphur, malathion, mineral oil
	Red scale	*Aonidiella aurantii*	Sucking insect causing defoliation, mottling and shrivelling of fruit, and withering of shoots and twigs	Dimethoate, DNOC, limesulphur, mineral oil, malathion
	Red spider mite	*Tetranychus* spp.	Silvering and mottling of leaves	Binapacryl, demeton-s-methyl, diazinon (not on yellow or green cultivars) dicofol, dimethoate, malathion, mineral oil, oxydemeton-methyl, sulphur, tetradifon
	Soft brown scale	*Coccus hesperidum*	Sucking insect on lower side of leaves causing stunting and sooty mould	Dimethoate plus oil
	Wooly aphid	*Eriosoma lanigerum*	Sucking insect covered with waxy threads causing galls	Demeton-s-methyl, diazinon (not on green or yellow cultivars), dimethoate, endosulfan, limesulphur, malathion, mineral oil, oxydemeton-methyl, phorate, vamidothion, use resistant rootstocks

Table 8 (*Contd.*)

Crop	Insect name Common	Scientific	Damage	Control
Apricot	Ants	— *See* Apple		
	Aphids	— *See* Apple		
	Australian bug	*Icerya purchasi*	Sucks sap and excretes honeydew	Mineral oil (malathion may be added when plants are in leaf)
	Bryobia mite	— *See* Apple		As apple, but delete lime-sulphur
	Grasshopper	— *See* Apple		
	Fruitfly	*Ceratitis capitata* *Pterandrus rosa*	Larvae feed inside fruit	Fenthion (do not use on alpha cultivar); malathion bait (100 g of 50 WP plus 7.5 kg sugar in 100l water); trichlorfon bait (50 g of 95 sp plus 6 g sugar in 100l water)
	Grey scale	*Diaspidiotus africanus*	Sucking insect causing discoloration of surrounding tissue	Parathion, dimethoate, mineral oil
	Leaf roller	Tortricidaceae	Small caterpillars which feed on new growth and bind leaves together	Azinophos-methyl
	Mealybug	— *See* Apple		Malathion, mineral oil
	Pernicious scale	— *See* Apple		As apple, but delete DNOC
	Red scale	— *See* Apple		Lime-sulphur, mineral oil, parathion
	Red spider mite	— *See* Apple		Same as apple but delete binapacryl; dimethoate

Table 8 (*Contd.*)

Crop	Insect name Common	Scientific	Damage	Control
Avocado	Avocado brown mite	*Oligonychus punicae* or *O. perseae*		Morestan 25 WP at 0.5–1 kg/ha formulation before fruit set
	Avocado weevil	*Heilipus lavri* *Conotrachelus aquacatae*		Cultural — remove foliage, twigs and fruit which house larvae from ground; chemical — methyl parathion 480 at 0.1–0.15%, azinphos-methyl EC 250 at 0.2–0.3%, lebaycid EC 500 at 0.1%, baythroid EC at 0.035–0.05%; control adults when they appear or when fruit is 3 cm diameter; repeat treatment every 2–3 wks.
	Black fly	*Alevrodes* sp.		Control as for avocado weevil
	Coconut scale	*Aspidiotus destructor*	Sucking insect causing discolored areas on leaves	Diazinon, malathion, parathion
	Helopeltis bug	*Helopeltis theivora*	Sucking insect which causes death of new shoots resulting in dieback and witches broom	Carbaryl, trichlorphon, phenthoate
	Leafhopper	*Metcalfiella monogramma*		Control as for avocado weevil

Table 8 (*Contd.*)

Crop	Insect name Common	Scientific	Damage	Control
Avocado (*contd.*)	Mites	Various	Feed primarily on undersides of leaves causing speckling	Dicofol, malathion
	Thrip			Control as for avocado weevil
	West African cocoa mealybug	*Planococcoides njalensis*	Sucking insect	Control attendant ants, malathion
	White fly	*Bemisia* sp.		Control as for avocado weevil
Banana	Banana root borer	*Cosmopolites sordidus*	Larvae tunnel in corms causing wilting and death of the newest leaves and breaking of stems	Dip new propagules in oil — 0.4% solutions of dieldrin, aldrin or chlordane; basal spray of aldrin, or dieldrin
	Pineapple mealybug	*Dysmicoccus brevipes*	Sucking insect feeding mainly on roots	Control attendant ants, dip propagules in methoxychlor, or chlordane
Breadfruit	Coconut scale	*See* Avocado		
Breadnut	*See* Breadfruit			
Carambola	Fruitfly	— *See* Apple		
Cashew	Citrus black fly	*Aleurocanthus woglumi*	Sucking insect which weakens tree and encourages sooty mould by exuding honeydew	Malathion, acephate
Cherimoya	*See* Annona			
Citrus	Ants	Formicidae	May feed on plants or tend homoptera pests	Chlordane, lindane, methoxychlor — brush or spray on trunk

Table 8 (*Contd.*)

Crop	Insect name Common	Scientific	Damage	Control
	Aphids	*Toxoptera* spp.	Sucking insect which distorts new growth and spreads viruses	Azinphos-methyl, dimethoate (not rough lemon, Villa franca lemon, Seville oranges, limes) triazophos (not lemons), malathion, mineral oil, monocrotophos
	Australian bug	—	*See apple*	Add malathion
	Bud mite	*Aceria sheldoni*	Causes fruit malformation, multiple branching, and brown markings on buds	Binapacryl, bromoprophylate, propargite chlorobenzilate, lime-sulphur
	Circular purple scale	*Chrysomphalus ficus*	Sucking insect feeding on fruit, branches, leaves	Azinphos-methyl, mineral oil
	Citrus black fly	*Aleurocanthus woglumi*	Sucking insect on leaf undersides	*See Annona*
	Flat mite	*Brevipalpus californicus*	Causes silvering of leaves	Bromopropylate, chinomethionat, dicofol propargibe
	Fruitfly	*Ceratitis capitata Pterandrus rosa*	Larvae feed on interior of fruits. Oviposition encourages secondary infection	Malathion bait (100g of 50 WP plus 7.5kg sugar in 100l water) trichlorfon bait (50g of 95 sp plus 6g sugar in 100l water)
	Fruit-piercing moths	Many	Nocturnal moths which pierce fruit rind allowing secondary infections	Pick fruit as soon as ripe
	Grey mite	*Calacarus citrifolii*	Causes concentric ring blotch dieback, and leaf drop	Bromopropylate, chlorobenzilate, dicofol, sulphur, propargite

Table 8 (*Contd.*)

Crop	Insect name Common	Scientific	Damage	Control
Citrus (*contd.*)	Heliothis bollworm	*Heliothis armigera*	Caterpillars hollow out buds	Endosulfan
	Mealybug	*Planococcus* spp. *Pseudococcus* spp. *Rastrococcus invadens*	Sucking insect which weakens plant and secretes honeydew	Control attending ants; malathion, methidathion plus mineral oil
	Mussel scale	*Lepidosaphes beckii*	Sucking insect causing curled leaves, mottled fruit and withered shoots	Parathion, malathion, dimethoate (see warning under psylla)
	Orange dog	*Papilio demodocus*	Caterpillar feeding on leaves of young trees. Early instars resemble bird droppings	Endosulfan, acephate, cypermethrin, B.T.
	Psyllid	*Trioza erytreae*	Sucking insects which feed on young leaves causing pock-marks. Transmit citrus-greening disease	Dimethoate (not for rough lemon, Villa Franca lemon, Seville oranges or limes), endosulfan, triazophos (not for lemons), malathion, mineral oil
	Red mite	*Panonychus citri*	Causing silvering and mottling of leaves	Azinophos-methyl, binapacryl, chinomethionat, dicofol, propargite malathion, triazophos, mineral oil, tetradifon
	Red scale	*Aonidiella aurantii*	Yellow-to-red cicular scale causing defoliation, shrivelling and mottling of fruit, and withering of shoots	Azinophos-methyl, dimethoate (not rough lemon, Villa Franca lemon, Seville oranges, or limes) malathion, mineral oil, methidathion,

Table 8 (*Contd.*)

Crop	Insect name Common	Scientific	Damage	Control
	Red spider mite	*Tetranychus* spp.	Small red mite causing silvering and mottling of leaves	Chinomethionat, dicofol, dimethoate (not for rough lemon, Villa Franca lemon Seville orange or lime), mineral oil, amitraz
	Rust mite	*Phyllocoptrata oleivora*	Cream-coloured mite causes severe silvering of leaves and fruit	Chinomethionat, chlorobenzilate, mancozeb, zineb
	Soft brown scale	*Coccus hesperidum*	Oval scale on undersides of leaves causing stunting and sooty mould	Azinophos-methyl, dimethoate (not for rough lemon, Villa Franca lemon, Seville orange, or lime), malathion, methidathion, mineral oil, triazophos
	Thrips	Thysanoptera	Small elongated insects which rasp leaf surfaces producing silvering	Malathion, parathion, tartar emetic, temephos, triazophos (not for lemon) dimethoate (*see warning above*) Chlordane, lindane, methoxychlor
Coconut	Ants and termites	Formicidae	Feed on young plants in nurseries and may tend aphids and scale	Chlordane, lindane, methoxychlor
	Aphids	*Cerataphis lataniae* and others	Sucking insects which feed on leaves and fruit thus weakening plants	Malathion, acephate
	Caterpillars	Various, especially *Parasa lepida*	Eat leaves causing defoliation	Malathion, carbaryl, B.T.

Table 8 (*Contd.*)

Crop	Insect name Common	Scientific	Damage	Control
Coconut (*contd.*)	Palm weevil	*Rhabdoscelus obscurus*	Larvae tunnel through the trunks	Lindane, chlordane
	Rhinoceros beetle	*Orocytes* spp.	Feeds on the apical bud	Mix methoxychlor or chlordane with sawdust or sand and place around the apical bud
	Coconut scale	*Aspidiotus destructor*	Sucking insects feed on leaves	Malathion, diazinon, parathion plus oil
Date	Date mite	*Oligonychus pratensis*	Webbing; scars distort and discolour fruit	Sulphur dust
	Date bug	*Asarcopus palmarum*	Concentrated between expanded parts of terminal leaves; oval to 3 mm × 1.5 mm with red eyes; infested tissue browns and shrivels; honeydew present	Nicotine sulfur dust or 40% nicotine sulfate at 1:500
	Red date scale	*Phoenicoccus marlatti*	Females 1.3 mm, oval, pink to red; rest on masses of cottony filaments; found behind fibre, at leaf bases, on fruit stems and roots	Malathion plus light oil
	Fig beetle	*Rotinis texana*	Large green beetle; punctures fruit with horn and feeds	None satisfactory
	Nitidulid beetle (dried fruit beetle)	Four spp.	Feeds on waste dates, then new crop; leads to mould attack	5% malathion 3 weeks before picking; sanitation
Durian Eugenia	Fruit borer Aphid	*Plagideicta mogniplaga* See Apple		

Table 8 (*Contd.*)

Crop	Insect name Common	Scientific	Damage	Control
Feijoa	Scale mealy bug	—	—	Phosmet 75% WP, plus 150 g oil/100 l water; carbaryl 80%
	Leaf roller caterpillar			
Fig	Long-horn beetle	Cerambycidae	Feeds on bark often girdling small limbs	Carbaryl
	Scale	Various	Sucking insects weaken plant by feeding on stems and leaves	Acephate, malathion
Granadilla	Weevils	Various	Feed on interior of fruit	Carbaryl
	Fruitfly	See Apple		
Grape	Fruitfly	See Apple		
	Mealybug	Planococcus kenyae	Sucking insect	Acephate, diazinon, malathion
	Bollworm	Heliothis armigera	Feed on leaves, shoots, flowers and berries	Acephate, carbaryl, diazinon, malathion, B.T.
	Fruitfly	See Apple		
	Fruit-piercing moths	See Citrus		
	Mealybug	Planococcus spp. Pseudococcus spp.	Sucking insect feeds on leaves and stems	Carbaryl, methiocarb Methadithion, omethoate, dimethoate fenchlorphos, formothion, carbaryl, methiocarb
	Phylloxera	Phylloxera spp.	Sucking insect feeds on roots	Resistant rootstocks
	Snoutbeetle	Phlyctinus callosus Eremnus cerealis E. setulosus	Feeds on young shoots and leaves and may later feed on fruit brunches	Carbaryl, methiocarb
	Vine blister mite	Eriophyes vitis	Feeds on buds and leaves causing distortion	Lime-sulphur, wettable sulphur or sulphur dust, methiocarb, endosulfan bromopropylate, chlorobenzilate

Table 8 (*Contd.*)

Crop	Insect name Common	Scientific	Damage	Control
Grapefruit	*See* Citrus			
Guava	Aphids — *see* Apple			
	False spider mite	*Brevipalpus phoenicis*	Feeding causes silvering of leaves	Dicofol, chlordimeform, monocrotophos
	Scale	*Coccus alpinus*	Sucking insects weaken plants and secrete honeydew	Control ants; malathion, azinphos, parathion, dimethoate, diazinon
	Hemispherical scale	*Saissetia coffeae*	Sucking insects weaken plants and secrete honeydew	Malathion, azinophos, parathion, dimethoate, mineral oil
	Mealybug	*Ferrisia virgata*	Sucking insect	Methyl parathion, methyl azinophos-methyl plus oil
	Cottony cushion scale	*Icerya purchasi*	Sucking insect	Malathion, parathion
	Cocoa thrip	*Selenothrips rubrotinctus*	Rasping insect which causes a silvering of a leaves	Malathion, lindane, dicrotophos, fenitrothion
Jackfruit	— *See* Breadfruit			
Jambolan	Bark eating caterpillar	*Indarbela quadrinotata*	Thick, ribbonlike silk webs with excreta hang from bark and main stems, especially near forks; below are zigzag galleries and entry holes of larvae	Sanitation; avoid overcrowding; insert iron spike into entry hole to kill larvae, soak swab in fumigant (petrol) and insert in holes, seal with mud; treat again and reseal if holes reopen
Kiwi fruit	Greedy scale	*Hemiberlesia rapax*	Sucking insect	Malathion, diazinon, parathion

Table 8 (*Contd.*)

Crop	Insect name Common	Scientific	Damage	Control
	Leaf roller	Tortidaceae	Caterpillars burrow through the skin into the pulp or make scars on the surface of fruit	Azinophos-methyl
	Mites	Various	Cause silvering of leaves	Diazinon, dicofol, dimethoate, malathion
	Tea thrips	*Heliothrips haemorrhoidalis*	Rasping insects cause silvering or browning of leaves	Fenitrothion
Lemon	— *See* Citrus			
Lime	— *See* Citrus			
Litchi	Fruitfly			
	Litchi moth	— *See* Apple *Argryroploce peltastica*	Eggs laid on fruit, larvae burrow into fruit	Cover fruit with bags. Insecticides not effective
Loquat	— *See* Apple			
Macadamia	Bark borer	*Salagena* spp.	Larvae feed on live bark occasionally girdling branches	Spraying not recommended. Control physically by inserting wire into passages and crushing
	Carob moth	*Spectrobates cerathoniae*	Larvae feed on nuts	Recommended cultivars are somewhat resistant
	False codling moth	*Cryptophlebia leucotreta*	Larvae feed on nuts	Recommended cultivars are somewhat resistant
	Green vegetable bug	*Nezara viridula*	Feeds on developing nuts causing nut drop and kernal staining	Malathion, acephate
	Litchi moth	*Argryroploce peltastica*	Larvae feed on nuts	Recommended cultivars are somewhat resistant

Table 8 (*Contd.*)

Crop	Insect name Common	Scientific	Damage	Control
Mandarin	— *See* Citrus			
Mango	Fruitfly — *see* Apple			
	Mango scale	*Diaspis cinnamonii* var. *mangifera*	Sucking insect attacks seedlings	Malathion plus oil
	Mango stone weevil	*Sternochetus mangifera*	Feeds on the mango seed destroying fruit in the process	Collect and destroy all fallen fruit weekly. Insecticides not recommended
	Mosquito-back	*Helopeltis* spp.	Sucking insects feed on new growth destroying growing points and causing witches broom	Carbaryl, phenthoate, trichlorphon
	Mealybug	*Rastrococcus invadens*	Sucking insect	Malthion plus oil, phenthoate, trichlorphon
Mangosteen	—			
Mulberry	—			
Naranjilla	Root-knot nematode	*Meloidogyne* spp.	Swellings on root	Fumigate, use rotations, resistant rootstocks
Olive	Root lesion nematode	*Pratylenchus musicola*	Attacks bark of root causing longitudinal black cracks; damage not serious	None effective
	Root-knot nematodes	*Meloidogyne* spp.	Knots develop on roots, particularly in sandy soils; older trees can be kept in economic production with proper cultural methods	None effective

Table 8 (*Contd.*)

Crop	Insect name Common	Scientific	Damage	Control
	Black scale	*Saissetia oleae*	Very widespread; defoliates and causes sooty mould; 3–5mm, brown to black nearly circular with 'H' pattern on backs of most	Light to medium oil in summer (1.66–2%)
	Olive scale	*Parlatoria olea*	Oval to circular, 1–1.5mm, grey with black exudate; deforms fruit in spring with dark spots which become conspicuous as fruit matures; spoiled for pickling and oil content drops; Also infest twigs and leaves	Pruning plus natural enemies (Persian strain of *Aphytis* hymenopterous parasite); or 454g parathion plus 6l (light to medium) summer oil in 380l water; spray late spring to early summer, 2 applications 10–15 days apart
	Oleander scale	*Aspidiotus hederae*	Similar appearance to olive scale; attacks fruit but parts of fruit beneath and surrounding body remain green; early attack may lead to pitting and deformation	As for olive scale
Orange	*— See* Citrus Citrus mealybug	*Planococcus citri*	Sucking insect	Azinophos, diazinon, dimethoate, malathion
Papaya	Mosquito back	*Helopeltis* spp.	Sucking insect may feed on fruit causing exudation of sap	Carbaryl
	Scale	Various	Sucking	Malathion

Table 8 (*Contd.*)

Crop	Insect name Common	Scientific	Damage	Control
Passionfruit				
Peach	Ants — *see* Apple			
	— *See* Granadilla.			
	Aphids — *see* Apple			
	Australian bug — *see* Apple			
	Black peach aphid	*Brachycaudus persicae*	Sucking insect attacks young foliage and bearing shoots reducing fruit set	Endosulfan, methamidophos, fenthion, diazinon, others
	Bryobia mite — *see* Apple			
	Elegant grasshopper	*Zonocerus elegans*	Feeds on leaves	Carbaryl
	Fruitfly — *see* Apple			
	Grey scale	*Diaspidiotus africanus*	Sucking insect causing discoloration of surrounding tissue	Parathion, Lime-sulphur
	Leaf roller	Tortricidae	Small caterpillars which feed on new growth and bind leaves together	Azinophos-methyl
	Mealybug	*Planococcus* spp. *Pseudococcus* spp.	Sucking insects	Dimethoate, malathion, mineral oil, parathion
	Pernicious scale	*Quadraspidiotus perniciosus*	Sucking insect causes discoloration of fruit and dieback of tree	Lime-sulphur, malathion, mineral oil, parathion
	Red scale — *see* Apple			
	Red spider mite — *see* Apple			
	White scale	*Pseudaulacaspis pentagona*	Sucking insect causes dieback	Lime-sulphur just before bud break
Pineapple	Black maize beetle	*Heteronychus orator*	Larvae feed on roots and stem	Chlordane, lindane

Table 8 (*Contd.*)

Crop	Insect name Common	Scientific	Damage	Control
	Pineapple mealybug	*Dysmicoccus boninsus*	Sucking insect attacks leaves, roots and fruits and spreads mealybug wilt	Dip propagules in methoxychlor before planting. Control ants; malathion, parathion
	Pineapple scale	*Diaspis bromeliae*	Sucking insect causes chlorosis and wilting	Malathion
Pitanga cherry				
Plums	Ant — *see* Apple			
	Aphid — *see* Apple			
	Australian bug — *see* Apple			
	Bryobia mite — *see* Apple			
	Elegant grasshopper — *see* Peach			
	Fruitfly — *see* Apple			
	Grey scale — *see* Peach			
	Mealybug — *see* Peach			
	Pernicious scale	*Quadraspidiotus perniciosus*	Sucking insect causes discoloration of fruit and dieback of tree	DNOC, lime-sulphur, malathion mineral oil, parathion
	Red scale — *see* Apple			
	Red spider mite — *see* Apple			
Pomegranate	Scale	Various	Sucking insect	Malathion plus oil
Quince	— *see* Apple			
Sapodilla	Fruitfly — *see* Apple			
Soursop	— *see* Annona			
Strawberry	Aphid	*Chaetosiphon fragaefolii*	Sucking insect transmits virus diseases	Dimethoate, azinphos-methyl, malathion

Table 8 (*Contd.*)

Crop	Insect name Common	Scientific	Damage	Control
Strawberry (*contd.*)	Red spider mite	*Tetranychus* spp.	Small mite causes silvering of leaves	Dimethoate, fenbutatin-oxide, propargite
Sweetsop	— *see* Annona			

Table 9
Methods of control of some tropical weeds

Species	Notes on mechanical control	Recommended herbicides
Acanthospermum hispidum (A)	Effective	Post: 2,4-D, MCPA, paraquat Pre : Simizine, fluometuron, dichlobenil, secbumeton, terbacil, diuron, pebulate
Ageratum conyzoides (A)	Effective	Post: 2,4-D, MCPA, 2,4,5-T, ametryne, paraquat, diquat, Pre: diuron glyphosate
Amaranthus spp. (A)	Effective	Most pre- and post-emergence herbicides
Bidens pilosa (A)	Effective	Post: 2,4-D, MCPA, paraquat, glyphosate Pre : Monuron, diuron, linuron, fluometuron, metobromuron, simazine, ametryne, metribuzin, terbacil, bromacil, alachlor, propanil
Boerhavia erecta (A)	Effective	Post: paraquat Pre : fluorodifen, chlorbromuron, metobromuron, linuron, fluometuron, terbatryne, diphenamid
Borreria spp. (P)	Slashing, hoeing, but may regrow	Post: glyphosate, metribuzin, dinoseb, paraquat Pre : linuron, diuron, fluometuron, neburon atrazine, desmetryne, lenacil
Cassia obtusifolia (A)	Effective	Post: paraquat, dinoseb, linuron, metribuzin, chloroxuron 2,4-D, 2,4,5-T, picloram chlorbromuron

Table 9 (*Contd.*)

Species	Notes on mechanical control	Recommended herbicides
Cassia obtusifolia (A) (*contd.*)		Pre : fluometuron, cynazine, atrazine, napropamide, diphenamid
		Post: paraquat, MCPA, 2,4-D
Cleome monophylla (A)	Effective	Pre : monuron, trifluralin, linuron, chlorbromuron, atrazine, prometryne, terbutryne, metribuzin, diuron oxadiazon
Commelina spp. (P)	Difficult as stem fragments root readily and are viable for prolonged periods	Post: 2,4-D, metribuzin, glyphosate
Cynodon dactylon (P)	Not effective, except during long periods of dry weather	Post: Dalapon, TCA, terbacil, bromacil, amitrole, asalam, glyphosate, fluazifop-butyl, sethoxydim
Cyperus esculentus (P)	Not effective, except repeated deep cultivation during prolonged dry periods	Post: bentazon, glyphosate
		Pre : EPTC, alachlor, metolachlor, dichlobenil, terbacil, bromacil
Cyperus rotundus (P)	Not effective except repeated deep cultivation during dry periods	Post: glyphosate, bentazon — repeated applications of both are necessary
		Pre : EPTC, alachlor, metolachlor, dichlobenil, terbacil, bromacil
Dactyloctenium aegyptium (A)	Effective	Post: Paraquat, glyphosate, propanil
		Pre : atrazine, terbatryne, simazine, pebulate, EPTC, nitralin, trifluralin, DCPA, diuron, fluometuron, prometryn, nitrofen
Digitaria scalarum (P)	Difficult to control by cultivation	Post: dalapon, glyphosate, amino-triazole
Eleusine indica (P)	Cultivation effective, hand pulling difficult	Post: propanil, paraquat, glyphosate sethoxydim, fluazifop-butyl.

Table 9 (*Contd.*)

Species	Notes on mechanical control	Recommended herbicides
Eleusine indica (P) (*contd.*)		Pre : asulam, pebulate, vernolate, butylate, benthiokarb, nitralin, trifluralin, isopropanil, dibutalin, alachlor, butachlor, propachlor, napropamide, diuron, linuron, norea, metobromuron, fluometuron, chlorbromuron, methabenzthiazuron, simazine, terbutryn, atrazine, prometryne, cyanazine, ametryne, metribuzin, dalapon
Imperata cylindrica (P)	Difficult — repeated cultivations necessary	Post: dalapon, sodium chlorate, amino-triazole, sodium chlorate, glyphosate (repeat applications), paraquat (repeat applications)
Leonotis nepetifolia (A)	Effective	Post: 2,4-D, MCPA Pre : linuron, terbutryne, alachlor, diuron, atrazine, ametryne, terbacil, bromacil
Nicandra physalodes (A)	Effective	Post: MCPB, 2,4-DB, 2,4-D, MCPA, dinoseb Pre : simazine, diuron
Oxalis spp. (P)	Not effective	Post: repeat applications of paraquat, glyphosate, amino-triazole, or dinoseb, NAA, triclopyr Pre : Trifluralin, dichlobenil, oxadiazon

Table 9 (*Contd.*)

Species	Notes on mechanical control	Recommended herbicides
Portulaca oleracea (A)	Effective, but stems will reroot on wet soils	Post: 2,4-D, MCPA, dinoseb, bentazon, paraquat, diquat Pre : linuron, metobromuron, chlorbromuron, methabenzthiazaron, simizine, atrazine, terbutryne, prometryne, cyanazine, trifluralin, nitralin, isopropanil, penoxalin, napropamide, alachlor, butylate, EPTC, chloramben, fluorodifen, nitrofen, methazole, oxydiazon
Rottboellia exaltata (A)	Cultivation effective, seeds viable for only 2–3 years in the soil so slashing before seed production will be very effective	Post: Difficult to control with herbicides, Paraquat, glyphosate, diuron/2,4-D Pre : trifluralin, oxadiazon
Setaria spp. (A)	Effective	Post: paraquat, dinoseb Pre : amiben, trifluralin
Sida spp. (P)	Effective	Post: glyphosate, 2,4-D ester, 2,4,5-T ester, dicamba, picloram, dinoseb/alachlor metribuzin Pre : diuron, fluometuron, prometryne, linuron
Sporobolus jaquemontii (P)	Hoeing	Post: TCA, dalapon, glyphosate Pre : Simazine

Table 9 (*Contd.*)

Species	Notes on mechanical control	Recommended herbicides
Striga spp. (A)	Hand weeding, hoeing to prevent seeding, since *Striga* is a parasite, most damage is done prior to emergence	Post: glyphosate Pre : trifluralin has some effect
Trianthema portulacastrum (A)	Effective if tap root is severed, stems may root in moist weather	Post: glyphosate, paraquat Pre : simazine, atrazine, prometryne, PCPA, EPTC, pebulate, diuron, fluometuron, chlorbromuron, alachlor, nitrofen, trifurolin, ametryne

A = Annual
P = Perennial

Table 10
Nutritional content of selected vegetables

Common name	Water (g)	Calories	Protein (g)	Fat (g)	Carbohydrate (g)	Fibre (g)	Calcium (mg)	Phosphorus (mg)	Iron (mg)	B-carotene equiv. (µg)	Thiamine (mg)	Riboflavin (mg)	Niacin (mg)	Ascorbic acid (mg)
Amaranth *Amaranthus* spp. (fresh leaves)	85	46	4.0	0.5	6	1.3	250	55	4.0	600	0.1	0.3	0.7	100
Beetroot *Beta vulgaris* (roots)	87	45	1.8	0.1	10	1.2	15	37	1.0	trace	0.02	0.03	0.3	3
Bottle gourd *Lagenaria siceraria* (fruit pulp)	93	21	0.5	0.1	5	0.6	443	34	2.4	25	0.03	—	0.4	10
Bunching onion *Allium fistulosum* (leaves)	90	36	1.8	0.5	6	1.0	40	—	3.0	328	0.05	0.10	0.5	50
Cabbage *Brassica oleracea* var. *capitata* (leaves)	90	23	1.5	0.2	5	1.0	24	28	0.5	18	0.1	0.7	0.7	40
Carrot *Daucus carota* (roots)	88	40	0.9	0.1	9	1.4	35	70	2.0	1800	0.05	0.05	0.6	8
Cassava *Manihot esculenta* (fresh leaves)	91	28	2.0	1.0	4	0.8	80	90	2.5	600	0.05	0.05	0.5	6
(fresh tuber)	62	149	1.2	0.2	35	1.1	68	60	1.9	30	0.04	0.05	0.6	31
Cauliflower *Brassica oleracea* var. *botrytis* (leaves)	87	43	4.0	0.3	6	1.5	100	—	2.0	600	0.15	0.20	1.0	80
Chick pea *Cicer arietinum* (seeds)	10	357	19.6	3.7	63	6.7	252	271	11.1	60	0.48	0.16	1.8	trace
Chinese cabbage *Brassica chinensis* (fresh leaves)	93	21	1.8	0.3	3	0.7	40	48	0.5	18	0.7	0.13	1.0	74
Chives *Allium schoenoprasum* (leaves)	92	27	2.7	0.6	4	0.7	83	41	0.8	—	0.10	0.06	0.5	32
Cho-Cho *Sechium edule* (fruit)	94	19	0.7	0.1	5	0.6	17	14	0.4	15	0.01	0.02	0.4	14
Cluster bean *Cyamopsis tetragonoloba* (fresh pods)	82	—	3.7	0.2	10	2.3	trace	—	5.8	198	—	—	—	49
Cock's comb *Celosia argentea* (leaves)	84	44	4.7	0.7	8	1.8	260	43	7.8	—	—	—	—	—
Cocoyam *Colocasia esculenta* (tubers)	73	102	1.8	0.1	23	1.0	51	88	1.2	trace	0.1	0.03	0.8	8
Cucumber *Cucumis sativus* (fruit)	95	15	0.8	0.1	3	0.8	13	30	0.5	trace	0.02	0.01	0.3	14

Table 10 (*Contd.*)
Nutritional content of selected vegetables

Common name	Water (g)	Calories	Protein (g)	Fat (g)	Carbohydrate (g)	Fibre (g)	Calcium (mg)	Phosphorus (mg)	Iron (mg)	B-carotene equiv. (µg)	Thiamine (mg)	Riboflavin (mg)	Niacin (mg)	Ascorbic acid (mg)
Dolichos bean *Lablab niger* (fresh pods)	90	34	2.0	1.0	6	1.0	50	96	1.4	120	0.08	0.12	0.5	20
Egg plant *Solanum melongena* (fresh fruit)	90	32	1.0	0.2	7	1.3	14	26	1.3	34	0.05	0.05	0.5	9
Endive *Cichorium endivia* (leaves)	92	21	1.5	0.2	4	1.1	60	36	1.2	—	—	—	—	—
French bean *Phaseolus vulgaris* (fresh pods)	88	36	1.9	0.2	4	0.8	35	48	0.8	120	0.08	0.12	0.5	20
Garlic *Allium sativum* (cloves)	66	122	7.0	0.3	25	1.1	26	109	1.2	—	0.03	0.08	0.4	7
Hot pepper *Capsicum frutescens* (fresh pods)	90	37	2.0	0.5	6	1.0	20	101	1.0	360	0.06	0.08	1.0	150
Jew's mallow *Corchorus olitorius* (leaves)	80	58	4.5	0.3	12	2.0	250	122	4.0	600	0.15	0.53	1.2	100
Kale *Brassica oleracea* var. *acephala* (leaves)	84	43	3.5	0.8	8	1.6	132	77	1.3	900	—	—	—	110
Kang Kong *Ipomoea aquatica* (leaves)	90	30	2.7	0.4	6	1.1	60	42	2.5	2865	0.09	0.16	1.1	47
Katuk *Sauropus and rogynous* (leaves)	81	58	48.0	0.9	11	—	51	86	2.7	6220	0.07	0.39	2.2	83
Kohl rabi *Brassica oleracea* var. *gongylodes* (leaves)	91	26	2.3	0.2	5	1.4	36	36	1.6	—	—	—	—	—
Leaf mustard *Brassica juncea* (leaves)	92	22	2.2	0.4	4	0.7	80	39	2.5	600	0.08	0.14	0.6	50
Leek *Allium ampeloprasum* var. *porrum* (leaves)	91	29	2.0	0.7	5	1.4	70	53	7.5	1827	0.07	0.11	0.8	29
Lesser yam *Dioscorea esculenta* (tuber)	74	102	1.5	0.2	24	0.6	12	35	0.8	—	0.10	0.01	0.8	15
Lettuce *Lactuca sativa* (leaves)	94	19	1.4	trace	3	0.5	35	21	1.0	180	0.1	0.1	0.4	15
Lima bean *Phaseolus lunatus* (seeds)	12	335	21.4	1.4	61	4.3	56	180	12.0	trace	0.33	0.16	2.1	1
Mung bean *Vigna radiata* (seeds)	11	341	22.9	1.2	62	4.4	105	330	7.1	55	0.53	0.26	2.5	4
Okra *Hibiscus esculentus* (fresh pods)	88	36	2.1	0.2	5	0.9	62	36	1.2	90	0.01	0.08	1.0	25
Onion *Allium cepa* var. *cepa* (bulbs)	87	48	1.5	trace	11	0.5	30	—	0.5	trace	0.04	0.02	0.3	10

Table 10 (*Contd.*)

Common name	Water (g)	Calories	Protein (g)	Fat (g)	Carbohydrate (g)	Fibre (g)	Calcium (mg)	Phosphorus (mg)	Iron (mg)	B-carotene equiv. (µg)	Thiamine (mg)	Riboflavin (mg)	Niacin (mg)	Ascorbic acid (mg)
	Representative values for nutrients per 100g of edible portion									Vitamin				
Parsley *Petroselinum crispum* (leaves)	85	49	3.7	0.6	7	1.4	200	47	8.0	1800	0.10	0.30	1.4	150
Potato, Irish *Solanum tuberosum* (tubers)	77	82	1.7	0.1	18	0.6	8	51	1.1	65	0.07	0.03	1.3	15
Pumpkin *Cucurbita moschata* (fruit)	90	36	1.0	trace	8	0.5	20	—	0.8	210	0.05	0.05	0.5	15
Radish *Raphanus sativus* var. *hortensis* (roots)	94	18	1.0	0.1	4	0.1	42	30	1.0	trace	0.03	0.03	0.3	25
Roselle *Hibiscus sabdariffa* var. *sabdariffa* (calyces)	86	44	1.6	0.1	11	2.5	160	60	3.8	285	0.04	0.06	0.5	14
Shallot *Allium cepa* var. *aggregatum* (leaves)	91	30	1.8	0.9	5	1.6	86	25	3.7	945	0.07	0.12	0.4	19
(bulbs)	81	67	1.9	0.3	15	0.7	36	45	0.8	trace	0.04	0.02	0.3	2
Snake gourd *Trichosanthes cucumerina* (fruit)	95	18	0.5	0.3	3	0.8	50	20	1.1	96	0.10	0.10	0.3	—
Sponge gourd *Luffa cylindrica* (fruit)	94	19	1.1	0.2	4	1.0	2	30	0.7	170	0.03	0.04	0.3	10
Squash *C. pepo* (fruit)	92	22	1.5	0.1	5	1.0	24	38	0.4		trace	0.17	0.8	
Sweet pepper *Capsicum annuum* (fruits)	94	23	0.9	0.1	3	0.8	4	25	0.7	180	0.06	0.08	1.0	140
Sweet potato *Ipomoea batatas* (tubers)	68	121	0.8	0.2	35	1.0	16	56	2.0	400	0.9	0.04	0.7	37
Tannia *xanthosoma sagitifolium* (tubers)	65	133	2.0	0.3	31	1.0	20	47	1.0	trace	0.1	0.03	0.5	10
Tomato *Lycopersicon lycopersicum* (fruits)	93	21	1.0	0.2	4	0.6	10	24	0.6	450	0.06	0.04	0.6	26
Water melon *Citrullus lanatus* (fruits)	93	22	0.5	0.1	5	0.4	8	9	0.3	250	0.04	0.05	0.1	8
Wax gourd *Benincasa hispida* (fruit)	92	28	0.7	trace	6	0.3	20	—	0.6	trace	0.04	0.03	0.6	15
Winged bean *Psophocarpus tetragonolobus* (seeds)	9	420	31.2	17.0	33	6.6	210	410	15.0	—	0.08	—	—	trace
Yam bean *Pachyrrhizus erosus* (tubers)	86	45	1.6	0.3	10	0.5	15	—	0.5	trace	0.03	0.02	0.2	20

Table 11
Transplanting response of selected vegetables

Good	Moderate	Poor[b]
Amaranth	Chinese cabbage[a]	Beetroot
Bologi[a]	Endive	Broad bean
Broccoli[a]	Indian spinach	Carrot
Cabbage[a]	Leek	Cucumber
Cauliflower[a]	Lettuce[a]	French or kidney bean
Cayote (Cho-cho)	New Zealand spinach	Garden pea
Egg-plant[a]	Onion[a]	Melon
Kale[a]	Parsley[a]	Pumpkin
Kohl-rabi[a]	Roselle	Radish
Pepper[a]	Shallots	Squash
Tomato[a]	Spinach	Sweet corn
	Water-leaf (Talinun)	Turnip

[a] Commonly sown in the nursery and transplanted
[b] Normally sown direct where they are to grow

Table 12

Vegetable seed sowing rates and growth period to maturity

Column 1: The amounts given represent sowing rates in either kg/ha or g/10m², the same figures apply to both rates
Column 2: Period to first harvest in days from sowing or transplanting (*t* indicates which are normally transplanted)

	Column 1 Seed sowing rates	Column 2 Days to maturity
Amaranth	1.5–2.0	30–50
Asparagus (t)	2–3	400–600
Beetroot	12–15	70–90
Bottle gourd (t)	4–5	70–90
Broad bean	70–100	100–150
Broccoli (t)	0.5	80–120
Brussels sprouts (t)	0.5–1.0	110–150
Cabbage (t)	0.5–1.0	70–90
Cape gooseberry (t)	—	80–100
Capsicum (t)	1–2	50–80
Carrot	5–6	70–85
Cassava (leaves)	—	50–70
Cauliflower (t)	0.5–1.0	60–120
Celery (t)	0.5–1.0	80–90
Celosia	1.5–2.0	30–40
Ceylon spinach (t)	10	55–70
Chick pea	25–30	90–150
Chinese cabbage (t)	0.5–1.0	50–80
Chives (t)	—	70–100
Cho-cho (t)	—	100–120
Cocoyam	—	240–300
Collard (t)	1	80–120
Corchorus (t)	5	40–60
Cowpea	12–15	50–70
Cucumber (t)	2.5–4.5	40–80
Dolichos bean	30–40	70–120
Egg plant (t)	0.5–0.8	80–120
Endive (t)	0.5	70–85
French bean (dwarf)	60–100	40–60
French bean (climbing)	25–60	70–90
Garden pea	50–75	60–80
Garlic (t)	—	90–120
Hot pepper (t)	1–2	80–100
Irish potato	—	90–120
Jack bean	25–30	100–120
Jerusalem artichoke	—	80–150
Kale (t)	1	50–85
Kang Kong	5–7	50–70
Kohl-rabi (t)	1.5–2.0	50–70
Leaf mustard (t)	1	50–70
Leek (t)	3–4	120–150
Lettuce (t)	0.5–1.0	60–85
Lima bean	50–80	80–110 (early cvs) / 180–210 (late cvs)
Melon, cantaloupe or sweet (t)	1.5–2.5	80–120
Mung Bean	5–7	50–70
New Zealand spinach (t)	15	80–100 (early cvs)
Okra	8–10	60–180

Table 12 (*Contd.*)

	Column 1 Seed sowing rates	Column 2 Days to maturity		Column 1 Seed sowing rates	Column 2 Days to maturity
Onion, bulb-forming (t)	4–5	100–140	Sweet melon (*see* Melon)		
Onion, bunching	4–5	60–120	Sweet pepper (*see* Capsicum)		
Parsley (t)	3–4	70–100	Sweet potato	—	80–200
Pumpkin (t)	7–8	80–140	Swiss chard (t)	8–12	50–60
Radish	10–12	30–50	Sword bean	25–30	90–150
Roselle	5.5–7.5	100–160	Tannia	—	240–300
Ridged gourd	3–5	40–70	Tomato (t)	0.5	70–100
Shallot	—	60–100	Turnip	3–4	60–70
Snake gourd (t)	4–6	50–70	Vegetable marrow (t)	3–4	80–100
Spinach (*see* Amaranth)			Watermelon	2.5–4.0	80–160
Squash (t)	7–8	80–140	Wax gourd	2–3	80–100
Sponge gourd	3–5	100–120	Winged bean	10–15	60–80
Sweet corn	10–20	65–95	Yam bean	—	150

Note:

1. The seed rates given in the above table are approximate, the actual amount of seed required will vary according to the spacings used, the viability of the seed, and other local factors

2. The days to first harvest given in the table above are also approximate. Variable factors affecting these figures include soil fertility, climatic conditions, cultivar characteristics, and the timely performance of suggested cultural operations used.

Table 13
Suggested fertilizer rates for vegetables

Vegetable	Nitrogen N (kg/ha) (g/10 m²)	Phosphorus P₂O₅ (kg/ha) (g/10 m²)	Potassium K₂O (kg/ha) (g/10 m²)
Leafy vegetables:			
lettuce, cabbage	224	224	224
spinach	168	112	168
Fruit vegetable crops:			
Tomatoes, melons, peppers	112	112	168
Root crops:			
sweet potatoes, carrots, dasheen	168	112	280
Legumes: beans, peas	56	84	56

Note: These rates are recommended as guidelines where specific recommendations or soil fertility data are lacking. Nitrogen should always be applied in two or three split applications at intervals of at least 14 days

Adapted from: M. H. MacVicar, G. L. Budges, and L. B. Nelson (eds) 1968, *Changing Patterns in Fertilizer Use* (Soil Science Society of America, Madison, WI)

Table 14

Pests of vegetables and suggested control measures

Crop	Pests	Scientific name	Description	Damage	Control
(1) Amaranth	Stink bug	*Aspavia armigera*	Spotted yellowish body, black head; 7–8mm long, 4–6mm across	Sucks leaf sap causing small clear spots	Malathion, permethrin, cypermethrin
(2)		*Cletus capensis*	Pale brown bug with white stripe on wings; 7–9mm long, 3mm across	Sucks sap from flowers, affecting seed production; feeds on shoots and leaves	Pyrethrum, malathion
(3)	Leaf caterpillar	*Hymenia recurvalis*	See 9		
(4)		*Psara bipunctalis*	Grey-green caterpillar, about 15mm long	Feeds on the leaves which it rolls within a web	Malathion, or hand picking
(5)	Stem borer	*Lixus trunculatus*	Adult beetle black, covered in down, 7–12mm long	Grub tunnels in the stem base, weakening plants	Burn debris, bromophos, carbaryl, lindane
(6) Asparagus bean	Spotted pod borer	*Maruca testulalis*	See 90		
(7)	Root-knot nematode	*Meloidogyne* spp.	See 119		
(8) Beetroot	Flea beetle	*Phylloreta* spp.	See 118		
(9)	Leaf caterpillar, beet web-worm	*Hymenia recurvalis*	Small smooth caterpillars up to 20mm	Feeds on leaves which may be rolled with a web	Bromophos, carbaryl, lindane
(10) Bitter gourd	Similar to Cucumber				
(11) Bottle gourd	Similar to Cucumber				

Table 14 (*Contd.*)

Crop	Pests	Scientific name	Description	Damage	Control
(12) Brussels sprouts	Similar to Cabbage and Cauliflower				
(13) Bunching onion	Cotton leafworm (See also Onion)	*Spodoptera littoralis*	See 34		
(14)	Beet armyworm	*Spodoptera exigua*	See 76.		
(15)	Onion thrips	*Thrips tabaci*	See 107		
(16) Cabbage	Cabbage moth	*Crocidolomia binotalis*	Yellow-green caterpillars with brown head, then brown to dark green with lighter stripes	Feed on lower leaf surfaces and growing point	Derris, pyrethrum, many organophosphates, *Bacillus thuringiensis* B.T.* cypermethrin
(17)	Aphid	*Aphis* spp.	See 117		
(18)	Diamond back moth	*Plutella xylostella*	See 35		
(19)	Cutworm	*Spodoptera littoralis*	See 34		
(20) Capsicum	Aphid	*Aphis* spp.	See 45		
(21)	Fruitfly	*Ceratitis capitata*	Black, yellow, and white fly with blue-red eyes, 5–6mm	Grubs tunnel in fruit introducing pathogens	Fenthion. Baits of malathion or trichlorphan in sugar solution
(22)	Root-knot nematode	*Meloidogyne* spp.	See 119		
(23)	Thrips	*Scirtothrips dorsalis*	Small (2–2.5 mm) yellow and black downy insects	Suck cell sap, resulting in distorted growth and underdeveloped fruit	Oxydemeton-methyl acephate,
(24) Carrot	Root-knot nematode	*Meloidogyne* spp.	See 119		

Table 14 (*Contd.*)

Crop	Pests	Scientific name	Description	Damage	Control
(25) Cassava	Red spider mite	*Tetranychus cinnabarinus*	*See 46*		
(26)	Tobacco white fly	*Bemisia tabaci*	White, four-winged fly	Larvae and adults suck leaf sap causing leaf fall. Virus vectors	dimethoate or pyrethrum
(27)	Variegated grasshopper	*Zonocerus variegatus*	3.5 cm, dark green with black, yellow and orange markings and an unpleasant smell	Feeds on leaves, especially of seedlings	Dieldrin, carbaryl sprays, aldrin or BHC baits
(28) Cauliflower	Aphid	*Aphis* spp.	*See 117*		
(29)	Cutworm	*Agrotis* spp.	Clay-coloured or blackish caterpillars, 25–40 mm	Cut off seedlings at ground and gnaw roots and leaves	Fenitrothion, diazinon, trichlorphan, tetrachlorvinphos; older larvae difficult to control
	(See also Cabbage)				
(30) Chick pea	Cutworm	*Agrotis* spp.	*See 29*		
(31)	Gram caterpillar	*Heliothis armigera*	*See 140*		
(32)	Root-knot nematode	*Meloidogyne* spp.	*See 119*		
(33) Chinese cabbage	Aphid	*Aphis* spp.	*See 117*		
(34)	Cutworm	*Spodoptera littoralis* (syn. *Prodenia litura*)	Pale green caterpillars, later brown with dark markings and yellow stripes; up to 50 mm	Feed on leaves and young stems just above soil level	BHC, diazinon methoxychlor

Table 14 (*Contd.*)

Crop	Pests	Scientific name	Description	Damage	Control
(35) Chinese cabbage (*contd.*)	Diamond back moth	*Plutella xylostella* (syn. *P. maculipennis*)	Green to brownish green caterpillars 10 mm long. Jump when disturbed *See* 118	Young caterpillars mine within the leaf, older ones feed on lower surface	Derris, pyrethrum, organophosphates, carbaryl, cypermethrin
(36) Chinese yam	Flea beetle	*Phyllotreta* spp.			
(37)	Yam shoot beetle	*Lilioceris livida*	Adult beetles are red-brown to black. The pink elongated eggs are laid in groups on leaves *See* 119	Adults and larvae feed on leaves and shoots	Carbaryl, dichlorvos, malathion
(38)	Root-knot nematode	*Meloidogyne* spp.			
(39)	Yam beetle	*Heteroligus meles*	Adult beetles 20–30 mm long. Brown or black with two conspicuous swellings on head. Larvae cream-white	Larvae feed on plant roots	Late planting. Dusting planting holes with BHC
(40)	Yam nematode	*Scutellonema bradys*	Minute soil organism	Feeds on tuber producing yellow lesions. Damage continues in storage	Clean planting material. Hot water dip 50–55°C for 40 minutes as pre-planting treatment for tubers
(41) Chives	Similar to Onion				
(42) Chives, Chinese	Similar to Bunching onion				
(43) Cluster bean	Aphid	*Aphis* spp.	*See* 45		

Table 14 (*Contd.*)

Crop	Pests	Scientific name	Description	Damage	Control
(44) Cock's comb	Similar to *Amaranthus*				
(45) Cocoyam	Aphid	*Aphis* spp.	2 mm long, green or black insects in clusters on shoots and stems	Suck cell sap causing leaf distortion and wilting	Disulfoton, pirimicarb phorate, menazon, demeton-s-methyl
(46)	Red spider mite	*Tetranychus* spp.	0.5 mm red or greenish spider-like insects on leaf underside	Suck cell sap, causing minute yellow spots on leaf	Dimethoate and other organophosphates
(47)	Root-knot nematode	*Meloidogyne* spp.	See 119		
(48)	Cutworm	*Spodoptera littoralis*	See 34		
(49) Cucumber	Aphid	*Aphis* spp.	See 117		
(50)	Epilachna beetle	*Epilachna* spp.	See 143		
(51)	Melon fruitfly	*Dacus cucurbitae*	Similar to *D. ciliatus* under Pumpkin, *see* 116		
(52)	Root-knot nematode	*Meloidogyne* spp.	See 119		
(53) Dolichos bean	Aphid	*Aphis* spp.	See 117		
(54)	Bollworm, gram caterpillar	*Heliothis armigera*	See 140		
(55)	Root-knot nematode	*Meloidogyne* spp.	See 119		
(56)	Spotted pod borer	*Maruca testulalis*	See 90		
(57) Egg-plant	Bollworm, corn earworm, gram caterpillar	*Heliothis armigera*	See 140		

Table 14 (*Contd.*)

Crop	Pests	Scientific name	Description	Damage	Control
(58) Egg-plant (*contd.*)	Cutworm	*Spodoptera littoralis* (syn. *Prodenia litura*)	*See* 34		
(59)	Epilachna beetle	*Epilachna hirta*	*See* 143		
(60)	Red spider mite	*Tetranychus* spp.	*See* 46		
(61)	Root-knot nematode	*Meloidogyne* spp.	*See* 119		
(62) Fennel	Rarely serious				
(63) French or kidney bean	Aphid	*Aphis* spp.	*See* 117		
(64)	Bean bruchid	*Acanthoscelides obtectus*	4.5mm long beetle. Curved, legless, off-white or yellow grub with brown head	The grub feeds in developing seeds	Carbaryl, diazinon lindane and pyrethrum dust
(65)	Epilachna beetle	*Epilachna* spp.	*See* 143		
(66)	Red spider mite	*Tetranychus urticae* and *T. telarius*	*See* 46		
(67) Garden pea	Cutworm	*Agrotis* spp.	*See* 29		
(68)	Pea weevil	*Bruchus pisorum*	5–6mm long black beetle, white speckled. Creamy white legless grub with brown head	Grub feeds on developing seeds	Seed fumigation (carbon disulphide or carbon tetrachloride)
(69)	Root-knot nematode	*Meloidogyne* spp.	*See* 119		
(70) Garlic	Similar to Onion				
(71) Hot pepper	Similar to *Capsicum annuum*				

Table 14 (*Contd.*)

Crop	Pests	Scientific name	Description	Damage	Control
(72) Irish potato	as for Tomato, plus:	Aphid	Aphis spp.	See 117	
	Cutworm	Agrotis ipsilon	See 29		
(73)					
(74) Indian or Ceylon spinach	Rarely serious. Root-knot nematode infestation may occur.	See 119			
(75) Jack bean; horse bean	Similar to Sword bean				
(76) Jute mallow; long-fruited jute	Beet armyworm	Spodoptera exigua	Caterpillar light green and striped, up to 50mm long	Young caterpillars feed on lower leaf surface, older larvae on entire leaf	Burn debris; B.T.* BHC, cypermethrin, parathion
(77)	Leaf beetle	Podagrica spp.	P. sjostedti is metallic blue with brown head and body; P. uniforma is shiny brown	Adults chew holes in leaves; larvae feed on roots	Tolerant cultivars
(78)	Root-knot nematode	Meloidogyne spp.	See 119		
(79)	Sweet potato butterfly	Acraea terpsichore	Caterpillar orange greenish-black and spiny. Butterfly orange	Caterpillar eats leaves under a web	Cypermethrin, permethrin, B.T.
(80) Kale or collard	Similar to other Brassicas, e.g. Cabbage, Cauliflower				

Table 14 (*Contd.*)

Crop	Pests	Scientific name	Description	Damage	Control
(81) Kang kong	Similar to Sweet Potato				
(82) Katuk	Root-knot nematode	*Meloidogyne* spp.	*See* 119		
(83) Kohl rabi	Similar to Cabbage and Cauliflower				
(84) Leaf mustard	Cabbage moth	*Crocidolomia binotalis*	*See* 14		
(85)	Cabbage budworm	*Hellula undalis*	Creamy-white, brown striped caterpillar (15mm)	Tunnels and feeds on leaves, stalks, growing points and shoots. Young plants killed, older ones spoiled	Dichlorvos, trichlorphon, malathion, carbaryl
(86) Leek	Similar to Onion				
(87) Lettuce	Aphid	*Aphis* spp.	*See* 45		
(88)	Root-knot nematode	*Meloidogyne* spp.	*See* 119		
(89)	White fly	*Bemisia tabaci*	*See* 26		
(90) Lima bean	Bean pod borer, spotted borer	*Maruca testulalis*	15mm caterpillars, yellow, greenish or reddish-white with dark spots	Feed on flowers, affecting pod development	Diazinon, lindane, carbaryl
(91)	Root-knot nematode	*Meloidogyne* spp.	*See* 119		
(92)	Stem-boring beetle	*Sagra adonis*	*See* 139		
(93)	Variegated grasshopper	*Zonocerus variegatus*	*See* 27		

Table 14 (*Contd.*)

Crop	Pests	Scientific name	Description	Damage	Control
(94) Melon	Similar to Cucumber and Pumpkin				
(95) Mung bean	Pod borer	*Heliothis* spp.	*See* 140		
(96)	Root-knot nematode	*Meloidogyne* spp.	*See* 119		
(97) Mustard	*See* Leaf mustard,				
(98) Okra	Cotton leaf caterpillar, cotton leafworm	*Xanthodes graellsi*	Dark green semi-looper caterpillar with a yellow stripe and black hairs; 30 mm	Feeds on and kills seedlings; strips leaves of older plants and damages pods	cypermethrin, carbaryl, malathion, mevinphos
(99)	Cotton stainer	*Dysdercus superstitiosus*	*See* 120		
(100)	Leaf-eating beetle	*Lagria villosa*	*See* 115		
(101)	Pink bollworm	*Pectinophora gossypiella*	Caterpillar at first white; later develops red banding	Feeds on flower buds and seeds	Carbaryl, cypermethrin, diazinon
(102)	Plant bug	*Mirperus jaculus*	Black bug, 22–25 mm long, 6–8 mm across	Feeds on stems and petioles, causing wilt	Control rarely necessary
(103)	Spiny bollworm	*Earias insulana* and *E. biplaga*	*See* 123	Punctures and deforms pods	Remove affected pods
(104) Onion	Beet army worm	*Spodoptera exigua*	*See* 76		
(105)	Onion fly	*Hylemya antigua*	Adult flies grey, 6 mm long. Eggs laid in soil	Larvae bore into stems and bulbs	Diazinon, ethion, fensulfothion

Table 14 (*Contd.*)

Crop	Pests	Scientific name	Description	Damage	Control
(106) Okra (*contd.*)	Stem and bulb nematode	*Ditylenchus dipsaci*	Minute soil pest	Feeds on leaves and bulbs. Seedlings killed, older plants stunted; leaf tips turn yellow	Long rotations, soil fumigation
(107)	Thrips	*Thrips tabaci*	1 mm, yellow-brown with dark bands	Feeds on cell sap, causing silvering, distortion and wilting	Malathion, diazinon, acephate, oxydemeton-methyl
(108) Parsley	Root-knot nematode	*Meloidogyne* spp.	*See 119*		
(109) Potato	Similar to Tomato				
(110)	Aphid	*Aphis* spp.	*See 45*		
(111)	Cutworm	*Agrotis* spp.	*See 29*		
(112) Pumpkin	Aphid	*Aphis* spp.	*See 117*		
(113)	Cucurbit leaf beetle	*Aulacophora* spp.	Yellow beetle. Grub orange to yellow	Adult beetle shreds leaves. Grub feeds on roots and fruit	Dichlorovos, trichlorphon, fenthion, thiodan, carbaryl
(114)	Leaf beetle	*Copa* spp.	*See 126*		
(115)	Leaf-eating beetle	*Lagria villosa*	Dark metallic brown, downy beetle, 12 mm long	Feeds on leaves	Malathion, carbaryl, cypermethrin
(116)	Lesser melon fly	*Dacus ciliatus*	Large brown fly 8–10 mm long. Larvae 10–12 mm	Larvae tunnel in the fruit, encouraging rot	Fenthion, malathion + sugar-baited sprays against adults
(117) Radish	Aphid	*Aphis* spp.	1.2 mm long, yellowish green to black, in colonies	Suck sap, causing leaf distortion and wilting	Disulfoton, pirimicarb, phorate, menazon, demeton-S-methyl, oxydemeton-methyl, or derris dust

Table 14 (*Contd.*)

Crop	Pests	Scientific name	Description	Damage	Control
(118)	Flea beetle	*Phyllotreta* spp.	1.5–3 mm jumping insect, metallic black	Feeds on seedling and young plant leaves, producing small holes	Seed dressing with BHC, sprays with derris or pyrethrum
(119)	Root-knot nematode	*Meloidogyne* spp.	Minute (about 1 mm) soil pests	Stimulates root gall formation; disrupts plant water supply	Crop rotation with a non-host plant. Resistant cultivars
(120) Roselle	Cotton stainer	*Dysdercus superstitiosus*	Reddish bug with white and black markings; 14–18mm long	Feeds on fruits and seeds	Carbaryl, cypermethrin
(121)	Flea beetle	*Podagrica* spp.	*See* Jute mallow, 76		
(122)	Root-knot nematode	*Meloidogyne* spp.	*See* 119		
(123)	Spiny bollworm	*Earias biplaga, E. insulana*	Fat, brown spiny caterpillar with yellow spots	Bores into unripe fruit and feeds on seeds	Carbaryl on early larvae, cypermethrin
(124) Shallot	Similar to Onion				
(125) Snake gourd	Cucurbit leaf beetle	*Aulacophora* spp.	*See* 113		
(126)	Leaf beetle	*Copa* spp.	5–6mm; *C. delata* is ivory-grey with yellow head and thorax	Adult feeds on leaves	Dichlorvos, trichlorphon, fenthion, carbaryl, endosulfan, cypermethrin
(127)	Leaf-eating beetle	*Lagria villosa*	*See* 115		
(128) Sponge gourd	Fruitfly	*Dacus* spp.	*See* 116		
(129) Sunset hibiscus	Rarely serious				

Table 14 (*Contd.*)

Crop	Pests	Scientific name	Description	Damage	Control
(130) Sweet corn	African army worm	*Spodoptera exempta* (syn. *Laphygma exempta*)	Pale green caterpillar, later brownish black with stripes	Feeds on the leaves, particularly young leaves and plants	Endosulfan, malathion, trichlorphon, carbaryl
(131)	Cotton earworm	*Heliothis armigera*	See 140		Remove crop debris.
(132)	Maize stalk borer	*Busseola fusca*	Larvae pink with blue spots	Caterpillars feed on leaves leaving upper surface intact	Apply malathion to leaf sheaths
(133)	Maize weevil	*Sitophilus oryzae*	Dark brown weevil 3.5–4mm. Off-white, plump, legless larva	Grub feeds within stored grain	Spray empty store with BHC, pirimiphos, or malathion. Grain can be mixed with malathion wettable powder
(134) Sweet potato	Root-knot nematode	*Meloidogyne* spp.	See 119		
(135)	Stem borer	*Megastes grandalis*	Adult moth dark brown with wingspan 25–30mm. Eggs laid in leaf axils	Caterpillars bore into stems and eventually feed on tubers	Carbaryl, chlorvinphos malathion or methomyl
(136)	Sweet potato hawk moth	*Agrius convolvuli*	Caterpillar up to 95mm, green or brown with a posterior horn	Eats leaves	Malathion, B.T.*, cypermethrin
(137)	Sweet potato weevil	*Cylas formicarius* and *C. brunneus*	6–8mm long black weevil with brown legs	Adult eats leaves and stem, larvae bore into tuber	Crop rotation. Use shoot tips as planting material
(138) Swiss chard	Similar to Beetroot				

Table 14 (*Contd.*)

Crop	Pests	Scientific name	Description	Damage	Control
(139) Sword bean	Stem-boring beetle	*Sagra* spp.	*S. carbunculus* is metallic blue with long hind legs, 20mm	Adult beetles defoliate young plants. White grubs tunnel in the stem causing galls	carbaryl, cypermethrin, diazinon
(140) Tomato	Bollworm, corn earworm, gram caterpillar	*Heliothis* spp.	Brown or greenish caterpillar with pale and dark stripes. Up to 40mm long	Feeds on fruit	Carbaryl, B.T.*, cypermethrin malathion or mevinphos
(141)	Cotton leaf roller	*Sylepta derogata*	Translucent green caterpillars with black heads	Feed on the leaf edges and roll them within a web	Carbaryl, cypermethrin; handpicking
(142)	Cutworm, click beetle	*Spodoptora littoralis* (syn. *Prodenia litura*)	See 34		
(143)	Epilachna beetle	*Epilachna* spp.	Oval beetle, red to yellowish brown with black spots, 6–8mm long. Yellow spine-covered larva	Feeds on leaves and gnaws stems of fruit	Malathion, phorate, carbaryl
(144)	Leaf hopper	*Empoasca* spp.	Pale green, wedge-shaped, 2.5mm long	Sucks sap on underside of leaves	Monocrotophos disulfoton, phosphamidon, formothion, dimethoate, and carbaryl
(145)	Mite	*Polyphagotarsonemus latus* (syn. *Hemitarsonemus latus*)	1.5mm, yellow spider-like pest	Sucks sap on underside of leaves causing distortion and mottling	Dicofol
(146)	Red spider mite	*Tetranychus* spp.	See 46		

Table 14 (*Contd.*)

Crop	Pests	Scientific name	Description	Damage	Control
(147) Tomato (*contd.*)	Root-knot nematode	*Meloidogyne* spp.	*See* 119		
(148) Turnip	Similar to Radish				
(149) Vegetable marrow	Similar to Pumpkin and Cucumber				
(150) Watermelon	Similar to Cucumber				
(151) Wax gourd	Aphid	*Aphis* spp.	*See* 117		
(152)	Fruitfly	*Dacus* spp.	*See* 116		
(153)	Root-knot nematode	*Meloidogyne* spp.	*See* 119		
(154) Winged bean	Similar to Lima bean				

*B.T. = Bacillus thuringiensis

Table 15

Diseases of vegetables and suggested control measures

Crop	Disease	Scientific name	Damage	Control
(1) Amaranth	Damping-off	*Pythium aphanidermatum*	Seedlings appear water-soaked at ground level and collapse	Preventative cultural control: avoid dense planting and overwatering. Dig in organic material to encourage antagonistic organisms
(2)	Leaf spot	*Cercospora beticola*	Small yellow or brown spots on leaves causing older leaves to fall	Copper salts, maneb, thiabendazole
(3)	Leaf and stem wet-rot	*Choanephora cucurbitarum*	Seedlings and weaker plants rot as they become covered in grey mould	Less sensitive cultivars. Cultural methods to promote vigorous growth
(4)	Leaf spot	*Cladosporium variabile*	Older leaves develop circular spots with yellow margins	Maneb, mancozeb plus daconil
(5) Asparagus bean	Anthracnose	*Colletotrichum lindemuthianum*	*See* 133	
(6)	Bean yellow mosaic virus		*See* 72	
(7)	Common bean mosaic virus		*See* 73	
(8) Beetroot	Leafspot	*Cercospora* spp.	*See* 33	
(9)	Bacterial soft rot	*Erwinia carotovora*	*See* 28	

Table 15 (*Contd.*)

Crop	Disease	Scientific name	Damage	Control
(10) Beetroot (*contd.*)	Downy mildew	*Peronospora parasitica*	Seedlings may be killed. Young plants develop irregular brown or white leaf spots with white mycelia on the lower surface	Bordeaux mixture, copper oxychloride, dithiocarbamates
(11)	Powdery mildew	*Albugo candida*	Raised white spots appear mainly on the upper leaf surfaces, resulting in stunted and distorted growth	Remove crop debris, use long rotations
(12) Bitter gourd	Similar to Cucumber			
(13) Bottle gourd	Similar to Cucumber			
(14) Broad bean	Broad bean rust	*Uromyces fabae*	Fawn-coloured pustules develop on both sides of the leaf	Not usually necessary
(15) Brussels sprouts	Similar to Cabbage and Cauliflower			
(16) Bunching onion	Similar to Onion			
(17) Cabbage	*Alternaria* or leaf blight	*Alternaria brassicae*	Brown-black spots with a yellow margin develop on older leaves, stems and petioles. Infected seed may result in 80–100% seedling death	Hot-water treatment of seed at 50°C for 25 min. Iprodione, Bordeaux mixture, sulphur dust, dithiocarbamates

Table 15 (*Contd.*)

Crop	Disease	Scientific name	Damage	Control
(18)	Black leg fungus	*Leptosphaeria maculans* (= *Phoma lingam*)	Infected seeds give rise to seedlings with red-brown spots on the leaves; seedlings may die. On older plants pale spots appear at the stem base and the root decays	Clean seed (hot water treatment or seed soak of thiabendazole, thiram, benomyl)
(19)	Black rot	*Xanthomonas campestris*	Blackening of the vascular tissue causing restriction of water supply. Yellow V-shaped spots on leaves	Seed disinfection. Crop rotation. Remove infected plants. Grow crops in the cool season
(20)	Cercospora leaf spot	*Cercospora brassicicola*	Small yellow, brown, or red-rimmed spots on leaves, from which the centres later fall away	Copper salts, maneb, thiabendazole
(21)	Club root	*Plasmodiophora brassicae*	Soil-borne infection which causes root galls, disturbing plant water supply and stunting growth *See 10*	Liming soil, PCNB compounds applied in planting hole
(22)	Downy mildew	*Peronospora parasitica*		
(23) Capsicum	Anthracnose	*Colletotrichum nigrum*	Dark sunken spots on the fruit. The disease enters through wounds	Clean seed. Crop rotation. Removing infected fruit
(24)	Bacterial wilt	*Pseudomonas solanacearum*	*See 136*	
(25)	Leaf spot, fruit rot	*Colletotrichum capsici*	Large grey areas bordered with black and dotted with black spores develop on lower leaves	Not generally necessary

Table 15 (*Contd.*)

Crop	Disease	Scientific name	Damage	Control
(26) Capsicum (*contd.*)	Powdery mildew	*Leveillula taurica*	Yellow spots on the upper surfaces of older leaves and white powder on the undersides	Less susceptible cultivars. Grow young crops apart from other host crops. Sulphur (may cause scorch in warm dry conditions)
(27)	Virus diseases	*See* 145–147		
(28) Carrot	Bacterial soft rot	*Erwinia carotovora*	Roots rot rapidly in storage and have an unpleasant smell. The soil-borne bacteria enter through wounds	Crop rotation. Store only undamaged roots in cool conditions
(29)	Leaf spot, leaf blight	*Alternaria dauci*	Irregular brown spots with yellow centres develop on older leaves late in the season	Clean seed. Dig in crop debris
(30) Cassava	Anthracnose	*Glomerella manihotis*	Spotting and distortion of leaves; death of young stems and cankers on older stems. Prevalent on young shoots after prolonged rain	Healthy planting material. Avoid planting in the rainy season
(31)	Bacterial blight	*Xanthomonas manihotis*	Developing shoots wilt and die with gummy exudate. Angular leaf spotting followed by rolling of leaves and defoliation	Use of resistant cultivars and healthy planting material. Removal of diseased shoots and six-month fallow. Disinfect knives used in pruning

Table 15 (*Contd.*)

Crop		Disease	Scientific name	Damage	Control
(32)		Brown steak virus		Mottling of leaves and streaking of stems	Healthy material. Remove infected plants
(33)		Leaf spot	*Cercospora* spp.	Small yellow or brown leaf spots which may join together and cause leaf fall	Copper salts, maneb, thiabendazole
(34)		Mosaic virus		Stunting and mottling of leaves	Healthy planting material. Resistant cultivars
(35)	Cauliflower	Root rot	*Sclerotium rolfsii*	*See* 98	
(36)		Black rot	*Xanthomonas campestris*	*See* 19	
(37)		Club root	*Plasmodiophora brassicae*	*See* 21	
(38)		Damping-off	*Pythium aphanidermatum* (= *P. ultimatum*)	*See* 1	
(39)		Stem rot	*Thanatephorus cucumeris* (= *Rhizoctonia solani*)	'Damping-off' of seedlings. In older plants the cortical tissue of the stem is destroyed	Cultural control: avoid deep planting and waterlogged soil, burn debris
		See also Cabbage diseases			
(40)	Chick pea	Fusarium root rot	*Fusarium oxysporum*	*See* 61	
(41)		Root rot	*Thanatephorus cucumeris*	*See* 39	
(42)		Wilt	*Verticillium albo-atrum*	Stunting of plants, leaves small, distorted and dark green. Internal stem discolouration normally grey-brown	Avoid infected soil, crop rotation, use of disease free and resistant cultivars
(43)	Chinese cabbage	Black rot	*Xanthomonas campestris*	*See* 19	
(44)		Cercospora leaf spot	*Cercospora brassicicola*	*See* 20	
(45)		Downy mildew	*Peronospora parasitica*	*See* 10	

Table 15 (*Contd.*)

Crop	Disease	Scientific name	Damage	Control
(46) Chinese yam	Anthracnose	*Glomerella cingulata*	Small brown spots, sometimes with yellow edge, on leaves and stems which cause blackening and withering	Removal of crop debris, resistant cultivars or maneb or benomyl plus propineb
(47)	Leaf mosaic		Leaf mottling, vein clearing, stunted and bushy growth. Transmission by mechanical means or aphids	Use of disease-free planting material and rogueing of infected plants
(48)	Tuber rot	*Botryodiplodia theobromae*	*See* 121	
(49)	Tuber rot	*Fusarium oxysporum*	*See* 148	
(50)	Tuber soft rot	*Thanatephorus cucumeris*	*See* 39	
(51) Chives	Similar to Onion			
(52) Chives, Chinese	Similar to Onion			
(53) Cluster bean	Anthracnose	*Colletotrichum* spp.	*See* 133	
(54)	Powdery mildew	*Erysiphe polygoni*	*See* 78	
(55)	Powdery mildew	*Leveillula* spp.	*See* 26	
(56) Cock's comb	Similar to Amaranth			
(57) Cocoyam	Leaf spot	*Phyllosticta* spp.	*See* 119	
(58)	Tuber rot	*Sclerotium rolfsii*	*See* 98	
(59) Cucumber	Cucumber mosaic virus		Green and yellow mottled leaves and fruit. Fruits often distorted and sometimes white	Control of aphid vector. Plant resistant cultivars. Remove infected plants
(60)	Downy mildew	*Pseudoperonospora cubensis*	*See* 112	

Table 15 (*Contd.*)

Crop	Disease	Scientific name	Damage	Control
(61)	Fusarium wilt	*Fusarium oxysporum*	Soil-borne disease infecting through root wounds and invading the vascular system. Seedlings may rot; older plants wilt, yellow and may die	Crop rotation. Resistant cultivars. Benomyl, carbendazim
(62)	Powdery mildew	*Erysiphe cichoracearum*	White powdery spots on upper and lower leaf surfaces, spreading from older to younger leaves. Fruits ripen prematurely and lack flavour *See 133*	Sulphur dust on sulphur-tolerant cultivars. Dinocap imazalil, benomyl
(63) Dolichos bean	Anthracnose	*Colletotrichum lindemuthianum*		
(64)	Common blight	*Xanthomonas phaseoli*	Small brown spots on leaves, stalks, pods and seeds; leaves tear and become ragged	Clean seed. Resistant cultivars
(65)	Halo blight	*Pseudomonas phaseolicola*	*See 74*	
(66)	Powdery mildew	*Leveillula taurica*	*See 26*	
(67) Egg-plant, garden egg	Bacterial wilt	*Pseudomonas solanacearum*	*See 136*	
(68)	Fruit rot	*Phytophthora parasitica*	Dark water-soaked spots on fruits. Mature fruits may rapidly brown and rot	Adequate spacing. Remove infected debris. Copper fungicides
(69)	Fusarium wilt	*Fusarium oxysporum*	*See 75*	Resistant cultivars. Long rotations. Maneb, zineb
(70)	Phomopsis rot	*Phomopsis vexans* (= *Diaporthe vexans*)	Blight and canker of stems, leaves and fruits	

Table 15 (*Contd.*)

Crop	Disease	Scientific name	Damage	Control
(71) French or kidney bean	Anthracnose	*Colletotrichum lindemuthianum*	*See 133*	
(72)	Bean yellow mosaic virus		Similar to common bean mosaic, but yellow mottling is more pronounced. Not seedborne.	Less susceptible cultivars
(73)	Common bean mosaic virus		Chlorotic, mottled and distorted leaves, stunted growth. Seed-transmitted, also insect vectors	Resistant cultivars
(74)	Halo blight	*Pseudomonas phaseolicola*	Brown water-soaked leaf spots develop chlorotic halo. Infected pods contain shrivelled seed. Reddish stem lesions can cause plant to collapse, slimy cream exudate	Clean seed. Crop rotation
(75)	Root rot	*Fusarium oxysporum*	Soil-borne infection which attacks the root vascular system. Root cortex and stem base blacken and rot. Plants are chlorotic and stunted	Crop rotation. Avoid infested land
(76) Garden pea	Ascochyta blight, leaf spot	*Ascochyta pisi*	Large sunken brown lesions on leaves and pods	Disease-free seed. Treat seed with benomyl. Burn crop debris. Minimum 3-year rotation

Table 15 (*Contd.*)

Crop	Disease	Scientific name	Damage	Control
(77)	Blight, leaf or pod spot	*Mycospaerella pinodes*	Numerous small purple-black spots develop on all parts of plant	Disease-free seed. 3-year crop rotation. Dig in crop debris
(78)	Powdery mildew	*Erysiphe polygoni*	Powdery areas usually on upper leaf surfaces only and on older leaves first	Dinocap, thiabendazole benomyl
(79) Garlic	Similar to Onion			
(80) Hot pepper	Similar to Capsicum			
(81) Indian spinach	Leaf spot	*Cercospora* spp.	*See* 33	
(82) Irish potato	Similar to Tomato			
(83) Jack or horse bean	Similar to Sword bean			
(84) Jute mallow	*See* Long-fruited jute			
(85) Kale or collard	Similar to Cabbage and Cauliflower			
(86) Kang kong	Similar to Sweet Potato			
(87) Kohlrabi	Similar to Cabbage and Cauliflower			
(88) Leek	Similar to Onion			
(89) Lettuce	Bottom rot	*Thanatephorus cucumeris* (= *Rhizoctonia solani*)	'Damping-off' of seedlings, older plants develop sunken brown spots on midribs and petioles of lower leaves which eventually rot	Cultural control: avoid deep planting and waterlogged soil, burn debris
(90)	Cercospora leaf spot	*Cercospora lactucae*	Small yellow or brown spots on leaves, eventually joining together and causing older leaves to fall	Copper salts, maneb, thiabendazole

Table 15 (*Contd.*)

Crop	Disease	Scientific name	Damage	Control
(91) Lettuce (*contd.*)	Downy mildew	*Bremia lactucae*	Pale green or yellow angular spots develop between leaf veins. White masses of spores on the lower surfaces which turn brown and eventually rot	Resistant cultivars, zineb, maneb, metalaxyl
(92)	Lettuce mosaic virus		Young affected plants are stunted with mottled leaves which may become distorted. Generally uneven growth.	Remove alternate host weeds. Use disease-free seed and resistant cultivars. Control aphids
(93)	Root rot	*Sclerotium rolfsii*	*See* 98	
(94) Lima bean	Common blight	*Xanthomonas phaseoli*	*See* 64	Use clean seed. 2–3-year crop rotation.
(95)	Downy mildew	*Phytophthora phaseoli*	Patches of white cottony growth develop on the pods which may dry and shrivel	Copper fungicides
(96)	Halo blight	*Pseudomonas phaseolicola*	*See* 74	
(97)	Root rot	*Thanatephorus cucumeris*	*See* 89	
(98) Long-fruited jute	Wilt	*Sclerotium rolfsii*	'Damping-off of seedlings; older plants may develop collar rot, yellow, wilt and die	Remove crop debris. Deep cultivate to bury sclerotia (resting spores)
(99) Melon	Similar to Cucumber and Pumpkin			
(100) Mint	Mint rust	*Puccinia menthae*	Yellow or brown pustules on stems and leaves, later turning dark brown	Bordeaux mixture, sulphur dust
(101) Mung bean	Anthracnose	*Colletotrichum lindemuthianum*	*See* 133	

Table 15 (*Contd.*)

Crop	Disease	Scientific name	Damage	Control
(102)	Collar rot	*Sclerotium rolfsii*	*See* 137	
(103)	Halo blight	*Pseudomonas phaseolicola*	*See* 74	
(104)	Leaf spot	*Cercospora* spp.	*See* 33	
(105)	Powdery mildew	*Erysiphe polygoni*	*See* 78	
(106)	Root rot	*Fusarium oxysporum*	*See* 75	
(107) Okra	Powdery mildew	*Erysiphe polygoni*	*See* 78	
(108)	Yellow vein mosaic virus		Yellow, mottled and distorted leaves	Remove infected plants
(109) Onion	Downy mildew	*Pseudoperonospora destructor*	Necrotic spots near leaf tips and blue-grey mildew in humid weather. Tops die back and growth of the bulb is affected	Clean seed. 4-year crop rotation. Remove crop debris. Maneb, zineb or nabam plus zinc sulphate
(110)	Onion smudge, anthracnose	*Colletotrichum circinans*	Black lesions on the outer scales of maturing bulbs, particularly at high temperatures	Resistant cultivars. Rapid drying of harvested crop
(111)	Smut, black soft rot	*Urocystis cepulae*	Only young plants are susceptible. Leaves show black spots, collapse and the plant dies. Black spore masses erupt near the base of bulb scales on surviving plants	Treat seed with thiram. Methyl bromide soil sterilization

Table 15 (*Contd.*)

Crop	Disease	Scientific name	Damage	Control
(112) Pumpkin	Downy mildew	*Pseudoperonospora* spp.	Bright yellow spots on upper leaf surfaces, blue-grey mould on lower surfaces. Fruit development and quality is impaired. In cucumbers affected plants may die	Grow young plants away from older crops. Adequate spacing. Avoid overhead irrigation. Dithiocarbamates and copper sprays. Some cultivars are copper-sensitive
(113)	Leaf and stem wet-rot	*Choanephora cucurbitarum*	*See 3*	
(114)	Leaf blight, brown spot	*Alternaria cucumerina*	Can defoliate plants and cause sunken spots on ripe fruit	Clean seed
(115) Radish	Downy mildew	*Peronospora parasitica*	*See 10*	
(116)	Leaf spot	*Cercospora brassicicola*	Small yellow or brown spots on the leaves. Centres may fall out as the spots enlarge	Copper salts, maneb, thiabendazole
(117) Roselle	Anthracnose	*Colletotrichum hibisci*	Sunken lesions or spots on calyces	Clean seed. Dithiocarbamates are recommended for other *Colletotrichum* spp. Adequate spacing. Remove infected fruits. Copper fungicides
(118)	Fruit rot	*Phytophthora parasitica*	Calyces develop white, cottonlike growths in patches, especially in humid conditions	
(119)	Leaf blight	*Phyllosticta hibisci*	Irregular-shaped spots on leaves enlarge and the dead tissue in the centre falls away	Not usually necessary

Table 15 (*Contd.*)

Crop	Disease	Scientific name	Damage	Control
(120) Shallot	Similar to Onion			
(121) Snake gourd	Storage or black rot	*Botryodiplodia theobromae*	A thick white mat of hyphae grows on the fruit surface and internal tissue decays	Avoid damage to fruits, store in well-ventilated conditions
	Other diseases similar to those of Cucumber			
(122) Sponge gourd	Downy mildew	*Pseudoperonospora cubensis*	*See* 112	
(123)	Powdery mildew	*Erysiphe cichoracearum*	*See* 62	
(124) Sweet corn	Corn smut	*Ustilago maydis*	Irregular white galls on leaves and ears, growing to a large size before releasing a mass of black spores	Less susceptible cultivars
(125)	Dry rot of ear, stalk or cob	*Diplodia macrospora*	Infected cobs have bleached sheathing leaves and whitish grey mould on the grains	Resistant cultivars
(126)	Maize rust	*Puccinia polysora*	Pustules occur on leaves and leaf sheaths, releasing orange-yellow spores	Resistant cultivars
(127)	Streak virus		Narrow chlorotic streaks on young leaves. Leaf-hopper vector (*Cicadulina* spp.)	Control of vector

Table 15 (*Contd.*)

Crop	Disease	Scientific name	Damage	Control
(128) Sweet potato	Black rot	*Ceratocystis fimbriata*	Growing plants and stored tubers affected. Young plants yellow and the underground stem blackens. Dark hollows form on the tubers which dry and shrivel. Toxins are produced which are not destroyed in cooking *See* 137	Resistant cultivars. Disease-free planting material. Crop rotation
(129)	Sclerotial wilt	*Sclerotium rolfsii*		
(130)	Soft rot	*Rhizopus nigricans* and *R. stolonifer*	Soft watery storage rots which normally occur only on damaged tubers	Avoid tuber damage. Cure tubers before storage. Destroy infected tubers
(131)	Storage or black rot	*Botryodiplodia theobromae*	Damaged tubers begin to develop a white mat of fungal threads which invade the internal tissues, causing decay	Avoid tuber damage. Store in cool, ventilated conditions. Treat surface of setts with benomyl or thiabendazole
(132) Swiss chard	Similar to Beetroot			
(133) Sword bean	Anthracnose, root rot	*Colletotrichum lindemuthianum*	All aerial parts are attacked especially the pods. Orange spots enlarge to form dark brown sunken areas	Clean seed. Resistant cultivars
(134) Tannia	Root rot	*Thanatephorus cucumeris*	*See* 89	
(135)	Tuber rot	*Sclerotium rolfsii*	*See* 98	

Table 15 (*Contd.*)

Crop	Disease	Scientific name	Damage	Control
(136) Tomato	Bacterial wilt	*Pseudomonas solanacearum*	Soil-borne bacteria which infect through the roots and invade the vascular tissue affecting plant water supply. Plants are stunted, wilt easily and normally die	Crop rotation. Resistant cultivars
(137)	Collar rot	*Sclerotium rolfsii*	'Damping-off' of seedlings and collar rot of older plants. Seedlings die rapidly, older plants turn yellow, wilt and die. White hyphae cover affected parts.	Remove infected plants and burn. Remove infected debris. Deep cultivate (30cm) to bury sclerotia (resting spores)
(138)	Cucumber mosaic virus		Slight leaf mosaic and a narrowing of the leaves ('fern leaf' symptom). Aphid vector	Remove and burn infected plants; resistant cultivars Control of vector.
(139)	Early blight	*Alternaria solani*	Brownish-black angular spots on leaves. Toxins secreted. Yields severely reduced	Remove debris. Hot-water treat seeds at 50°C for 25 min. Crop rotation. Carbamates and chlorothalonil give some control, daconil
(140)	Grey leaf spot	*Stemphylium solani*	Irregular brown to black spots with grey centres on leaves and stems. Leaves yellow and fall. Yields significantly reduced	Resistant cultivars. Spray seed bed with dithiocarbamates such as maneb

Table 15 (*Contd.*)

Crop	Disease	Scientific name	Damage	Control
(141) Tomato (*contd.*)	Late blight	*Phytophthora infestans*	Necrotic spots on leaves become brown and water-soaked. White downy patches form on lower surfaces and the leaves fall. Large green-brown blotches may occur on fruits	Semi-resistant cultivars. Remove infected debris. Maneb, copper fungicides, daconil
(142)	Leaf mould	*Cladosporium fulvum*	Older leaves develop yellow spots which turn brown and enlarge causing leaf fall. Beige to purple spores form on the under-sides. Petioles, flowers and stems may also be affected	Resistant cultivars. Mancozeb
(143)	Septoria leaf spot	*Septoria lycopersici*	Water-soaked spots with grey centres, black borders and distinct yellow haloes occur on oldest leaves first, causing defoliation	Maneb, zineb
(144)	Target leaf spot	*Corynespora cassicola*	Dark brown spots with darker concentric rings on leaves. Yields severely reduced.	Mancozeb
(145)	Tomato bunchy top virus		Bunching of terminal leaves	Remove and burn infected plants
(146)	Tomato double streak virus		Mottling of leaves, streaks on stem and petioles, leaf fall, distorted fruits, stunted growth. Mechanical transmission	Remove and burn infected plants. Control vector

Table 15 (*Contd.*)

Crop	Disease	Scientific name	Damage	Control
(147)	Tomato mosaic virus		Leaves become mottled with raised dark green patches and young leaves may be distorted	Soil sterilization, resistant cultivars and prevention of mechanical transmission from the hands of workers who smoke
(148)	Tomato wilt	*Fusarium oxysporum*	Soil-borne fungus. Enters through root wounds and invades the plant vascular system, producing toxins. Leaves wilt, yellow and the plant normally dies	Crop rotation. Resistant cultivars. Prevent root damage
(149) Turnip	Similar to Cabbage and Cauliflower			
(150) Vegetable marrow	Similar to Cucumber and Pumpkin			
(151) Watermelon	Similar to Cucumber and Pumpkin			
(152) Winged bean	Similar to Lima bean			
(153)	Orange gall or false rust	*Synchytrium psophocarpi*	Small light green watersoaked pimples mainly on lower leaf surfaces, but also on petioles, vines and pods. Galls turn orange as spores form. *See 73*	Remove infected debris. Copper fungicides
(154) Yam bean	Mosaic virus			

Table 16
Suggested storage conditions for fresh fruit

Fruit	Relative humidity (%)	Temperature (°C)	Expected storage life (days)
Akee	—	—	—
Annona	85−90	41	42
Apple	90	−4−4	90−240
Apricot	90	0	7−14
Avocado	85−90	4−13	14−28
Banana (Cavendish, green)	85−90	7−14	21−28
Breadfruit	—	—	—
Breadnut	—	—	—
Canefruit	90−95	0	14−21
Carambola	—	—	—
Cashew apple	85−90	0−2	35
Cherimoya	See Annona		
Coconut	80−85	0−2	28−56
Date (dry types)	Same as moisture content of dates	15.5	95
Date (soft, invert sugar types)	Same as moisture content of dates	−17.5	365
Durian	85−90	4−5	14−21
Feijoa	—	—	—
Fig	85−90	−0.6−0	7−10
Granadilla (purple)	85−90	6−7	21
Grape	90−95	−0.6−0	14−180
Grapefruit	85−90	6−7	48−84
Guava	85−90	8−10	14−35
Jackfruit	85−90	11−13	42
Kiwi	—	0	90
Lemon	85−90	6−7	42−84
Lime	85−90	11−13	49
Litchi	85−90	2	42−84
Loquat	—	0	14−21
Mandarin	85−90	6−7	42
Mango	85−90	6−8	14−42
Mangosteen	85−90	4−6	49
Mulberry	—	—	—
Olive	85−90	7.5−10	28−42
Orange	88−92	4−6	35−42
Papaya	85−90	8−10	14−28
Passionfruit	See Granadilla		
Peach	90	−0.6−0	14−28
Pear	90−95	−2−−0.6	60−110
Persimmon	90	−1	90−120
Pineapple (25% yellow)	85−90	4−7	7−14
Pitanga cherry	—	—	—
Plum	90−95	−0.6−0	14−28
Pomegranate	90	0	14−28
Quince	90	−0.6−0	60−90
Sapodilla	—	—	—

Table 16 (*Contd.*)

Fruit	Relative humidity (%)	Temperature (°C)	Expected storage life (days)
Soursop	*See* Annona		
Strawberry	90−95	0	5−7
Sweetstop	*See* Annona		
Tree tomato	—	—	—

Table 17
Suggested storage conditions for fresh vegetables

Vegetable	Relative humidity (%)	Temperature (°C)	Expected storage life
Asparagus	95	0–2	2–3 weeks
Beetroot[1]	95	0–4.4	3–5 months
Broccoli	90–95	0	10–14 days
Brussels sprouts	90–95	0	3–5 weeks
Cabbage (early)	98	0	3–6 weeks
Carrot[2]	90–95	0	4–5 months
Cauliflower	90–95	0	2–4 weeks
Chinese cabbage	90–95	0	1–2 months
Chives	90–95	0	3 days
Collard	90–95	0	10–14 days
Cucumber	90–95	8.7–10	10–14 days
Egg-plant	90	8.7–10	7 days
Endive	90–95	0	2–3 weeks
French bean	90–95	4.4–8.7	7 days
Garlic (dry)	65–70	0	6–7 months
Hot pepper (dried)	60–70	10–26.7	6 months
Irish potato			
early crop[4]	90	4.4	4–5 months
		10	3 months
late crop[5]	90	3.3–4.4	5.8 months
Jerusalem artichoke	90–95	−5–0	2–5 months
Kale	90–95	0	10–14 days
Kohlrabi	90–95	0	14–28 days
Leek	90–95	0	1–3 months
Lettuce (head)	95	0	1–3 months
Lima bean	90–95	38 shelled 5–7 unshelled	about 7 days
Melons:			
watermelon[6]	80–85	4.4–10	2–3 weeks
cantaloupe[7] (muskmelon)	85–90	2.2–4.4	15 days
other	85–90	8.7–10	2–6 weeks
Mustard	90–95	0	10–14 days
Okra	90–95	8.7–10	7–10 days
Onion, green	90–95	0	3 days
Onion, dry[8]	65–70	0	1–8 months depending on cultivar
Parsley	90–95	0	2 months
Peas[9]	90–95	0	7–14 days
Peppers[10]	90–95	8.7–10	2–3 weeks
Pumpkin	50–75	10–12.8	5–12 weeks
Radish[11]	90–95	0	3–4 weeks without tops 1–2 wks with

Table 17 (*Contd.*)

Vegetable	Relative humidity (%)	Temperature (°C)	Expected storage life
Squash, winter	50−75	10−12.8	5−12 weeks
Sweet corn	90−95	0	4−8 days
Sweet potato[12]	85−90	12.8−15.6	4−6 months
Swiss chard	90−95	0	10−14 days
Tomato[13]			
mature green	85−90	13.9−15.6	7−14 days
ripe	85−90	8.7−10	3−7 days
Turnip	90−95	0	10−14 days
Vegetable marrow	90	0−4.4	3−4 days
	90	8.7−10	2 weeks

1. Remove tops before storage, otherwise storage will be limited to 10−14 days
2. This recommendation is for mature carrots without tops; immature carrots should be topped also but will only keep for between four and six weeks
3. Storage at 0−10°C is also acceptable and retards red colour loss
4. Storage at 4.4°C necessitates a curing period of between four and five days at 15.6−21.1°C before storage
5. Curing period of 10−14 days at 8.7−15.6°C and high relative humidity necessary prior to storage
6. Flavour is improved and flesh colour intensified by holding one week at room temperature. Certain cultivars reportedly are able to be kept for between two and three months at 2.8−3.9°C
7. This is for three-quarters to the 'full slip' stage of ripeness (hard ripe); if the melons are 'full slip', 0−2.2°C can be used for between five and 14 days
8. Must be cured prior to storage by allowing them to dry in the field after digging, or in a barn until the necks are tight and the outside layers of the bulbs dry and papery
9. Can be stored an extra week if covered in crushed ice; unshelled peas maintain better quality than shelled ones
10. Packaging in plastic or other moisture-conserving film will lengthen storage life by one week
11. Winter or black radishes require 0°C and 90−95% relative humidity and can be held for between two and four months
12. Cure prior to storage by holding at 29.4°C and 85−90% relative humidity for between four and seven days
13. The riper the tomato, the lower the storage temperature the fruit will tolerate without damage

Source: United States Department of Agriculture Handbook, Number 66, *The Commercial Storage of Fruits, Vegetables, Flowers and Nursery Stocks*, US Government Printing Office, 1968

Glossary

Agar	An inert substance derived from seaweed which is used to solidify media used in tissue culture.
Air layering	Propagation technique involving wounding a stem then wrapping it with moist rooting medium and polythene until roots form
Alkaline	Having a high pH
Amend	To mix material such as organic matter into soil to improve its quality
Annual	A plant which completes its life cycle from seed to seed in a single year
Aphids	Small, soft-bodied insects which suck plant sap causing distortion of the leaves and new growth
Apical	Top or terminal
Apomictic seed	Seed which develops from the ovary tissue of a mother plant without fertilization taking place
Aromatic	Having a strong smell
Asexual	Not sexual, normally used in reference to vegetative propagation, i.e. by cuttings, budding, grafting
Assimilate	Absorb
Astringent	Having a taste which causes the mouth to pucker
Auxin	A plant hormone
Axillary	Growing where a leaf joins a stem (the axil)
Balled and wrapped	A technique for moving trees in which the tree is dug from the ground with a ball of soil attached and wrapped in hessian or other material
Bareroot plant	Nursery plant which has been dug from the field while dormant with no soil around the roots.
Basin irrigation	Watering system in which a plant is enclosed by a ridge to form a round basin into which irrigation water is applied
Bear	Produce fruit
Berry	A simple fruit with pulpy flesh and no stone
Biennial	A plant which completes its life cycle from seed to seed in two years
Blanching	The technique of covering the leaves of plants with soil or other material to exclude light

468

Bloom	A flower, or the waxy or hairy covering on a fruit
Bole	Swollen base of a palm tree trunk
Box ridging	System in which a box-shaped ridge is formed around the base of a plant to control erosion and aid irrigation
Brackish	Slightly salty, due to a high level of sodium, chlorine, or other alkaline elements
Bramble	A fruit which is a member of the *Rubus* genus
Break dormancy	Passing of a plant from dormancy to active growth
Broadcast(ing)	A method of applying either fertilizer or seeds evenly on the surface of the soil
Budding	Form of vegetative propagation in which a bud from the scion is implanted in a rootstock
Budding tape	Special tape used to protect a union during graft healing
Budwood	Stems cut from a tree to use as scions in budding
Bunchy top	A virus disease affecting papayas and several other plants
Calyx	Green cap at the top of many fruits, e.g. strawberries and kiwis
Cambium layer	Group of rapidly dividing cells which produce xylem and phloem
Cane	Long shoot of a grape, berry or other vining crop
Capsule	A dry fruit, formed from more than one carpel, which opens to allow the seeds to disperse
Carpel	One of the sections of a compound ovary
Cellophane	Thin, clear plastic which can be moulded around fruits to keep them fresh
Central leader	Main upright shoot of a tree
Central leader form	Fruit tree training form in which there is a main upright shoot and fruit borne on branches spaced down the trunk
Chilling requirement	Minimum number of hours of cold which a plant must receive while dormant before it will begin active growth
Chlorosis	Yellowing of plant leaves, often due to a mineral deficiency
Climacteric	The maximum respiration rate of a fruit which is reached just before full ripening
Clingstone	Term referring to peach cultivars in which the seed does not separate easily from the flesh
Clone	Vegetatively propagated offspring of a plant which are the same genetically as the parent
Compost	Brown organic material which results from the decay of plant refuse
Compound (leaf)	A leaf blade which is divided into distinct segments; sometimes these are separate leaflets (pinnae)
Container stock	Young plants growing in containers
Copper naphthenate	A wood-preserving chemical
Copra	Dried coconut product from which oil is derived
Corm	The swollen, food-storing base of a stem, growing underground.
Cormel	A small corm
Corolla	A general term for all the petals of a flower
Cover crop	A grass or legume forage crop which prevents erosion in an orchard
Crown	The top of a pineapple, or the point at which the stem joins the roots of a plant

Creosote	Petroleum-based, wood-preserving liquid
Crotch	The place at which a branch joins a tree trunk
Cultivar	A plant variety which has arisen as a result of intentional hybridization or selection and is only found in cultivation (*see* Variety)
Cutting	Piece of a plant which is rooted to form an offspring genetically identical to the mother plant
Cyme	Flattened flower cluster
Damping off	Disease which kills seedlings before or just after they emerge
Deciduous fruits	Fruits which lose all their leaves for a yearly dormant period
Deficiency	A lack of, as in nutrient (mineral) deficiency in plants, leading to physiological disorders
Dicotyledon	The larger of the two subclasses of flowering plants
Dioecious	Having male and female flowers on separate plants
Diurnal	Daily, usually applied to a day-night cycle such as temperature change
Division	Propagation technique involving separating a plant into several pieces and replanting
Dooryard fruit	Fruit species mainly grown in small numbers around homes but not commercially
Dormancy/Dormant	Not growing actively
Drip irrigation	Watering system using water applied at a slow rate through tiny tubes
Drupe	One-seeded fruit with a hard seed coat, soft flesh and thin skin
Dwarfing rootstock	A rootstock which causes the scion to grow to less than normal size
Ecological	Related to studies of plants and their environment
Eelworm	(*see* Nematode)
Electronic leaf	Mechanical device used with a mist system to detect when the cuttings are dry and turn on the mist
Elliptic	Egg-shaped
Elongated	Longer than wide
Endocarp	The innermost layer of flesh surrounding the seed in a fruit
Endosperm	The part of a seed which stores and supplies energy to the young seedling during germination
Epidermis	'Skin' of a leaf or fruit
Ethylene	A gas which causes ripening of fruits, ageing of plant parts, and other effects
Evergreen	Always having leaves
Eye	The bud of a tuber, or the small opening which is the site of pollination on a fig
Fertilization	The fusion of the male gamete (pollen grains) with the ovule or egg cell (in the carpel) to form a new generation
Field stock	Young plants growing in the field of a nursery
Flood irrigation	Watering by allowing water to flow over a field to reach all the plants
Follicle	Dry, one-celled, many-seeded fruit which opens along one suture when mature
Freestone	Term referring to peach cultivars in which the seed separates easily from the flesh

Fruitlet	One section or segment of a fruit, as a pineapple
Fumigation	Treatment of soil or crops with gas to kill micro-organisms and other pests
Fungicide	A chemical used to control fungal diseases
Furrow irrigation	Watering system in which water runs in shallow channels past the roots of plants
Gall	Abnormal swelling on a stem or root
Genus	Group of related species
Geotrophic	Growing towards the ground. Also called positively geotrophic. Negatively geotrophic means growing away from the ground
Girdling	Removing a strip of the outer bark from a stem or a tree trunk
Glabrous	Smooth, not hairy
Gland	An organ or cell which forms specific substances, known as secretions
Grafting	Technique by which two plants are joined together to form one
Grafting wax	Special wax used to protect a union during graft healing
Greenhouse	Plant-growing structure covered with glass, clear plastic, or similar material
Greenwood bud	Bud for grafting taken from an immature, succulent portion of a plant
Growth regulator	Chemical used to influence plant growth or reproduction, for example, to encourage flowering
Gummosis	Continual oozing of sap from a wounded plant, often a tree trunk
Hand	Cluster of bananas. Many hands are found on each stalk
Hand pollination	Pollination of a plant by man using a small brush to transfer pollen from one plant to another
Hand pruner	Small, scissor-like tool for cutting small branches, secateur
Hardwood cutting	Cutting made from a woody portion of a plant
Hardening	Process of gradually adjusting a tender plant to normal outdoor growing conditions
Heading	Type of pruning which removes part of a branch just above a bud, causing side buds to grow below the cut
Heating cables	Special electric wires which are buried in the soil to warm it for seed germination or rooting cuttings
Hedgerow	Method of planting by placing plants close together in a row so that when they mature they grow to touch one another
Herbaceous (plant)	A non-woody plant which has no permanent above-ground parts
Herbicide	A chemical used to kill weeds
Hermaphrodite	Having both male and female reproductive parts
Heterozygous	Containing different genes in the pair for at least one characteristic, and hence unable to breed true.
Hilum	The scar of attachment of a seed to the ovary or carpel wall
Host plant	A plant infected by a disease or infested with pests
Hotbed	Small plant-propagating structure heated by rotting manure or electricity

Husk	Dry outer covering of a seed or fruit
Hybrid	An offspring resulting from the sexual union of two genetically different parent plants
IBA	Indolebutyric acid, a plant hormone
Immune	Not affected by a particular pest of disease organism
Incompatible	Two plants not capable of fertilizing each other or of being grafted together
Inert	Not chemically reactive
Infestation	Attack of a plant by a disease or insect pest
Inflorescence	Cluster of flowers
Ironstone	Hard rock layer rich in iron which can restrict soil drainage, plinthite
Juvenility/juvenile period	Earliest period of plant growth when the plant is unable to reproduce sexually
Lamina	The blade or expanded portion of a leaf or petal
Lanceolate	Leaf shape which tapers to a point at both ends
Larva	Immature form of an insect, frequently wormlike in shape
Lateral	Growing to the side
Laterite	Red, crusted soil layer which restricts soil drainage, ironstone, plinthite
Latex	White plant sap, normally sticky
Layering	Propagation technique involving wounding of a plant stem and then covering the wound with a rooting medium to encourage root formation
Lesion	Spot or wound on a leaf or the bark of a plant
Liner	Plant seedling or rooted cutting planted in the field after germination of the seed in the nursery
Lopping shears	Large, scissor-type tool used for cutting medium-sized branches
Marcottage	Air layering
Margin	The edge of a leaf
Medium (pl. media)	A soil substitute
Methyl bromide	Soil-sterilizing gas
Micronutrients	Elements such as zinc which are needed by plants in very small amounts
Micropyle	Small opening on a seed though which water and air pass
Mildew	Fungus disease characterized by white patches of mycelia on plant leaves
Minor fruit	Not grown on a large scale
Mist system (intermittent mist system)	Spray nozzles and a timeclock used to keep cuttings moist during rooting
Mites	Very small members of the spider family which damage plants by sucking sap
Modified central leader form	Tree-training system in which fruit is borne on branches spaced evenly up the tree trunk and the leader of the tree is removed
Monocotyledon	Member of the plant subclass Monocotyledonae characterized by parallel leaf veins and lack of woody tissue

Monoecious	Having separate male and female flowers together on one plant
Mottle	Yellow discoloration in an irregular pattern
Mound layering	Propagation technique involving piling soil around a shrubby plant to encourage roots to form at the bases of the stems
Mulch	Any material laid on the soil surface to decrease erosion, conserve water, and reduce weed growth
Mycelium (pl. mycelia)	Thread-like body of a fungus which invades the host plant
Mycoplasmas	Extremely minute, bacteria-like organisms which cause many plant diseases
Mycorrhizae	Soil fungi which live in association with plant roots and are beneficial to them
Nematicide	A chemical used to control nematodes (eelworms)
Nematode	A wormlike, often microscopic organism which seriously damages many plants. Nematodes frequently live in the soil but may invade many parts of the growing plant
Nocturnal	Active at night
Node	Place on a stem where a leaf is or was attached
Nodule	A swelling, usually on the roots of plants belonging to the legume family caused by nitrogen fixing bacteria
Non-receptivity	Inability of a stigma to receive pollen
Nucellor seed	Seed develops from the nucellus of the mother plant without fertilization.
Nursery bed	Prepared soil area used for germinating seeds or growing lines.
Nutrient	An essential food material of plants, usually mineral
Nymph	The young stage of an insect, resembling the adult in many ways but either without wings or with wings incompletely developed
Obovate	Oval-shaped but with a tip wider than the base
Off-shoot	Small plant growing from the base of a larger parent plant
Open-pollinated	Non-restricted pollination in which a plant is allowed to be pollinated naturally, not a hybrid
Organic matter	Decomposed plant or animal refuse
Ovary	The female part of a flower which swells to become the fruit
Ovate	Egg-shaped
Palatable	Good tasting
Palmately lobed	Leaves with sections (lobes) like the shape of a human palm
Panicle	Flower cluster formed by irregular branching
Papain	Enzyme from papaya fruit which decomposes proteins
Pappus	The modified hair-like perianth whorl of plants in the Compositae (e.g. lettuce) family
Parthenocarpic	Describes fruits formed without pollination
Pathogen	Micro-organism which causes a disease
Pedicel	Slender stalk that supports a fruit
Peduncle	Main flower stalk
Peeper	Small banana shoot which has not yet formed leaves
Pendulous	Hanging
Perennial	An individual plant which continues to live from year to year
Pericarp	The ripened walls of a plant ovary which surrounds the seed

Perlite	Small pieces of expanded volcanic rock used in media
Pesticide	A chemical, usually manufactured, used to kill insects and other pests. Includes insecticides, herbicides, fungicides and nematicides
Petal	The often brightly coloured and prominent part of a dicotyledonous flower
Petiole	Leaf stalk
pH	The degree of alkalinity or acidity of a substance, based on the hydrogen-ion concentration present
Photoperiod (ism)	The response of plants to the relative duration of day and night, i.e. short- and long-day flowering plants.
Physiological disorder	A poor state of plant health, normally induced by the deficiency of an essential element or nutrient, sometimes by a shortage of water
Phytotoxic	Poisonous to plants
Pinnately compound	Leaf shape in which the leaf is composed of smaller leaflets on both sides of a central stem
Plinthite	A highly weathered soil type which changes irreversibly to hardpan upon alternate wetting and drying
Pod	A carpel containing seeds, usually applied to legumes but sometimes to other plants such as okra
Pollen	The male (micro) spores of seed-bearing plants which transfer the male chromosomes to the female stigma. They are formed in the anther which is the upper portion of the stamen
Pollinator	Plant or cultivar needed for fertilization and fruit set on another plant on the same species
Polyembryonic	Containing more than one embryo
Polyethylene (polythene)	A thin, flexible plastic sheeting used for various purposes such as mulching and soil sterilizing
Pome fruits	Tree fruits of the Rosaceae family which have many central seeds in each fruit
Precocious	Bearing fruit at a young age
Predator	An animal which feeds on other animals
Propagation	Reproduction
Propagating frame	Small, low structure used for raising seedlings or rooting cuttings and having a removable top made of plastic or glass
Propagule	A part of a plant which can be used for reproduction
Prune	Dried plum
Pruning	Selective removal of plant parts to improve plant health, fruit yield, etc.
Pseudostem	A type of growing shoot arising from an underground rhizome consisting of overlapping petioles
Psyllid	Sucking insect which transmits citrus virus diseases and causes galls to develop on the leaves
Pubescent	Having hair
Puree	Food product with a paste-like consistency
Quincunx	Planting pattern in which trees are planted in each of four corners of a square and an additional tree is planted in the middle
Raceme	Spike-shaped flower cluster which continues to grow during opening and opens from the bottom to the top

Ratoon crop	The harvest produced from the suckers of an original planting of pineapples or bananas
Receptable	End of the flower stalk on which the flower parts are borne
Reflexed	Curving backward
Renewal spur	Shortened cane of a grapevine from which the following season's canes will grow
Residual pesticide	A chemical used for controlling pests which can remain active for a long time
Resistant	Able partially to withstand a pest, disease organism, or adverse environmental condition
Rest	Period of dormancy required before active growth will begin
Rhizome	A thickened horizontal stem, growing at or just below the soil surface
Rind	A thick skin of a fruit, usually citrus
Ringbarking	*see* Girdling
Root cutting	Cutting made from a section of root
Rootstock	Seedling or cutting used as the root system of a grafted plant
Rooting hormone	Chemical used to encourage faster rooting or encourage rooting of more cuttings
Rosette	Plant form in which all leaves arise from a very short stem
Runner	Above-ground stem used as a means of vegetative self-propagation, stolon
Saline	Salty
Sap	Liquid which flows from some plants and fruits when wounded
Scab	Fungus disease which causes raised brown spots
Scaffold branches	The main branches of a fruit tree
Scale	An insect pest which attaches itself in one spot to feed on a plant and then develops a waxy shell for protection
Scarification	Scratching the outer coat of a seed to encourage germination by increasing the rate of water intake
Scion	The bud or cutting used as the new shoot system of a grafted plant
Secateurs	*see* Hand pruner
Self-sterile	Incapable of self-fertilization
Self-fruitful	Capable of self-fertilization
Semi-hardwood cutting	Cutting made from newly matured stem with some bark
Sepal	The outermost part of a flower, normally enclosing the flower bud in the early stage of development
Serrate	Refers to the leaf edges which are toothed
Sessile	Without a stem or petiole
Sett	A portion of a tuber used to propagate from large tubers and bearing one or more dormant buds or 'eyes' or small bulbs
Shadehouse	Plant-growing structure used to shelter plants from excess sunlight
Slip	Shoot growing below the fruit of a pineapple
Softwood cutting	Type of cutting made from tender, new growth
Species	Closely related groups of plants which are subdivisions of a genus
Spike	A simple flowering axis, with the flowers attached directly to the main axis, without pedicels (flower stalks)
Spore	A microscopic reproductive body, of fungi or bacteria, often produced in great numbers. Some spores can resist

	extremely unfavourable conditions
Spur	Short branch of a tree with short internodes on which fruit is borne
Stamen	The male reproductive portion of a flower consisting of a filament which supports the anther
Stigma	Female part of the flower which receives pollen located above the style
Stolon	*see* Runner
Stone fruits	Members of the Rosaceae family which have only one seed per fruit
Stool layering	*see* Mound layering
Stratification	Chilling of seeds under moist conditions to promote germination
Stub	Short piece of branch
Style	Part of the female reproductive system of the flower, the style is between the stigma and the ovary on the pistil
Subsoil	Layers of soil found under the topsoil
Succulent	Used to describe leaves of a plant which are enlarged to store water
Sucker	An abnormally vigorous, vertical growing shoot; watersprout
Sunscald	Injury to a plant due to excess sunlight
Surface roots	Plant roots growing in the upper soil layers
Surfactant	A chemical added to a pesticide to make the spray stick to and cover the leaves better
Suture	Point at which two edges are joined
Symbiosis	A close and mutually beneficial combination of organisms of different species which live in close association
Systemic (insecticides)	Insecticides which are taken in via plant roots and/or leaves and transported throughout the plant system; this makes the tissues of the plant poisonous to sucking insect pests
T-bud and inverted T-bud	Grafting technique involving a T-shaped cut in the rootstock with the scion consisting of a single bud
Tap root	Single main root that grows vertically down through the soil
Tendril	A thin climbing extension or modification of a leaf, petiole, or stem
Terminal	Growing at the tip of a stem
Terracing	Erosion control method used for steep hillsides, involving making flat steps on which crops are planted
Tiller	A shoot arising from the axil of a lower leaf, as in many members of the Gramineae (grass family)
Thinning	Removing one or more young fruits from a cluster to allow the remaining ones to mature. Also a type of pruning which removes an entire branch at its origin resulting in a decrease in plant density
Tip layering	A propagation technique in which lower branches are bent to the soil, wounded and covered with soil until rooting occurs
Topography	Land slope
Topsoil	The uppermost soil layer
Top working	Changing a tree from one cultivar to another using grafting
Toxin	Poison

Trace nutrients	*see* Micronutrients
Transpiration	Loss of water vapour from a plant
Transplanting	Transferring seedlings from a seed-bed or container to a bed or ridge where they will mature
Trellis	A wood or wood and wire support for a vine
Trifoliate	Three leaflets forming a single leaf, as in many legumes
Tuber	An underground swollen food reserve, formed from a stem
Umbel	A flowering shoot with a flattened top; all the minor branches of the umbel arise from one point
Union	The point at which a stock and scion are joined together
Unisexual	With only stamens or pistil (style, stigma, and ovary)
Variety	A subdivision of a species which has arisen from natural hybridization (*see* Cultivar)
Vase form	Fruit tree-training form in which the main fruit-bearing branches all originate within 30−60 cm of each other on the main trunk
Vector	The organism or object which spreads a disease from one plant to another, for example, insect vector
Vermiculite	Pieces of expanded mica used in rooting media
Viability	Ability of seeds to germinate
Watersprouts	Vigorous, non-productive shoots which grow vertically
Weak union	Graft site which does not heal completely
Windbreak	A fence or rows of shrubs or trees established to protect newly planted crops from the prevailing wind
Witches broom	The growth of many short twigs at one point on a branch

References

Abbot, J. C. (1970) *Marketing Fruit and Vegetables*, FAO, Rome

Agnoli, M. and Guileani, F. (1977). *Cashew Cultivation*, Florence

Akamine, E. K. (1976). 'Problems in Shipping Fresh Hawaiian Tropical and Subtropical Fruits', *Acta Horticulturae*, **57**, 151−161

Atherton. J. G. and Rudich, J. (1986). *The Tomato Crop*. Chapman, Hall, London

Bammi, R. K. and Randhawa, G. S. (1968). 'Viticulture in the Tropical Region of India', *Vitis*, **7**, 124−9

Chandler, W. H. (1964). *Evergreen Orchards*, Lea and Febeger, Philadelphia

Chen, H. F. and Yong, M. S. (1980). *Malaysian Fruits in Colour,* Tropical Press SDN, BHD.

Cobley, L. S. and Steele, W. M. (1976). *An Introduction to the Botany of Tropical Crops*, Longman, London

Collins, J. L. (1960). *The Pineapple*, Leonard Hill, London

Cook, A. A. (1975) *Diseases of Tropical and Subtropical Fruits and Nuts*, Hafner (Macmillan), New York

Coursey, D. B. (1967). *Yams*, Tropical Agriculture Series, Longman, London

Coursey, D. G. *et al.* (1976). 'Recent Advances in Research on Post-harvest Handling of Tropical and Subtropical Fruits', Lima, ISHS, 135−143

Epenhuijsen, D. W. (1974). *Growing Native Vegetables in Nigeria* FAO, Rome

Eckert, J. W. (1978). 'Post-harvest Diseases of Citrus Fruits', *Outlook on Agriculture*, **9** (5), 225−32

El Baradi, T. A. (1975). 'Guava', *Abstr. on Trop. Agr.*, **1** (3), 9−16

FAO (1968). *Food Composition Table for Use in Africa*, FAO and US Dept. Health, Education and Welfare, Bethesda, Maryland

Gangolly, S. R., Singh, R., Katyal, S. L. and Singh, D. (1957). *The Mango*, Indian Council for Agricultural Research, New Delhi

Garner, R. J. (1967). *The Grafter's Handbook*, London

Garner, R. J. *et al.* (1976). *The Propagation of Tropical Fruit Trees*, FAO/CAB, Commonwealth Agricultural Bureaux, Farnham Royal

George, R. A. T. (1985). *Vegetable Seed Production*. Longman, London

Giesberger, B. (1972). 'Climate Problems in Growing Deciduous Trees in the Tropics and Subtropics', Trop. Abstr., **27**, 1−8

Grubben, G. J. H. (1975). *The Cultivation of Amaranth as a Tropical Leaf Vegetable*, Communication 67, Department of Agricultural Research, Royal Tropical Institute, Amsterdam

Grubben, G. J. H. (1977) *Tropical Vegetables and their Genetic Resources*, FAO Rome

Grubben, G. J. H. (1978) *Vegetable Seeds for the Tropics*, Bulletin 301, Department of Agricultural Research, Royal Tropical Institute, Amsterdam

Haarer, A. E. (1964) *Modern Banana Production*, Leonard Hill, London

Hartmann, H. T. and Kester, D. E. (1983). *Plant Propagation*, Prentice-Hall, Englewood Cliffs, New Jersey

Hill, D. S. (1975). *Agricultural Insect Pests of the Tropics and Their Control*, Cambridge Univ. Press, London

Irvine, F. R. (1969) *West African Crops*, Oxford University Press, London

Kay, D. E. (1973). *Root Crops*, Crop and Product Digest No. 2, Tropical Products Institute, London

Kay, D. E. (1979). *Food Legumes*, Crop and Product Digest No. 3, Tropical Products Institute, London

Kranz, J., Schmutterer, H., and Kock, W. (1978). *Diseases, Pests and Weeds in Tropical Crops*, John Wiley, Chichester, England

Lever, R. J. A. W. (1969) *Pests of the Coconut Palm, FAO, Rome*

Martin, F. W. (ed). (1984). CRC Handbook of Tropical Food Crops. CRC Press, Boca Raton, Florida, U.S.A.

Morton, J. F. (1987). *Fruits of Warm Climates*. Creative Resource Systems, Inc., North Carolina, USA

Nagy, S. and Shaw, P. E. (1980). *Tropical and Subtropical Fruits*. AVI Publishing, Westport, Connecticut.

National Academy of Sciences (1975). *Underexploited Tropical Plants with Promising Economic Value*, Washington DC

Onwueme, I. E. (1978). *The Tropical Tuber Crops*, John Wiley, Chichester, England

Onwueme, I. E. (1978). *Pest Control in Bananas* (1971). PANS Manual No. 1, London

Opeke, L. K. (1982). *Tropical Tree Crops*. John Wiley, Chichester.

Peirce, L. A. (1987). *Vegetables*. Wiley, New York

Philips, K. and Dahleen, M. (1985). *Market Fruits of South-East Asia*. South China Morning Post. Hong Kong.

Phillips, T. A. (1968) *An Agricultural Notebook*, Longman, London

Platt, B. S. (1962). *Table of Representative Values of Food Commonly Used in Tropical Countries*, Medical Research Council, Spec. Rep. Series no. 302, HMSO, London

Purseglove, J. W. (1968) *Tropical Crops: Dicotyledons*, Longman, London

Purseglove, J. W. (1972) *Tropical Crops: Monocotyledons*, Longman, London

Reuther, Walter. (1973). *The Citrus Industry Vol. I, II, III, IV*, University of California, Berkeley.

Ruck, H. C. (1975). *Deciduous Fruit Tree Cultivars for Tropical and Sub-Tropical Regions*, Commonwealth Agricultural Bureau, Farnham Royal, England.

Samson, J. A. (1970). 'Rootstocks for Tropical Fruit Trees', *Tr. Abstr.*, **25**, 145–151

Samson J. A. (1977). 'Problems of Citrus Cultivation in the Tropics', *Span*, **20**, 127–9

Samson J. A. (1980). *Tropical Fruits*, Longman, London

Sauls, J. W. *et al.*, (1976). *Proceedings of the First Tropical Fruit Short Course: The Avocado*, Univ. of Florida, Gainesville

Shalatin, G. (1973). 'New Approaches to Grape Growing in the Tropics; Grape Vine Training and Pruning Studies in Kenya', *Fruits*, **29**, 375–383.

Simmonds, N. W. (1966). *Bananas*, Longman, London

Singh, R. N. (1962). *The Mango*, Leonard Hill, London

Terra, G. J. A. (1966). *Tropical Vegetables*, Communication No. 54e, Royal Tropical Institute, Amsterdam

Tindall, H. D. (1965). *Fruits and Vegetables in West Africa*, FAO, Rome

Tindall, H. D. (1968). *Commercial Vegetable Growing*, Tropical Handbook Series, Oxford Univ. Press, London

Tindall, H. D. (1983). Vegetables in the Tropics. Macmillan Press, London and Basingstoke

Waithaka, J. H. G. and Puri, D. K. (1971). 'Recent Research on Pineapple in Kenya', *World Crops*, **23**, 190–192

Wardlaw, C. W. (1961). *Banana Diseases*, Longman, London

Williams, C. N. and W. Y. Chew. (1979). *Tree and Field Crops of the Wetter Regions of the Tropics*, Longman, London

Winkler, A. J. (1962). *General Viticulture*, Univ. of California, Berkely

Yamaguchi, Mas. (1983). *World Vegetables*. AVI, New York

Index